T0203724

Climate Change and Animal Health

This benchmark publication assembles information on the current and anticipated effects of climate change on animal health. It empowers educators, managers, practitioners, and researchers by providing evidence, experience, and opinions on what we need to do to prepare for, and cope with, the largest threat ever to have faced animals on this planet. With expert contributors from across the globe, the text equips the reader with information and the means to develop sustainable adaptation or mitigation actions.

After introducing animal health in a climate change context, chapters look at specific animal health impacts arising from climate change. The book concludes with suggestions on teachable and actionable ideas that could be used to mobilize concepts provided into education or advocacy.

This book was written amid the COVID-19 pandemic and in the face of ever-increasing reports of on-the-ground, real-life climate impacts. Large-scale wildfire and ocean heat waves killed unprecedented numbers of animals, while droughts in some areas and floods in others displaced thousands of livestock and made food scarce for even more. Climate change is real, and it is here. How we respond will have profound implications for people, biodiversity, welfare, conservation, societies, economies, and ecosystems.

Today's veterinary educators are awakening to the need to adapt and train a new generation of animal health professionals who can understand and plan for climate change, and this book is an essential resource.

CRC One Health One Welfare

Learning from Disease in Pets: A 'One Health' Model for Discovery
Edited by Rebecca A. Krimins

Animals, Health and Society: Health Promotion, Harm Reduction and Equity in a One Health World
Edited by Craig Stephen

One Welfare in Practice: The Role of the Veterinarian
Edited by Tanya Stephens

Climate Change and Animal Health
Edited By Craig Stephen and Colleen Duncan

For more information about this series, please visit https://www.routledge.com/CRC-One-Health-One-Welfare/book-series/CRCOHOW

Climate Change and Animal Health

Edited by
Craig Stephen and Colleen Duncan

CRC Press
Taylor & Francis Group
Boca Raton London New York

CRC Press is an imprint of the
Taylor & Francis Group, an **informa** business

Front cover image: Cows who survived Hurricane Florence, stranded on a porch, surrounded by flood waters. North Carolina, USA. Photograph by Jo-Anne McArthur, weanimalsmedia.org

Back cover image: A fox watches on while escaping a forest fire which is destroying wild natural habitat. Photograph by solarseven via Shutterstock.

First edition published 2023
by CRC Press
6000 Broken Sound Parkway NW, Suite 300, Boca Raton, FL 33487-2742

and by CRC Press
4 Park Square, Milton Park, Abingdon, Oxon, OX14 4RN

CRC Press is an imprint of Taylor & Francis Group, LLC

© 2023 Craig Stephen and Colleen Duncan

ISBN: 978-0-367-71202-0 (hbk)
ISBN: 978-0-367-71201-3 (pbk)
ISBN: 978-1-003-14977-4 (ebk)

DOI: 10.1201/9781003149774

Typeset in Times
by MPS Limited, Dehradun

Contents

Editors

Craig Stephen is a veterinarian and epidemiologist who has worked at the interface of human, animal, and environmental health for 30 years. His work evolved from finding and describing emerging environmental threats around the globe to helping build the circumstances that permit interspecies and intergenerational health equity. Dr Stephen has held a variety of One Health leadership positions, including serving as the founding president and director of the Centre for Coastal Health, the scientific director of the Animal Determinants of Emerging Diseases Research Network, the scientific director of the British Columbia Occupational and Environmental Health Network, a Canada research chair in Integrating Human and Animal Health and, most recently, the founding president of a new think tank striving to "future-ready" animal health professionals. He edited and co-wrote the books *Animals, Health, and Society: Health Promotion, Harm Reduction, and Health Equity in a One Health World* and *Wildlife Population Health*. He co-edited *One Health: The Theory and Practice of Integrated Health Approaches*. Dr Stephen has more than 200 peer-reviewed and technical reports. He currently operates a One Health and EcoHealth practice while retaining clinical professorships at the School of Population and Public Health (University of British Columbia) and the School of Veterinary Medicine (Ross University).

Colleen Duncan is personally and professionally committed to sustainability and education. Dual-specialized in both veterinary anatomic pathology (ACVP) and epidemiology (PhD, ACVPM), she has worked in both diagnostics and research on a wide range of species and diseases. Dr. Duncan is on the veterinary faculty at Colorado State University and affiliated with the Colorado School of Public Health, the One Health Institute, and the School of Global Environmental Sustainability. Her current efforts include the study of animal health impacts associated with climate change, the protection of animal health from environmental harms, and the identification of ways to minimize the environmental impact of veterinary care.

Preface

Imagine a future where animal health is protected and promoted and the unanticipated or undesired effects of climate changes can be prevented, mitigated, or adapted to. Then ask, what do we need to know, and what must we do to get there? Those questions were the motivating factor for this book. An esteemed set of authors has come together to help us think about the challenges climate change is presenting to animal health and to provide evidence, experience, and opinions on what we must do to address the largest threat animals on this planet have ever faced. This book is our contribution to the CRC One Health One Welfare series that focuses on the interconnections between human well-being, animal health and welfare, and the environment, with emphasis on interdisciplinary collaboration and solutions.

Each group of authors was selected based on their interest and expertise in specific topic areas highly relevant to humanity's efforts to enable evidence-based action on key subjects driving climate change actions for animal health. The goal is to help people who are responsible for animal health education, policies, and practices gain knowledge and confidence about the factors and decisions that shape their responses to climate change. We asked all authors to not only summarize existing evidence, experience, and opinions on their subject matter, but also to focus on key points for action and learning.

This book as written between 2021 and 2022, amid the COVID-19 pandemic and in the face of ever-increasing reports of on-the-ground, real-life climate impacts. Large-scale wildfires spread across massive swaths of the northern and southern hemisphere, killing unprecedented numbers of animals. Ocean heat waves similarly killed billions of near-shore creatures. Droughts in some areas, and floods in others, displaced thousands of livestock and made food scarce for even more. Climate change was real, and it was here. Despite this reality, the scholarship, advocacy, and policies dealing with ways to mitigate and adapt to the effects of climate change on animal health remained rare in comparison to the discourse on its effects on people and human communities. This book makes a start at collecting information for those interested in and dedicated to confronting climate change for animal health and the benefits that brings ecosystems and societies. We have organized this material into three sections:

1. Overview of climate change and animal health (Chapters 1–5): Starting with an overview of climate change science, these first chapters include an introduction on how to think about health in a climate change context, how health will influence patterns of diseases, and how it will affect the determinants of health. This part of the book also introduces the idea of climate change action.
2. Specific animal health impacts arising from climate change (Chapters 6–10): These chapters build on the introductory chapters, providing more details on potential effects and further insights into the

unique nature of infectious, noninfectious, and social impacts induced by climate change on animals and society.

3. Action to protect animal health from threats posed by climate change (Chapters 11–17): Authors of these chapters provide insights and advice on areas of action ranging from international policy to personal involvement. This section includes advice on how to maintain hope in the face of grave challenges from climate change and shares lessons learned from those who have led bottom-up changes. The concluding chapter aims to extract lessons for action shared across all the chapters.

The evidence and experience on climate change and animal health is mounting and expected to grow even faster as the effects of climate change become more persistent, pervasive, and impactful. The editors sincerely hope that this book will provide a jumping off point for all readers to take some action.

Contributors

Julia E. Baak
Department of Natural Resource Sciences
McGill University
Québec, Canada

Guillaume Belot
World Organisation for Animal Health (OIE)
Paris, France

Tianna Brand
World Organisation for Animal Health (OIE)
Paris, France

Chris G. Buse
Faculty of Health Sciences
Simon Fraser University
Burnaby, Canada

Maud Carron
World Organisation for Animal Health (OIE)
Paris, France

Emily S. Choy
Department of Natural Resource Sciences
McGill University
Québec, Canada

Stéphane de la Rocque
World Organisation for Animal Health (OIE)
Paris, France

Michelle Dennis
College of Veterinary Medicine
University of Tennessee
Knoxville, Tennessee, USA

François Diaz
World Organisation for Animal Health (OIE)
Paris, France

Michael A. Drebot
Science Reference and Surveillance Directorate
National Microbiology Laboratory
Public Health Agency of Canada
Winnipeg, Canada

Colleen Duncan
College of Veterinary Medicine and Biomedical Sciences
Colorado State University
Fort Collins, Colorado, USA

Ahmed. H. El Idrissi
Independent Consultant

John E. Elliott
Environment and Climate Change Canada
Gatineau, Canada
and
Pacific Wildlife Research Centre & Department Canada Avian Research Centre
Delta, Canada
and
Faculty of Land and Food Systems
Delta, Canada

Kyle H. Elliott
Department of Natural Resource Sciences
McGill University
Québec, Canada

Christa A. Gallagher
School of Veterinary Medicine
Ross University
Saint Kitts and Nevis, West Indies

Gregory Graff
Department of Agricultural and
    Resource Economics
Colorado State University
Fort Collins, Colorado, USA

Keith Hamilton
World Organisation for Animal
    Health (OIE)
Paris, France

Bethany Jackson
Murdoch University
Perth, Australia

Rose M. Lacombe
Department of Natural Resource
    Sciences
McGill University
Québec, Canada

L. Robbin Lindsay
One Health Division
Science Reference and
    Surveillance Directorate
National Microbiology Laboratory
Public Health Agency of Canada
Winnipeg, Canada

Jeremy S. Littell
U.S. Geological Survey
Alaska Climate Adaptation
    Science Center
Fairbanks, Alaska, USA

Carrie McMullen
Ontario Veterinary College
University of Guelph
Ontario, Canada

Sophie Muset
World Organisation for Animal
    Health (OIE)
Paris, France

Benjamin Nordbrook
Department of Agricultural and
    Resource Economics
Colorado State University
Fort Collins, Colorado, USA

Nick H. Ogden
Public Health Risk Sciences Division
Scientific Operations and Response
    Directorate
National Microbiology Laboratory
Public Health Agency of Canada
Winnipeg, Canada
and
Groupe de recherche en épidémiologie
    des zoonoses et santé publique
Faculté de médecine vétérinaire
Université de Montréal
Quebec, Canada

Jane Parmley
Ontario Veterinary College
University of Guelph
Ontario, Canada

Julio Pinto
Food and Agriculture Organization
    of the United Nations (FAO)
Rome

Delia Grace Randolph
International Livestock Research
    Institute (ILRI)
Nairobi, Kenya

Cheryl Sangster
Aboyne, Scotland

Will Sander
College of Veterinary Medicine
University of Illinois
Urbana–Champaign, Illinois, USA

Andrew Seidl
Department of Agricultural and
    Resource Economics
Colorado State University
Fort Collins, Colorado, USA

Diego S. Silva
Sydney School of Public Health
University of Sydney
Sydney, New South Wales, Australia

Maxwell J Smith
School of Health Studies
Faculty of Health Sciences
Rotman Institute of Philosophy
Western University
Ontario, Canada

Katie Steneroden
Center for Food Security & Public Health
College of Veterinary Medicine
Iowa State University
Ames, Iowa, USA

Craig Stephen
McEachran Institute
Alberta, Canada

Tim K. Takaro
Faculty of Health Sciences
Simon Fraser University
Burnaby, Canada

Jimmy Tickel
Institute for Infectious Animal
    Diseases
Texas A&M
College Station, Texas, USA

Simone Vitali
Zoo and Wildlife Veterinary Services
    Western Australia
Australia

Chadia Wannous
World Organisation for Animal
    Health (OIE)
Paris, France

1 An Introduction to Current Climate Projections and Their Use in Climate Impacts Research

Jeremy S. Littell

CONTENTS

KEY LEARNING OBJECTIVES

- Be able to define climate and climate change, and have basic knowledge of the components of the Earth's climate system.
- Recognize the uses of climate models in climate science and the uncertainties that apply to the projections made with them.
- Distinguish between forecasts and scenarios.

DOI: 10.1201/9781003149774-1

IMPLICATIONS FOR ACTION

- Evaluating the likelihood of a given climate impact on animal health or management strategies requires consideration of the main sources of climate projection uncertainties.
- Adaptation requires consideration of global-to-regional contexts of climate changes and impacts, but also adaptive capacity, including governance.

INTRODUCTION – WHAT IS CLIMATE AND WHY DOES IT MATTER?

At its simplest, climate is the expected weather conditions in some place for some time. One definition is "the statistics of weather," including the mean of and variation in atmospheric conditions and their physical effects at the Earth's surface on variables such as temperature, precipitation, humidity, wind, and so on. Most people have an intuitive expectation of the annual and seasonal variation in weather in the places they have spent time, and so they can grasp the difference between *weather* (the conditions right now, today, tomorrow) and *climate* (the conditions averaged over longer periods, their variability). For purposes of this book, a good definition of *climate change* is:

> *"a change in the state of the climate that can be identified (e.g. by using statistical tests) by changes in the mean and/or the variability of its properties and that persists for an extended period, typically decades or longer. Climate change may be due to natural internal processes or external forcings such as modulations of the solar cycles, volcanic eruptions and persistent anthropogenic changes in the composition of the atmosphere or in land use."* [1]

Greenhouse gases (GHGs, such as CO_2 – carbon dioxide, CH_4 – methane, water vapor, and many others) affect the way infrared radiation (heat) behaves in the atmosphere. Changes in their concentrations do not much affect how light enters the Earth system, but they do affect how infrared radiation (essentially heat) is transmitted or absorbed. They change the thermal balance of the Earth system by increasing heat absorption. Greenhouse gas concentrations are one kind of climate *forcing* – the global temperature responds to changes in GHGs. The effect on global temperature has been proportional to the increase in GHGs since the industrial revolution and the more rapid increases in the 20th century.

Global temperature[1] has increased more than +0.8°C since the middle of the 20th century (Figure 1.1, [2]). For the most recent decade (2011–2020), the increase in mean global surface temperature since 1850–1900 is assessed to be +1.09°C (90% uncertainty level +0.95 to +1.20°C) (Figure 1.2, [3]).

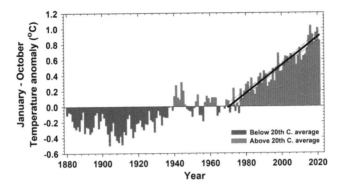

FIGURE 1.1 Global land and ocean temperature anomalies from 1880–2021, for the year-to-date available data (January–October) with 1970–2021 trend. Annual average values represent difference from the 20th century average (horizontal temperature anomaly line at 0.0), with blue bars indicating lower than long-term average and red bars indicating higher. The last year with average temperature lower than the long-term average was 1976. 21 of the warmest 22 years in the record were 2001–2021. The trend in the anomalies from 1970 to 2021 is about +0.18°C/decade, representing a warming of around +0.9°C for this period. Data: NOAA NCEI [4,5].

The recently released Intergovernmental Panel on Climate Change (IPCC) Sixth Assessment Report (AR6) concluded, "It is unequivocal that human influence has warmed the atmosphere, ocean and land. Widespread and rapid changes in the atmosphere, ocean, cryosphere and biosphere have occurred" [6]. Packed into that statement is the assertion that the temperature changes at a planetary scale have cascading effects and feedbacks that translate to impacts through many pathways, starting with the physical climate system and rippling through the planetary hydrologic, ecological, and marine systems. These notable global changes and their climatic descriptions, however, veil a complexity that matters a great deal for any real-world application of climate change information. The overarching drivers are no less important, but the variation among both physical and human geographic contexts results in a wide range of climatic changes and climate impacts that are regionally distinct. Preparation for, and adaptation to, the plausible impacts therefore varies with changes, climate system sensitivity, and impacts. This is no less true in considering the impacts of climate change on animal health and its subsequent impact on welfare, conservation, economies, ecosystems, and societies – climate is one (albeit an overarching and far-reaching one) of many in a complex, tangled set of stressors.

The goal of this chapter is, first, to briefly discuss the basics of climate science required as foundation for subsequent chapters that deal with specifics of climate impacts and effects and, second, to present consensus future *projections* – which describe plausible future climate scenarios – and their uncertainties. For these purposes, it may be useful to think less globally and more locally, starting from the decision context(s) of those concerned with specific impacts on animal health

(a) Change in global surface temperature (decadal average) as reconstructed (1–2000) and observed (1850–2020)

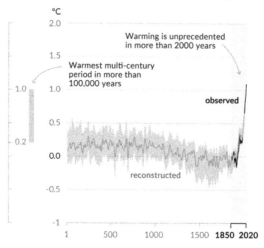

(b) Change in global surface temperature (annual average) as observed and simulated using human & natural and only natural factors (both 1850–2020)

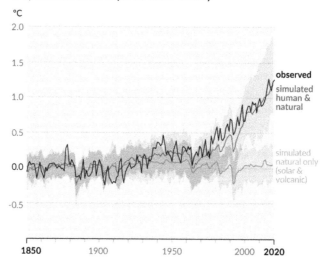

FIGURE 1.2 Changes in global surface temperature relative to 1850–1900. Figure source: SPM1, [6]. Panel (a): (b) Changes in global surface temperature reconstructed from paleoclimate archives (solid gray line, 1–2000) and from direct observations (solid black line, 1850–2020), both relative to 1850–1900 and decadally averaged. The vertical bar on the left shows the estimated temperature (very likely range) during the warmest multi-century period in at least the last 100,000 years, which occurred around 6500 years ago during the current interglacial period (Holocene). The Last Interglacial, around 125,000 years ago, is the next most recent candidate for a period of higher temperature.

at regional to continental scales so that decision-making and adaptation can account for this variation and be most effective.

CLIMATE CHANGE PAST AND IMPLICATIONS FOR NAVIGATING CLIMATE CHANGE FUTURE

Before we discuss what the future may hold, it may be useful to understand the historical context. Understanding the climate changes that have shaped animal populations and their adaptive capacity to change requires a much longer-term history of climate than that in Figure 1.1. There have been periods in the distant geologic past where the climate was radically different than now, for a range of reasons including different continental/ocean configurations, different atmospheric composition, and different orbital configurations of the Earth in relation to the sun, such as the seasonal timing of the Earth's closest approach to the sun. In-depth treatment of these differences is beyond the scope of this chapter because current ecological and human contexts, such as this book's focus on animal health and climate change, are the products of more recent climate variability. For example, the current climate is now warmer than at any time in at least the last 2000 years, very likely[2] the last 6500 years, and likely much longer – 100,000 years (Figure 1.2).

The early historical record also overlaps with a period when global climate was emerging from a period called the "Little Ice Age," from roughly the late 1400s to the mid-19th century, when northern hemisphere temperatures were cooler than now and preceding centuries. Over longer time scales, millennial climate variability since the Last Glacial Maximum (21,000 years ago) and during the interglacial epoch of the last ~12,000 years (the Holocene) includes the climates that exerted evolutionarily selective pressures on species, accompanied human migrations and cultural developments, and defined the landscapes in which we live and work.

This longer-term Holocene global climatic history is also the longest plausible climate with which we have collective experience, though climate proxies (geological and ecological records related directly to climate) allow reconstructions of past climates hundreds of millions of years into the past.

These past variations are good examples of the climatic variability known to occur within the climate system. The recent climate, and especially the future

---

These past warm periods were caused by slow (multi-millennial) orbital variations. The gray shading with white diagonal lines shows the very likely ranges for the temperature reconstructions. Panel b, bottom): Observed changes in global surface temperature over the past 170 years (black line) relative to 1850–1900 and annually averaged, compared to CMIP6 climate model simulations (see Box SPM.1, [6]) of the temperature response to both human and natural drivers (brown), and to only natural drivers (solar and volcanic activity, green). Solid colored lines show the multi-model average, and colored shades show the very likely range of simulations. Used with permission.

climate we expect, is quite different from the climate in which modern eco-systems and human institutions formed, and we therefore cannot rely on what we know from the paleoclimatic record as a guide to the future. However, the variability in the climate system – its capacity for persistent conditions over years or decades, its ability to produce extremes we have not previously ex-perienced, its tendency to reorganize rapidly due to both external and internal drivers – is something we should continue to expect.

PRIMER ON CLIMATE CHANGE SCIENCE – CLIMATE PROJECTIONS AND SCENARIOS

Since 1990, developments in climate change science have been routinely sum-marized globally about twice a decade by the IPCC, and a history of the de-velopment of this science can be found in the Working Group One (WG1) reports, the most recent of which appeared in 2021 ([7], AR6 WG1 report). In the late 1990s, the United States initiated its own National Climate Assessment (NCA) and most recently published NCA4 in 2017/2018, with the NCA4 Climate Science Special Report [8] providing a U.S.-focused report similar to IPCC WG1. The NCA5 is currently underway and expected to be published in 2023. These authoritative reports summarize the state of climate science and scientific advances since previous reports. The impacts and responses to the changes in the climate system are described in other related work (e.g. IPCC Working Group 2 and 3, and NCA Impacts, Risks, and Adaptation in the United States, [9]). For more information than can be described here, these reports offer the breadth and depth of climate science research.

For the more practical purpose of introducing climate change projections as background for the rest of this volume, it may be useful to consider three broad fields of climate change science that underlie climate impacts research, climate-informed decision making, and adaptation. The first is climate measurement and observation – the direct observation, spatial interpolation, and summary of his-torical climate variation and current climate trends. The second, called "**detection and attribution**" research, relies on the first as foundational drivers of other changes. Detection and attribution are, in some ways, like a sort of climate for-ensics. This field involves determining the causes (including different forcings) of climatic changes or events, quantifying climatic influences on other processes, and determining what part of that process's variation is due to climate change and what part is due to other factors. The third is climate modeling, which is focused on simulating the climate system to improve understanding of how the whole system works and responds to changes in its drivers and components, and developing projections of how global climate could change in the future. Historical ob-servation, current detection and attribution, and future climate modeling rely on over a century of research on the climate system. The forcings and *feedbacks* (processes that increase or decrease the rates of other processes) that interact to affect global-to-local climate are shown in relation to other components of the climate systems in Figure 1.3. To attribute and explain the changes in Figure 1.1,

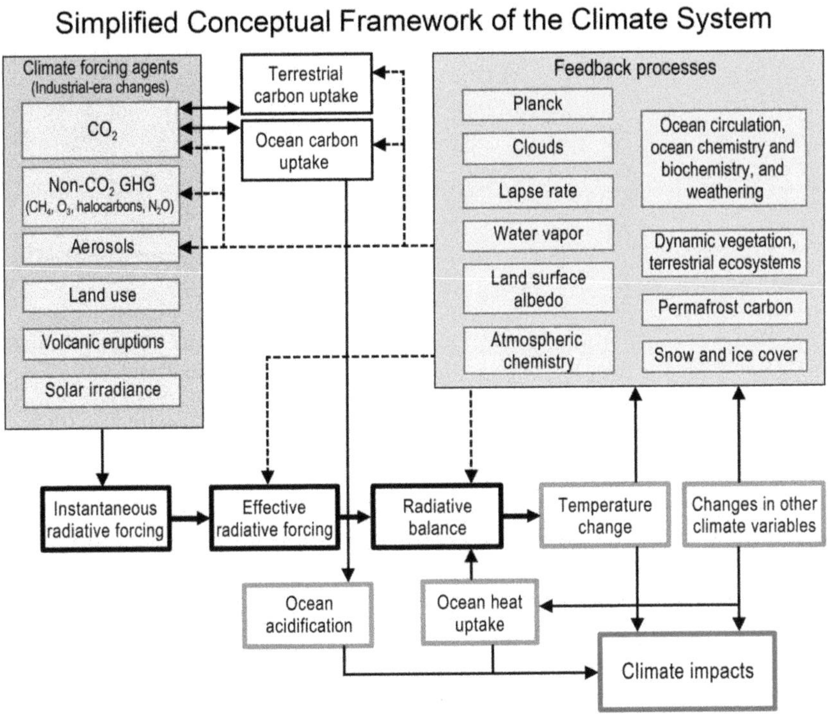

Simplified Conceptual Framework of the Climate System

FIGURE 1.3 The climate system is complex, and the pathways by which climate drives impacts of importance to people depend on which processes one considers. Simplified conceptual modeling framework for the climate system as implemented in many climate models (Chapter 4: Projections). Modeling components include forcing agents, feedback processes, carbon uptake processes, and radiative forcing and balance. The lines indicate physical interconnections (solid lines) and feedback pathways (dashed lines). Principal changes (blue boxes) lead to climate impacts (red box) and feedbacks.

Source: [10] – NCA4 CCSR Chapter 2, adapted from [11].

for example, and the cascading effects through global ecosystems requires an integrated systems approach like that described in Figure 1.3. Projecting accurately the potential futures of the climate thus requires developing models that correctly integrate these processes in a quantitative framework.

Global climate models (GCMs) are mathematical representations of the climate system and its components. Modern climate models are complex simulations of the atmosphere, ocean, land surface, and other components, such as sea ice, the hydrologic cycle, the carbon cycle, and their interactions. The GCMs run on large computers required to handle the computations involved in simulating multiple layers in the atmosphere, multiple depths in the ocean, and a global grid of cells (approximately 100 km resolution, but this varies considerably among models). A century of development has led to increasingly realistic simulations

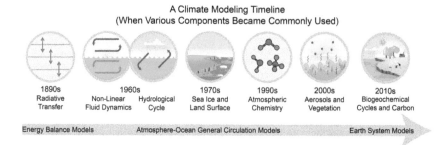

FIGURE 1.4   Different aspects of the climate system were introduced into global climate models at different times as models evolved toward increasing complexity.

Source: [13] - NCA4 CCSR Chapter 4.

by incorporating more components in Figure 1.3, such as sophisticated atmospheric chemistry and biogeochemical cycles (Figure 1.4). Recent generations of climate models include Earth system models (ESMs), indicating they incorporate and simulate many processes beyond atmosphere-ocean climate dynamics, such as land surface – atmosphere carbon exchange (e.g. [12]).

Climate models are capable of skillfully simulating the historical observed climate (Figure 1.1) without the need to "tune" the models with observed climate. In other words, models simulate the climate known to have occurred (see Figure 1.2b for a global example) given the known forcings and conditions. Comparing climate model projections with subsequent climate responses has also shown that, in general, climate models perform well (e.g. [14]). This assessment gives us greater confidence they can simulate future climate, although these simulations do not provide the probabilistic context of weather forecasts and have notable limitations in skill, especially for higher spatial and temporal resolution (e.g. [15]).

The GCMs' (including ESMs') construction differs from model to model, especially in their parameterization of processes that are still areas of active research, such as cloud physics and aerosol feedbacks. These differences in model construction result in a range of simulated climatic changes at global-to-regional scales, in part because the exact sensitivity of the current climate to forcing can be estimated but is uncertain even though the range of likely and unlikely fundamental sensitivities is well constrained. One result is that some models project faster and/or greater warming while others project less warming or slower warming for the same forcing, and other climatic changes that result are also different. Another is that models handle internal climate dynamics differently, resulting in differences in simulation of specific processes.

The IPCC Fifth Assessment Report (AR5) used several dozen GCMs and ESMs from climate modeling research groups around the world, collectively run under a common set of conditions under the auspices of the Fifth Coupled Model Intercomparison Project (CMIP5). These models provided projections of future climate changes common in climate impacts research since the mid 2010s.

CMIP6, with climate model outputs coordinated for IPCC AR6, includes a similar number of climate models and has become available more recently.

The future GHG forcings used to drive these models, however, are uncertain – they depend on policy choices and economic forces that will unfold between now and the end of the 21st century (or beyond). To develop greenhouse gas emissions for input as forcings into climate models, **emissions scenarios** must be constructed that represent the range of plausible global socioeconomic and policy trajectories. It is important to note that these scenarios are not forecasts or predictions – they are each a possible outcome from the time of their development into the futures they describe. The CMIP5/AR5 generation of climate model outputs used greenhouse gas forcings consistent with Representative Concentration Pathways (RCPs, e.g. [16]). Four RCPs were commonly used, and their names are derived from the end-of-21st century mean radiative forcing at the top of the troposphere (the layer of the atmosphere that extends from the Earth's surface to the lower boundary of the stratosphere), in W/m^2: RCP 2.6 (GHG mitigation, very low forcing), 4.5 (early GHG stabilization, medium forcing), 6.0 (later GHG stabilization, medium forcing), and 8.5 (very high GHG baseline, high forcing) [17]. The CMIP6/AR6 generation of climate model outputs used similar forcings consistent with Shared Socioeconomic Pathways (SSPs, [18,19]) and named after similar radiative outcomes: SSP1-2.6, SSP2-4.5, SSP3-7.0, and SSP5-8.5. In addition, an even lower scenario, SSP1-1.9 [20] corresponds to limiting GHG emissions to remain below global temperature increase of 1.5°C as in the Paris Agreement [21]. For each of these RCP or SSP scenarios, the climate projections generated by a group of GCMs/ESMs can be compared (called an **ensemble**), and the mean and range compared to simulated historical references to obtain a range of plausible climate changes. Figure 1.5 [22] shows a comparison for three CMIP5/RCP ensembles with three CMIP6/ SSP ensembles in terms of global temperature responses relative to the late 19th century, as well as to the more recent period 1986–2005. CMIP6 projections are proving to be warmer than CMIP5, especially for higher emissions, due at least in part to higher mean climate sensitivity that probably stems from cloud feedbacks and cloud-aerosol feedbacks [23]. Figure 1.5 also shows the slightly lower ranges of warming at 2100 that would be expected if the projections were constrained to historical warming trends and thus consistent with previous generation (CMIP5) assessments [24].

Downscaling is the process of taking coarser resolution GCM/ESM output and using other information to increase the resolution (decrease the size of grid cells) for which information is projected or summarized. Downscaling methods can be classified on a continuum from very simple statistical methods to dynamical methods more like weather simulation. Statistical methods rely on statistical relationships between gridded historical observed climate and climate model simulated historical climate, whereas dynamical methods rely on regional weather simulation models to interpret global climate model output. Downscaling approaches vary with the objective of the research – factors, such as whether finer spatial averages are needed or if extremes are required,

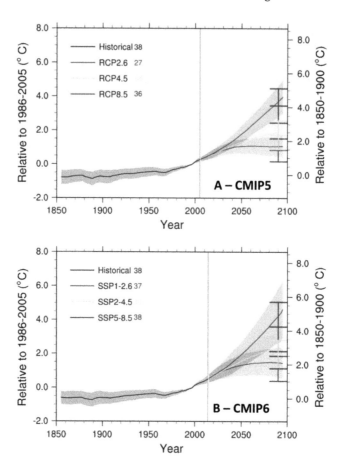

FIGURE 1.5   CMIP5 (top) and CMIP6 (bottom) global surface temperature responses for comparable SSP and RCP scenarios. Bold trajectory lines indicate mean temperature scenarios, shaded ranges indicate uncertainty across models. Note that the uncertainty range values are not directly comparable to the right (late 19th century) anomaly axis – they are referenced to the left (recent past) axis. Bold whisker plots indicate temperature ranges at 2100 derived by constraining model projected trends to observed warming trends [24]. This constraint is a form of tuning. Number references after scenario legends indicate number of models in each ensemble. Figure: [22], Modified from Fig. 8. Permission to reproduce figure under Creative Commons Attribution 4.0 - https://www. earth-system-dynamics.net/policies/licence_and_copyright.html.

determine the most appropriate approaches. In general, statistical methods are computationally faster but have limitations derived from historical observation density (number of long-term weather stations) and quality (completeness of record). Dynamical methods are computationally more intensive but provide physically consistent output of (usually) more variables and are better able to directly simulate the changes in extremes that might result from changes in climate. A full treatment of downscaling is beyond the scope of this introduction, but regional climate services and boundary organizations are a good place to find regionally downscaled climate projections for specific purposes.

The combination of multiple emissions scenarios and multiple climate models results in a wide range of global climate outcomes, and these translate quickly to *uncertainty* in future climate, especially at local-to-regional scales where the magnitude, direction, and timing can vary considerably depending on which emissions scenario – climate model(s) are considered. Based on the scientific knowledge of the sources of uncertainty in climate projections, there are strategies to incorporate uncertainty directly into climate impacts modeling. Hawkins and Sutton [25] describe three principal sources of uncertainty in climate model projections (essentially climate variability, model construction and sensitivity, and emissions differences) that vary in their relative contribution through time into the 21st century. It is important to note that total uncertainty increases with time from the present, even though these relative contributions from the three sources change. Interannual to decadal climate variability has a larger amplitude in some regions of the planet than others, but this variability contributes to both regional and global climate variation on these timescales. In the near term, these variations are large compared to the GHG-forced trend in the climate system, and so natural climate variability is the largest source of uncertainty in the near term, roughly a couple of decades. Uncertainty is also proportionally larger for smaller areas considered and in regions with higher decadal climate variability in the historical record. Projection uncertainty due to climate model differences begins to exceed uncertainty due to climate variability several decades (roughly 30–50 years from model output) into the future. This is because the climate models' cumulative responses to forcing begin to exceed the magnitude of interannual climate variability by that time. Finally, emissions scenario differences begin to dominate in the later decades of the 21st century (or more than about 50 years from model output). The mean residence time of CO_2 in the atmosphere is several decades, so projections do not diverge much due to emissions before the mid 21st century, but realized near-term differences in GHG emissions result in larger climate differences late in the 21st century. It has been argued [26,27] that downscaling of GCM/ESM output adds a fourth source of uncertainty, with the additional contributions depending on the strategy employed for model subsampling and the method(s) used for finer resolution interpolation.

For most climate impacts research, the combination of multiple climate models and multiple emissions scenarios results in a range of climate futures. Resources and capacity, however, often limit the number that can be considered for drivers of climate impact models for other physical processes, species distributions, or health outcomes. Considerations for choosing and using climate model projections for impacts modeling or other use-inspired research [27–29] suggest a combination of strategies to address these sources of uncertainty. First, in the near term, climate variability can be expected to be larger than the total change in mean climate due to forcings for the same period, so planning for historical climate variability plus the changes in climate trend is an initial step. Next, these strategies suggest using as many climate models as is practical, ideally with some objective method for choosing models based on skill (or lack of skill) if a subset is required. For scenario planning that is less concerned with skill and equally or more concerned with the range of risks associated with the plausible outcomes, bracketing the "average" outcome with models that project more and less change than the ensemble mean provides a range tailored to higher (risk averse) and lower (risk tolerant) vulnerability or impact. The strategies also suggest using multiple emissions scenarios (recently, RCPs or SSPs) with choices depending on time frame of required decision or adaptation as well as risk framing. Here, if the required decision is in the near term, where climate variability and climate model uncertainty contribute most to total uncertainty, the distinction between lower, medium, and higher emissions is less important because they do not diverge until later in the 21st century. Maximum contrast with historical conditions would be with a higher emissions scenario, and lower emissions scenarios may overlap considerably with recent conditions. Finally, if resources allow, multiple downscaling methods can provide useful insights for applications in specific regions or for specific processes, but even if capacity to produce these is not limiting, evaluating the similarities and differences among them and translating them to practical use requires considerable effort. It can be tempting to try to find the "most likely" plausible future and use it as a single scenario, but the uncertainties in climate, modeling, and emissions indicate this approach is less ideal than it might seem.

FUTURE GLOBAL AND REGIONAL PROJECTIONS

International scientific consensus (e.g. [6,7]) is that the Earth's climate is warming and that it is very likely that GHGs have been the main driver of that warming since at least 1979. Despite the current state of climate modeling, however, the climate futures discussed above are projections – not forecasts, at least in part because assigning probabilities to climate model simulations is hindered by the nature of the scenarios that drive them. Perhaps even more importantly, climate variability arises from forcings such as volcanic eruptions, solar variability, and internal dynamics, but the variability simulated by climate models will not ne-cessarily match the real climate system in timing of these events. The interannual increases and decreases in simulated climate variables therefore cannot be used as forecasts. The statistics of those variations will be reasonable for skillful models, but the seasonal to decadal timing may not match. For some decisions and uses, it is sufficient to plan on continued increases in global temperature, but for most real-world applications, much more specific information is required, and the un-certainties described in section 3 can be addressed to develop a range of plausible outcomes given current scientific capabilities. It is worth noting that the differ-ences between CMIP3 (mid-2000s) and CMIP5 (2010s) model skill (and differ-ences between SRES and RCP scenarios) were sufficiently small that the resulting climate projections could be considered from the same general distribution [30]. There is also considerable overlap between CMIP5 and the newer CMIP6. CMIP5 (or even CMIP3) projections are not necessarily obsolete, but characterizing them in the context of the newer expected ranges of projections will help bridge real-world decision making until CMIP6 projections and research based on them be-come more common. This section provides an overview of global changes expected under the most recent (CMIP6/AR6) projections and some related regional differentiation of climate impacts pathways. It also sketches some of the mechanisms that tend to be associated with ecological impacts, noting some of the scientific limitations in using versions of those derived from climate models.

Relative to 1961–1990, future 21st century global climate is unequivocally projected to warm [6]. These changes are not equally distributed, however, with the land surface being virtually certain to warm faster than the ocean surface and the Arctic being virtually certain to warm more than the global land surface – there is high confidence in Arctic warming being greater than double the global rate (Figure 1.6, Table 1.1).

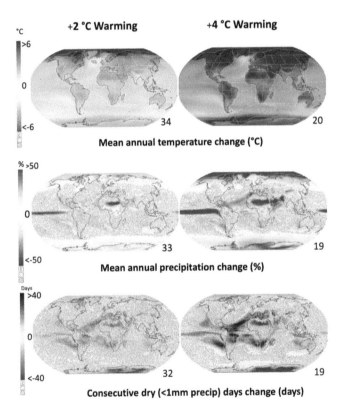

FIGURE 1.6 CMIP6/AR6 ensemble average projections for changes in mean annual temperature (top), mean annual precipitation (middle), and consecutive dry days (bottom) for a +2°C global warmer future (left) and a +4°C global warmer future, relative to 1850–1900. Lower right numbers indicate CMIP6 climate models in ensemble. Changes vary in robustness: unmarked color means the change is large compared to historical variability; left-leaning hatching indicates no robust change or signal; cross-hatching indicates conflicting signals. Regions indicated correspond to IPCC WG1 regions.

Source: [31]. Maps created with IPCC WG1 Interactive Atlas (https://interactive-atlas.ipcc.ch/).

TABLE 1.1

Median and 10th–90th confidence ranges for annual temperature and precipitation changes for Global, Northern, and Equatorial regions by SSP for mid- and late-21st century, relative to 1961–1990

		Annual Temperature change (°C)		Annual Precipitation change (%)	
		2041–2060	2081–2100	2041–2060	2081–2100
Global	SSP1-2.6	+1.6 (+1.1 to +2.1)	+1.7 (+1.1 to +2.4)	+3.1 (+2.1 to +4.7)	+3.7 (+2.3 to +5.7)
	SSP2-4.5	+1.8 (+1.4 to +2.4)	+2.7 (+2.0 to +3.3)	+3.0 (+1.9 to +4.7)	+4.8 (+3.0 to +7.4)
	SSP5-8.5	+2.3 (+1.7 to +3.0)	+4.5 (+3.4 to +5.7)	+3.5 (+2.0 to +5.2)	+7.0 (+4.4 to +10.3)
Northern	SSP1-2.6	+3.5 (+1.8 to +5.5)	+3.9 (+2.1 to +6.2)	+11.5 (+6.5 to +17.3)	+12.8 (+7.7 to +21.6)
	SSP2-4.5	+4.1 (+2.5 to +6.2)	+5.8 (+4.0 to +8.0)	+12.5 (+8.0 to +17.8)	+18.1 (+11.1 to +25.8)
	SSP5-8.5	+4.9 (+3.3 to +7.2)	+9.3 (+6.3 to +12.4)	+15.1 (+9.5 to +22.1)	+29.4 (+18.7 to +42.7)
Equatorial	SSP1-2.6	+1.6 (+1.2 to +2.2)	+1.6 (+1.1 to +2.4)	+2.7 (−0.5 to +5.5)	+3.1 (+0.4 to +5.9)
	SSP2-4.5	+1.8 (+1.4 to +2.4)	+2.6 (+2.0 to +3.4)	+2.4 (−0.7 to +5.2)	+3.6 (−0.3 to +7.0)
	SSP5-8.5	+2.3 (+1.8 to +2.9)	+4.7 (+3.3 to +6.4)	+2.3 (−0.8 to +5.5)	+3.8 (−2.3 to +10.5)

Source data: [31]

Northern = NWN, NEN, GIC, NEU, RAR, and Arctic Ocean; Equatorial = SCA, NWS, CAR, WAF, NEAF, SAH, ARP, SAS, SEA. See Figure 1.7 for a map of regional acronyms. Numbers of models in ensembles for (temperature, precipitation) respectively: SSP1-2.6(33, 31); SSP2-4.5(34, 32); SSP5-8.5(34, 33).

Table 1.1 shows mid- and late-21st century projected global changes for annual temperature and precipitation by SSP scenario globally and for contrasting northern and equatorial groups of IPCC WG1 regions. Precipitation changes are not as easily summarized, although climate models have higher agreement that annual precipitation increases are likely to very likely larger (in terms of percent change) in the high latitudes and equatorial regions than elsewhere. Precipitation decreases are likely in the mid-latitudes, particularly in and around the southeast Pacific, north central Atlantic and Mediterranean, southeast Atlantic, and southeast Indian Ocean (Figure 1.6). Whether these changes are large compared to historical variability is also highly variable. Many impacts for health or managed resources are not driven as much by annual averages as by seasonal changes, derived variables such as dry or wet spells, extremes, and other mechanisms that vary more with geography and physical climate. As an example, the change in maximum annual consecutive dry days is also shown in Figure 1.6. Changes in dry-spell duration generally, but not exactly, follow mean changes in precipitation. This illustrates another of the IPCC statements in the AR6 Summary For Policymakers, *"Many changes in the climate system become larger in direct relation to increasing global warming. They include increases in the frequency and intensity of hot extremes, marine heatwaves, and heavy precipitation, agricultural and ecological droughts in some regions, and proportion of intense tropical cyclones, as well as reductions in Arctic sea ice, snow cover and permafrost."* [6]

Figure 1.7 shows five common combinations of climate changes evident in CMIP6 projections for different regions of the world from IPCC AR6 WG1 [3]. In all cases, the temperature increases are notable relative to historical variability, but the impacts vary with changes in average and extreme precipitation, and these are sometimes at odds with each other. For example, some of the tropics (northern and northeast South America, eastern South Africa, all of Western Australia) and temperate regions (Mediterranean, western North America, northern Central America) fall into the "hotter and drier and, in some regions, wetter extremes." Preparing for a wider range of extremes may be warranted even if the trend in mean climate suggests a significant trend in one direction. Each of these regions has an accompanying general narrative of these changes, quantitative scenarios exist for each region, and downscaled projections are rapidly becoming available at the time of this writing. However, the detailed expression of uncertainty required for decision making is not immediately evident from such figures. It would be interesting, and perhaps useful, if such maps could rapidly be distilled into an estimate of the areas of the world most vulnerable to climate related harms. The problem with doing so is that large changes in physical climate alone do not necessarily equate to vulnerability. There are also many potential indices of vulnerability (economic, extinction/species, sociocultural), and a consensus on appropriate ranking of such impacts would be even more elusive than model consensus on changes in the physical climatology. Perhaps a way around this conundrum is to consider the mechanisms driving risk in different geographies, ecosystems, and economies around the planet. Climate

**Hotter and wetter extremes,
in some regions more precipitation
or fire weather**
*Examples: NWN, NEN, ENA, SSA, WCE, EEU, WSB,
ESB, RFE, WCA, ECA, EAS, NZ, GIC, RAR*

**Hotter, in some regions
wetter extremes
or more precipitation**
*Examples: PAC, SES, SAH, WAF,
CAF, MDG, ARP, SEA*

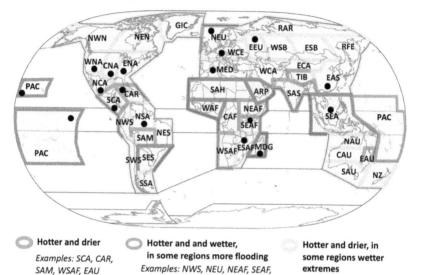

Hotter and drier
*Examples: SCA, CAR,
SAM, WSAF, EAU*

**Hotter and and wetter,
in some regions more flooding**
*Examples: NWS, NEU, NEAF, SEAF,
TIB, SAS*

**Hotter and drier, in
some regions wetter
extremes**
*Examples: WNA, NCA, SWS,
NSA, NES, MED, ESAF, N/C/SAU*

● **Increase in Tropical cyclone intensity or severe winds**

FIGURE 1.7 Changes in climatic impact-drivers will happen everywhere, but unique combinations of changes will affect each region. IPCC WG1 regions are grouped into five clusters, each one based on a combination of projected mid-21st changes for global 2°C in climatic impact drivers relative to 1960–2014. All coastal regions except North East North America (NEN) and Greenland/Iceland (GIC) will be exposed to at least two among increases in relative sea level, coastal flooding, and coastal erosion. Adapted from Figure TS22a, IPCC AR6 WG1 [3].

change results in increased ***exposure*** everywhere for outcomes with high ***sensitivity***, but which ones, how they change, and who bears the cost varies with impacts pathway and region (see, for example, the IPCC AR5 WGII report Part B, [32]). ***Adaptive capacity*** (in animals, people, their societies/economies) ultimately determines the impacts of increased exposure, what vulnerabilities exist, and whether adaptation that can decrease these vulnerabilities is within reach. Anywhere with a high rate of change and less adaptive capacity, the threat and risk are higher than places with only a high rate of change. Moreover, high rates of change may be systemically related to decreased adaptive capacity. High rates of change coupled with lower access to resources in lower or middle income countries or regions could result in higher vulnerability to many of the changes in regional climates that result from global climate changes described above.

Vulnerability to changes in physical climate may also be greater in specific change circumstances, such as where thresholds drive important processes and are resistant to climate changes until they shift rapidly to new conditions. This could apply equally to cryosphere (ice, snow, permafrost) changes that are robust to seasonal warming until conditions warm enough that they melt/thaw, or to marine systems where, once crossed, species with thermal restrictions or dissolved oxygen requirements are suddenly restricted or even killed. Similarly, and counterintuitively, slower changes in climate that repeatedly cross thresholds of temperature may be more impactful than rapid changes that cross a threshold briefly and keep changing, at least from a perspective of landscape change. For example, it is one thing to deal with a rapid and irreversible change in habitat conditions (say, a forest or woodland shift to shrub or grassland after disturbance), entirely another to deal with chronic disturbance that exceeds the historical conditions people and species adapted to but does not immediately result in wholesale ecosystem change. Contrasting RCP or SSP scenarios may result in such differences in the future, and planning for, much less managing in anticipation of them, is difficult given current scientific uncertainty. Where current vulnerability and divergent future trajectories can result in completely novel systems with which people have relatively sparse experience, the term "deep uncertainty" applies. Lempert et al. [33] define deep uncertainty as,

> *"when experts or stakeholders do not know or cannot agree on: (1) appropriate conceptual models that describe relationships among key driving forces in a system; (2) the probability distributions used to represent uncertainty about key variables and parameters; and/or (3) how to weigh and value desirable alternative outcomes."*

This definition is a technical way of describing what may seem evident to many grappling with climate impacts, namely that the changes are driven by global processes, but the adaptation to them must necessarily be local, or regional, to accommodate those stakeholders and their experience with the parts of a system that can be weighed and valued. This is no less true when considering the tremendous range of animal health outcomes around the world, and the contexts and stress histories that interact with climate changes to determine future conditions. Projecting trends, impacts, and risks associated with animal health and disease will require using climate science much as described in this chapter, but using it in planning, adaptation, and decision making will require collaboration and coproduction efforts that include this global-to-local texture in exposure and adaptive capacity.

ACKNOWLEDGMENTS

Any use of trade, firm, or product names is for descriptive purposes only and does not imply endorsement by the U.S. Government. Two anonymous reviewers provided helpful comments on an earlier version of the manuscript. I acknowledge the World Climate Research Programme, which, through its

Working Group on Coupled Modelling, coordinated and promoted CMIP5 and CMIP6. I thank the climate modeling groups for producing and making available their model output, the Earth System Grid Federation (ESGF) for archiving the data and providing access, and the multiple funding agencies who support CMIP6 and ESGF.

NOTES

1 Globally averaged temperature includes data from both land surface and ocean, but whether the ocean data are sea surface temperature or marine air temperature and methods for combining them vary. There is generally good agreement among these approaches, but see [2] for a discussion, and Vose et al. [34] for discussion relevant to data in Figure 1.1.
2 The IPCC uses standardized likelihood framing to indicate probabilities. "Very likely" indicates the evidence in favor of the statement suggests a 90–100% probability, and falls between "virtually certain," or 99–100% probability and "likely", or 66–100% probability. Low-likelihood indicators are inversely proportional, for example, "unlikely" is 0–33% probability Mastrandrea et al. [35].

REFERENCES

1. IPCC. Annex I: Glossary [Matthews, J.B.R. (ed.)]. In: Masson-Delmotte V, Zhai P, Pörtner HO, Roberts D, Skea J, Shukla PR, Pirani A, Moufouma-Okia W, Péan C, Pidcock R, Connors S, Matthews JBR, Chen Y, Zhou X, Gomis MI, Lonnoy E, Maycock T, Tignor M, Waterfield T, editors. Global Warming of 1.5°C. An IPCC Special Report on the impacts of global warming of 1.5°C above pre-industrial levels and related global greenhouse gas emission pathways, in the context of strengthening the global response to the threat of climate change, sustainable development, and efforts to eradicate poverty [Internet]. Geneva, World Meteorological Organization; 2018 [Cited 2021 Nov 12]. Available from: https://www.ipcc.ch/site/assets/uploads/sites/2/2019/06/SR15_AnnexI_Glossary.pdf
2. Gulev, S.K., Thorne, P.W., Ahn, J., Dentener, F.J., Domingues, C.M., Gerland, S., Gong, D., Kaufman, D.S., Nnamchi, H.C., Quaas, J., Rivera, J.A., Sathyendranath, S., Smith, S.L., Trewin, B., von Shuckmann, K. and Vose, R.S., 2021. Changing State of the Climate System. In: Masson-Delmotte, V., Zhai, P., Pirani, A., Connors, S.L., Péan, C., Berger, S., Caud, N., Chen, Y., Goldfarb, L., Gomis, M.I., Huang, M., Leitzell, K., Lonnoy, E., Matthews, J.B.R., Maycock, T.K., Waterfield, T., YelekçI, O., Yu, R. and Zhou, B. eds. *Climate Change 2021: The Physical Science Basis. Contribution of Working Group I to the Sixth Assessment Report of the Intergovernmental Panel on Climate Change.* Cambridge, United Kingdom and New York, NY, USA: Cambridge University Press, (In Press).
3. Arias, P.A., Bellouin, N., Coppola, E., Jones, R.G., Krinner, G., Marotzke, J., Naik, V., Palmer, M.D., Plattner, G-K, Rogelj, J., Rojas, M., Sillmann, J., Storelvmo, T., Thorne, P.W., Trewin, B., Achuta Rao, K., Adhikary, B., Allan, R.P., Armour, K., Bala, G., Barimalala, R., Berger, S., Canadell, J.G., Cassou, C., Cherchi, A., Collins, W., Collins, W.D., Connors, S.L., Corti, S., Cruz, F., Dentener, F.J., Dereczynski, C., Di Luca, A., Diongue Niang, A., Doblas-Reyes, F.J., Dosio, A., Douville, H., Engelbrecht, F., Eyring, V., Fischer, E., Forster, P., Fox-Kemper, B., Fuglestvedt, J.S., Fyfe, J.C., Gillett, N.P., Goldfarb, L.,

Gorodetskaya, I., Gutierrez, J.M., Hamdi, R., Hawkins, E., Hewitt, H.T., Hope, P., Islam, A.S., Jones, C., Kaufman, D.S., Kopp, R.E., Kosaka, Y., Kossin, J., Krakovska, S., Lee, J-Y, Li, J., Mauritsen, T., Maycock, T.K., Meinshausen, M., Min, S-K, Monteiro, P.M.S., Ngo-Duc, T., Otto, F., Pinto, I., Pirani, A., Raghavan, K., Ranasinghe, R., Ruane, A.C., Ruiz, L., Sallée, J-B, Samset, B.H., Sathyendranath, S., Seneviratne, S.I., Sörensson, A.A., Szopa, S., Takayabu, I., Treguier, A-M, van den Hurk, B., Vautard, R., von Schuckmann, K., Zaehle, S., Zhang, X. and Zickfeld, K., 2021. Technical Summary. In: Masson-Delmotte, V., Zhai, P., Pirani, A., Connors, S.L., Péan, C., Berger, S., Caud, N., Chen, Y., Goldfarb, L., Gomis, M.I., Huang, M., Leitzell, K., Lonnoy, E., Matthews, J.B.R., Maycock, T.K., Waterfield, T., Yelekçi, O., Yu, R. and Zhou, B., eds. *Climate Change 2021: The Physical Science Basis. Contribution of Working Group I to the Sixth Assessment Report of the Intergovernmental Panel on Climate Change.* Cambridge, United Kingdom and New York, NY, USA: Cambridge University Press.

4. NOAA National Centers for Environmental Information. Climate at a Glance: Global Time Series [Internet]. [Sep 2021, cited 2021 Sep 28]. Available from: https://www.ncdc.noaa.gov/cag/

5. Menne, M.J., Williams, C.N., Gleason, B.E., Jared Rennie, J. and Lawrimore, J.H., 2018. The global historical climatology network monthly temperature dataset, version 4. *Journal of Climate, 31*(24), pp. 9835–9854.

6. Intergovernmental Panel on Climate Change, 2021. Summary for Policymakers. In: Masson-Delmotte, V., Zhai, P., Pirani, A., Connors, S.L., Péan, C., Berger, S., Caud, N., Chen, Y., Goldfarb, L., Gomis, M.I., Huang, M., Leitzell, K., Lonnoy, E., Matthews, J.B.R., Maycock, T.K., Waterfield, T., Yelekçi, O., Yu, R. and Zhou, B. eds. *Climate Change 2021: The Physical Science Basis. Contribution of Working Group I to the Sixth Assessment Report of the Intergovernmental Panel on Climate Change.* Cambridge, United Kingdom and New York, NY, USA: Cambridge University Press.

7. Intergovernmental Panel on Climate Change, 2021. Climate Change 2021: The Physical Science Basis. Contribution of Working Group I to the Sixth Assessment Report of the Intergovernmental Panel on Climate Change. In: Masson-Delmotte, V., Zhai, P., Pirani, A., Connors, S.L., Péan, C., Berger, S., Caud, N., Chen, Y., Goldfarb, L., Gomis, M.I., Huang, M., Leitzell, K., Lonnoy, E., Matthews, J.B.R., Maycock, T.K., Waterfield, T., Yelekçi, O., Yu, R. and Zhou, B. eds. *Climate Change 2021: The Physical Science Basis. Contribution of Working Group I to the Sixth Assessment Report of the Intergovernmental Panel on Climate Change.* Cambridge, United Kingdom and New York, NY, USA: Cambridge University Press.

8. U.S. Global Change Research Program, 2017. Climate Science Special Report. In: Wuebbles, D.J., Fahey, D.W., Hibbard, K.A., Dokken, D.J., Stewart, B.C. and Maycock, T.K. eds. *U.S. Global Change Research Program.* Washington, DC, USA: p. 470. doi: 10.7930/J0J964J6.

9. U.S. Global Change Research Program, 2018. Impacts, Risks, and Adaptation in the United States: Fourth National Climate Assessment, Volume II. In: Reidmiller, D.R., Avery, C.W., Easterling, D.R., Kunkel, K.E., Lewis, K.L.M., Maycock, T.K. and Stewart, B.C. eds. *U.S. Global Change Research Program.* Washington, DC, USA: p. 1515. doi: 10.7930/NCA4.2018.

10. Fahey, D.W., Doherty, S.J., Hibbard, K.A., Romanou, A. and Taylor, P.C., 2017. Ch. 2: Physical drivers of climate change. In: Wuebbles, D.J., Fahey, D.W., Hibbard, K.A., Dokken, D.J., Stewart, B.C. and Maycock, T.K. eds. *Climate Science Special Report: Fourth National Climate Assessment*, Volume I. Available from: https://science2017.globalchange.gov/chapter/2/

11. Knutti, R. and Rugenstein, M.A.A., 2015. Feedbacks, climate sensitivity and the limits of linear models. *Philosophical Transactions of the Royal Society, A373*, p. 20150146. 10.1098/rsta.2015.0146

12. Lawrence, D.M., Hurtt, G.C., Arneth, A., Brovkin, V., Calvin, K.V., Jones, A.D., et al., 2016 Sep. The land use model intercomparison project (LUMIP) contribution to CMIP6: Rationale and experimental design. *Geoscientific Model Development*, 9(9), pp. 2973–2998.

13. Hayhoe, K., Edmonds, J., Kopp, R.E., LeGrande, A.N., Sanderson, B.M., Wehner, M.F. and Wuebbles, D.J.. Chapter 4: Climate Models, Scenarios, and Projections. In: Wuebbles, D.J., Fahey, D.W., Hibbard, K.A., Dokken, D.J., Stewart, B.C. and Maycock, T.K. eds. *Climate Science Special Report: Fourth National Climate Assessment*, Volume I. Washington, DC, USA: U.S. Global Change Research Program, pp. 133–160. doi: 10.7930/J0WH2N54.

14. Hausfather, Z., Drake, H.F., Abbott, T. and Schmidt, G.A., 2020. Evaluating the performance of past climate model projections. *Geophysical Research Letters*, 47(1), pp. e2019GL085378-e2019GL085378.

15. Palmer, T. and Stevens, B., 2019. The scientific challenge of understanding and estimating climate change. *Proceedings of the National Academy of Sciences*, 116(49), pp. 24390–24395.

16. Moss, R.H., Edmonds, J.A., Hibbard, K.A., Manning, M.R., Rose, S.K., Van Vuuren, D.P., et al., 2010. The next generation of scenarios for climate change research and assessment. *Nature 2010 463:7282*, 463(7282), pp. 747–756.

17. van Vuuren, D.P., Edmonds, J., Kainuma, M., Riahi, K., Thomson, A., Hibbard, K., et al., 2011. The representative concentration pathways: An overview. *Climatic Change*, 109(1), pp. 5–31.

18. Riahi, K., van Vuuren, D.P., Kriegler, E., Edmonds, J., O'Neill, B.C., Fujimori, S., et al., 2017. The Shared socioeconomic pathways and their energy, land use, and greenhouse gas emissions implications: An overview. *Global Environmental Change*, 42, pp. 153–168.

19. Gidden, M.J., Riahi, K., Smith, S.J., Fujimori, S., Luderer, G., Kriegler, E., et al., 2019. Global emissions pathways under different socioeconomic scenarios for Use in CMIP6: A dataset of harmonized emissions trajectories through the end of the century. *Geoscientific Model Development*, 12, pp. 1443–1475. 10.5194/gmd-12-1443-2019.

20. Rogelj, J., Popp, A., Calvin, K.V., Luderer, G., Emmerling, J., Gernaat, D., et al., 2018 Mar. Scenarios towards limiting global mean temperature increase below 1.5°C. *Nature Climate Change 2018 8:4*, 8(4), pp. 325–332.

21. United Nations Treaty Collection [Internet]. Paris agreement 2016. Cited 30 Sep 2021. Available from: https://treaties.un.org/Pages/ViewDetails.aspx?src=IND&mtdsg_no=XXVII-7-d&chapter=27&clang=_en

22. Tebaldi, C., Debeire, K., Eyring, V., Fischer, E., Fyfe, J., Friedlingstein, P., et al., 2021. Climate model projections from the scenario model intercomparison project (ScenarioMIP) of CMIP6. *Earth System Dynamics*, 12, pp. 253–293. 10.5194/esd-12-253-2021, 2021.

23. Meehl, G.A., Senior, C.A., Eyring, V., Flato, G., Lamarque, J.F., Stouffer, R.J., et al., 2020 Jun. Context for interpreting equilibrium climate sensitivity and transient climate response from the CMIP6 earth system models. *Science Advances [Internet]*, 6(26). Available from: https://www.science.org/doi/abs/10.1126/sciadv.aba1981

24. Tokarska, K.B., Stolpe, M.B., Sippel, S., Fischer, E.M., Smith, C.J., Lehner, F. and Knutti, R., 2020. Past warming trend constrains future warming in CMIP6 models. *Science Advances*, 6, pp. eaaz9549. 10.1126/sciadv.aaz9549, 2020.

25. Hawkins, E. and Sutton, R., 2009. The potential to narrow uncertainty in regional climate predictions. *Bulletin of the American Meteorological Society*, *90*(8), pp. 1095–1108.
26. Wootten, A., Terando, A., Reich, B.J., Boyles, R.P. and Semazzi, F., 2017. Characterizing sources of uncertainty from global climate models and down-scaling techniques. *Journal of Applied Meteorology and Climatology*, *56*(12), pp. 3245–3262.
27. Terando, A., Reidmiller, D., Hostetler, S.W., Littell, J.S., Jr. TDB, Weiskopf, S.R., et al., 2020. Using Information From Global Climate Models to Inform Policymaking—The role of the U.S. Geological Survey. *Open-File Report [Internet]*. Available from: 10.3133/ofr20201058.
28. Littell, J.S., McKenzie, D., Kerns, B.K., Cushman, S. and Shaw, C.G., 2011. Managing uncertainty in climate-driven ecological models to inform adaptation to climate change. *Ecosphere*, *2*(9), pp. art102–art102.
29. Snover, A.K., Mantua, N.J., Littell, J.S., Alexander, M.A., Mcclure, M.M. and Nye, J., 2013. Choosing and using climate-change scenarios for ecological-impact assessments and conservation decisions. *Conservation Biology*, *27*(6), pp. 1147–1157.
30. Knutti, R. and Sedláček, J., 2013. Robustness and uncertainties in the new CMIP5 climate model projections. *Nature Climate Change*, *3*(4), pp. 369–373.
31. Gutiérrez, J.M., Jones, R.G., Narisma, G.T., Alves, L.M., Amjad, M., Gorodetskaya, I.V., Grose, M., Klutse, N.A.B., Krakovska, S., Li, J., Martínez-Castro, D., Mearns, L.O., Mernild, S.H., Ngo-Duc, T., van den Hurk, B. and Yoon, J.H., 2021. Atlas. In: Masson-Delmotte, V., Zhai, P., Pirani, A., Connors, S.L., Péan, C., Berger, S., Caud, N., Chen, Y., Goldfarb, L., Gomis, M.I., Huang, M., Leitzell, K., Lonnoy, E., Matthews, J.B.R., Maycock, T.K., Waterfield, T., YelekçI, O., Yu, R. and Zhou, B., eds. *Climate Change 2021: The Physical Science Basis. Contribution of Working Group I to the Sixth Assessment Report of the Intergovernmental Panel on Climate Change*. Cambridge, United Kingdom and New York, NY, USA: Cambridge University Press. (In Press) Interactive Atlas available from Available from http://interactive-atlas.ipcc.ch/
32. Intergovernmental Panel on Climate Change, 2014. Climate Change 2014: Impacts, Adaptation, and Vulnerability. Part B: Regional Aspects. In: Barros, V.R., Field, C.B., Dokken, D.J., Mastrandrea, M.D., Mach, K.J., Bilir, T.E., Chatterjee, M., Ebi, K.L., Estrada, Y.O., Genova, R.C., Girma, B., Kissel, E.S., Levy, A.N., MacCracken, S., Mastrandrea, P.R. and White, L.L. eds. *Contribution of Working Group II to the Fifth Assessment Report of the Intergovernmental Panel on Climate Change*. Cambridge, United Kingdom and New York, NY, USA: Cambridge University Press, p. 688.
33. Lempert, R.J., Popper, S.W. and Bankes, S.C., 2003. *New Methods for Quantitative, Long-Term Policy Analysis [Internet]*. 1st ed. RAND Corporation. Available from: http://www.jstor.org/stable/10.7249/mr1626rpc
34. Vose, R.S., Huang, B., Yin, X., Arndt, D., Easterling, D.R., Lawrimore, J.H., et al., 2021. Implementing full spatial coverage in NOAA's global temperature analysis. *Geophysical Research Letters*, *48*(4), pp. e2020GL090873.
35. Mastrandrea, M.D., Field, C.B., Stocker, T.F., Ebi, K.L., Frame, D.J., Held, H., et al., 2010. Guidance Note for Lead Authors of the IPCC Fifth Assessment Report on Consistent Treatment of Uncertainties IPCC Cross-Working Group Meeting on Consistent Treatment of Uncertainties Core Writing Team [Internet]. *Intergovernmental Panel on Climate Change*. Available from: https://www.ipcc.ch/site/assets/uploads/2017/08/AR5_Uncertainty_Guidance_Note.pdf

2 Overview of Climate Change and Animal Health

Craig Stephen

CONTENTS

KEY LEARNING OBJECTIVES

- Be able to explain how health is the product of interacting individual, social, and environmental assets, deficits, and threats.
- Understand how the socio-ecological model of health can help to explore and identify multiple opportunities to work across the continuum of care in health management planning.
- Be able to describe the major themes of health-climate interactions for animal health, including the impacts of animal health on human community resilience.

DOI: 10.1201/9781003149774-2

IMPLICATIONS FOR ACTION

- Animal health management will need to involve a transdisciplinary team linking multiple types of knowledge producers and knowledge users to reduce vulnerability and enhance resilience by managing the determinants of health rather than only reacting to problems after they emerge.
- Prevailing uncertainties about the effects of climate change on animal health at a local level (due to data gaps and the complexity of climate-health interactions) requires attention and investment into building general resilience against multiple and unknown threats by protecting the determinants of animal health.
- Healthy animals are an important contributor of social, economic, and natural capital that lead to climate change resilience and therefore must be incorporated into climate change adaptation and resilience planning and programs.

WHAT IS ANIMAL HEALTH?

Society faces the challenge of preventing adverse health consequences and finding health benefits that will result from climate change. A series of questions arise as soon as we start planning how to manage animal health impacts from climate change. For whom are we managing animal health? Is it for the benefit of an individual animal, the persistence of a species, the protection of the services and values animals bring society, or the protection of people from animal-related threats? By what means, at what cost, and for how long do we act to protect animal health in the face of rapid and somewhat unpredictable effects of climate change? Which health threats do we worry about? To answer these questions, we first need to know what we mean by animal health and understand if others see health in the same way.

Optimal health seems to be an almost universal goal. Healthy individuals, healthy communities, and healthy ecosystems feature prominently in government policies, climate change action plans, and personal goals. However, polices, plans, and goals rarely clearly describe how to recognize health. We are more accustomed to describing when health is absent or when circumstances are not conducive to health and welfare. Efforts to develop a shared vision for animal health action can quickly be derailed when there is no clarity or consistency on how to recognize when the goal of health has been met.

Is Health in the Eye of the Beholder?

The definition of health is the subject of centuries-old debate. The combinations of capacities, attributes, and attitudes that define a healthy state have varied over time and between cultures, disciplines, and subdisciplines. Many pieces of biomedical, sociological, ecological, and individual knowledge can be gathered for the same animals to define health, but there is no guarantee that people will assemble the components in the same way. This allows more than one way to describe health, even for the same animals. For example, social conflicts and disagreements over health have been one of the biggest impediments to salmon aquaculture securing its social license to operate (1). Proponents see aquaculture contributing to healthy communities and ecosystems by reducing exploitation of wild stocks and contributing to food security and employment. Sophisticated veterinary care and animal husbandry are said to protect the salmon from the rigours of life in the wild and reduce the frequency and impacts of disease. Opponents see the potential transmission of infectious disease to free-ranging wild animals, crowding of fish in confined spaces that lead to disease, habitat alterations, and imposition on aboriginal rights combining to create an unhealthy situation.

Gunnerson (2) described five categories of health in the veterinary literature: (i) health as normality; (ii) health as biological function; (iii) health as homeostasis; (iv) health as physical and psychological well-being and (v) health as productivity including reproduction. These seemingly objective categories are subjective or context specific in that someone must establish thresholds and criteria for normality, well-being, or acceptable productivity. The combination of factors that lead to homeostasis will vary with species and environments, as will the individual, population, and ecosystem measurements we use to judge if an animal is healthy.

Our views of animal health are closely tied to how we regard the animal. For example, it would be rare to find a person who deemed their pet house cat to be terminally unhealthy because it was unable to produce offspring, but it would be common practice to cull a commercial dairy cow that was unable to conceive a calf and produce milk. We may work hard to protect peregrine falcons that use urban spaces as nesting sites but mount large-scale efforts to kill urban pigeons in these same places. Social factors influence our attitudes toward animals, including (i) the extent to which we are responsible for harm to them; (ii) the extent to which the harmed animals are under our stewardship; (iii) the severity of the problems that cause harms; and (iv) cultural and economic factors, including the popularity of the species (3). Regulations, legislation, and expectations for animal health, therefore, differ based on social values and uses of the species, whether farm animals, pets, laboratory animals, wild animals, or pests.

This chapter explores four ways animal health has been described. The first two look at health as a measurable biological state that is reflected either by an animal's diseases status or a degree of functioning the determines the extent of physical, social, and mental well-being. The next two viewpoints see health as a cumulative effect of capacities provided by individual, social, and environmental factors that provide the needs for daily living, allow the animals (or population) to cope with the stressors of daily living, and meet human expectations for how the animals exist and persist based on scientific or social criteria. Those two viewpoints see health not as an objective reality, but rather as a combination of socially interpreted factors and circumstances that allow animals to meet human expectations for them.

ANIMAL HEALTH AS THE ABSENCE OF DISEASE

A disease-centric preoccupation in animal health is pragmatically reasonable because disease-related outcomes, like epidemics, food safety threats, and trade barriers demand responses from managers. Disease events have the potential for significant economic, ecological, and social impacts and, therefore, can be the subject of significant regulatory and public attention. However, the definition of health as the absence of disease lags 80 years behind modern concepts of human health, does not reflect the focus on production outcomes found in livestock herd health, and is inconsistent with emerging models of wildlife health. Using absence of disease as the health standard does not recognize that a population can be deemed healthy (often based on measures of abundance, productivity, public safety, or profitability) but still have individual members that harbor disease-causing agents or are diseased. An absence of disease standard often does not define the threshold of dysfunction, disruption, or infection when an animal changes from being healthy to diseased along the clinical course from exposure to death or recovery. This standard most often refers only to infectious and parasitic diseases and does not consider in legislation noninfectious disorders such as neoplastic, endocrine, or nutritional diseases . Finally, this perspective defines health by what is dysfunctional and unacceptable rather than based on positive attributes, thus focusing attention on the means to treat and prevent hazard, risks, and harms, rather than protecting assets for health and well-being. This often leads to reactive programs to prevent or recover from an adverse event rather than proactive investment in keeping animals well.

Defining health based on the presence or absence of a selected disease or etiologic agents does not reflect health as living organisms experience it. Any living being is simultaneously challenged by multiple stressors, vulnerabilities, and threats and is benefiting from multiple assets. Its health is the combination of all these variables. Health, as experienced by a living organism, is a cumulative effect, not the absence of a single contributor, like a disease.

HEALTH AS A COMPLETE STATE OF WELL-BEING

The World Health Organization declared in 1948 that human health was not merely the absence of diseases but rather a state of complete physical, mental, and social well-being. This definition was applauded for recognizing the co-existing physical, mental, and social domains of health and the need to look at factors both internal and external to an organism to understand its health. The animal health world tended to adopt this perspective by adding the concept of welfare to the purview of health regulations or organizations (4). Animal welfare and health are allied concepts. Some people conceive health and welfare as normal functioning and freedom from disease. Others conceive welfare more as a sense of coherence between the capacity to identify, benefit, and use resources to deal with stress and the reality of current living conditions (5). This parallels the viewpoint that animal welfare is compromised when adaptations possessed by the animal make an imperfect fit to the challenges it faces in the circumstances in which it lives (6). Health and welfare, thus, both imply successful biological function, positive experiences, and freedom from adverse conditions.

Conceiving health as a complete state of physical, mental, and social well-being has been criticized on three fronts. First, the ideal of complete well-being in each of those domains would be hard to achieve and even harder to recognize under normal conditions. This means it would be hard for any animal to be deemed healthy. Second, it unintentionally medicalized health practice by expanding the scope of factors for medical practice to measure. This allowed health professionals to categorize more individuals as unhealthy most of the time because they were unable to attain full satisfaction of all physical, social, and environmental determinants of health (7). Third, our capacity to understand animal mental well-being is still in its formative stage, limiting our ability to assess all aspects of social and mental well-being.

HEALTH AS CAPACITY TO COPE WITH LIFE

The Ottawa Charter for Health Promotion (8) and the concept of salutogenesis introduced in the 1970s–1980s, emphasized health as the capacities and resources needed to adapt to, respond to, or control life's challenges and changes. Instead of asking why something is sick, this conception of health asks why an individual, group, or population stays well despite stressful situations and hardships (9). This approach assesses health status and inequities over the lifespan at the population level. It captures not only adverse outcomes like disease, but also the positive determinants of health. These determinants are provided by the individual and collective factors and conditions that enable an individual or population to be healthy. For people, these include income and social status, employment and working conditions, education and literacy, childhood experiences, physical environments, social supports and coping skills, healthy behaviors, access to health services, biological and genetic endowment, gender, and culture (10). Determinants of health, whether in domestic or wild animal populations, include individual level factors (e.g. genotype, demographics, immune function, body condition, behavior, social status), population-level factors (e.g. density, social structure, husbandry, predation), and environmental factors (e.g. climate, season, habitat quality, food, and water security), as well as socioeconomic and cultural factors (e.g. policies, cultural practices, values and beliefs, anthropogenic causes of mortality) (11) (Figure 2.1). Each theme of determinants is comprised of an interacting subset of factors, circumstances, or conditions. The specific variables used to describe each factor vary with the unique needs for each species or population.

This approach to health helps identify factors that affect an individual's or population's vulnerability before harms occur, without needing to rely on the

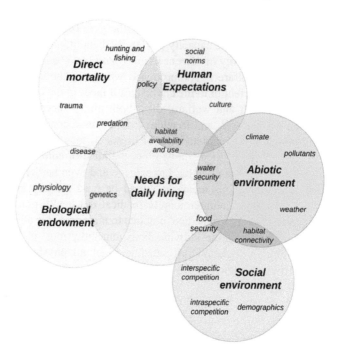

FIGURE 2.1   The determinants of health model for wildlife developed by Wittrock et al. (11) shows six main themes of internal and external factors determining wildlife and fish health (circles) and provides examples of individual factors within each theme that combine to establish the health status of an animal or population based on the cumulative effect of its assets and stressors.

occurrence of death or disease to signal the need for action. It not only helps conceive of the breadth of factors that increase susceptibility to disease, but also directs attention to the major drivers of population declines and extinctions found outside of the realm of pathogenesis, such as habitat loss, over exploitation, and climate effects (11). For example, the limitations on the ability to meet the needs for daily living, deal with stressors, and meet human expectations for a community of fish in a coral reef may be more at risk from reef destruction due to ocean warming, overfishing, and introduction of alien species than from any fish disease. A fish health program that fails to address the non-disease determinants of health will not be addressing the main factors limiting the capacity of those fish to flourish, free of obstacles, in a way that conforms with their evolved needs and our social and scientifical expectations.

A Combined Perspective of Health

Health exists on a continuum ranging from optimal health and ending with death. Health management, therefore, requires a continuum of care (Figure 2.2). Death, at one end of the continuum, is a part of life and its occurrence is not a failure of

CONTINUUM OF CARE

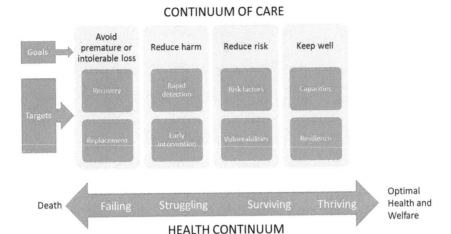

FIGURE 2.2 Matching the continuum of animal health care with the health continuum.

health care. But premature, excessive, or inhumane death are to be avoided. Optimal health and welfare, at the other end of the continuum, is the unimpaired flourishing, free of obstacles, to live in a way that conforms with opportunities, capacities, and social and scientific expectations. In between, we find animals with inadequate access to determinants of health; those being exposed to risk factors that may transition into disease, deformity, disability, or reduced productivity; those afflicted by the latter, and those struggling to survive in the face of life's challenges. The continuum of animal health care parallels these stages. Climate change health actions should match where we find an animal on the health continuum along its life course (Figure 2.2). The relative investment made in treating problems or aiding in recovery versus promoting and protecting circumstances that ensure access to the determinants of health will be dictated by program objectives, population status, social priorities, and available knowledge and resources. Health stewardship in the face of climate change means promoting a continuum of care that will prevent anticipated impacts, resist unanticipated impacts, and ensure recovery without persistent and irreversible harms (12).

Animal health is the balance between the assets that the determinants of health provide and the deficits that hazards and challenges present. It can be thought of as the consequence of reciprocal causation unfolding at multiple individual and environmental levels of influence (13). An integrated model of health (Figure 2.3) accommodates the need to: (i) incorporate a wider suite of social, biological, and ecological information into health assessment and management, where needed and appropriate; (ii) consider a wider variety of health management options; (iii) be more alert to unanticipated impacts on health outcomes due to relationships between determinants of health and risk factors; (iv) make better use of multifaceted management strategies along the continuum of care; and (v) consider the effectiveness of health management from a wider perspective.

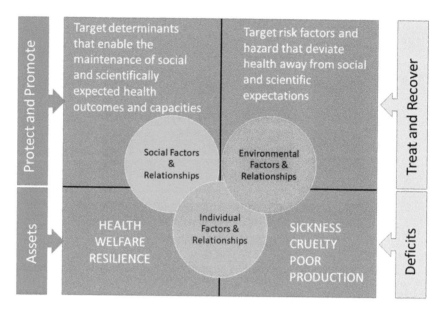

FIGURE 2.3 An integrated model of health that considers the interplay between deficits that deviate animals away from a desired health status with assets for health provided by the determinants of health. The balance of health assets and deficits are determined by the interplay of social, environmental, and individual factors.

The idea that health is affected by the interaction between the characteristics of the individual, the population, and the physical, social, and political environments is known as the socio-ecological model of health. The socio-ecological model emphasizes that health is a product not only of individual factors but also on the multilevel, interacting factors that influence behaviors and decisions that affect the vulnerability to threats and resilience in a changing world. Take for example the relationship between climate change, poverty, and the spread of Japanese encephalitis in mountainous areas of Nepal (based on 14–16) (Figure 2.4). This zoonotic

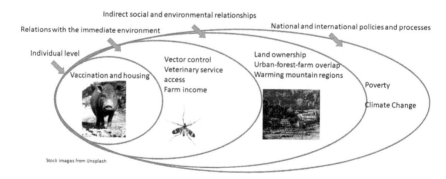

FIGURE 2.4 A simplified example of the socio-ecological model of health. This illustration of spread of Japanese encephalitis in Nepal shows the multiple individual, social, and environmental factors interacting in a nested hierarchy of causal relationships.

mosquito-borne disease is a viral infection of birds, which spills over into pigs, which then serve as an amplifying host that increase human risk of exposure. Swine and human vaccination, as well as farm housing to reduce swine and human mosquito exposures, can control this disease. But in Nepal, many pig farmers are extremely poor. They often lack money to build barns to house pigs and homes for their families inside away from mosquitoes or to regularly vaccinate their animals or family. Policies and cultural legacies that prevent poor farmers from owning land creates a quasi-nomadic population of farmers that move from lot to lot in peri-urban areas. When those farms are near rice paddy fields and forests that allow ready mixing of birds, mosquitoes, pigs, and people, infection risks can increase. The lack of veterinary infrastructure compounds problems in creating farmer awareness of options for disease control. Poverty can limit access to veterinary services and mosquito avoidance materials such as window screens. As climate change advances and land-use policies change, pig farming is expanding into new areas, higher in the mountains. The warming climate allows mosquitos to also move into previously colder areas, thereby spreading the disease into new regions. This simplified example illustrates how multiple, interacting, and embedded levels of causes affect the course of a disease and a community's ability to manage it. Ecological, climatological, individual, and social factors all play some role in the spread and impact of Japanese encephalitis in this circumstance. The interplay of health literacy, income, rural development planning, access to health care for people and animals, vector ecology in the face of climate change, and poverty reduction all affect the likelihood that a disease control program that is effective, locally feasible, acceptable, and sustainable can be successfully conceived and implemented.

CLIMATE CHANGE AND HEALTH

The idea that health is linked to climate predates written history. Climate disaster and seasonal trends in health conditions, such as starvation, allergies, and infections, are recounted in most cultures from far back in time until today (17). Our current dilemma is that climate warming in this century is not only happening faster than in the past but is going to exceed natural multidecadal fluctuations (18). This will challenge the ability of individuals, populations, food production systems, and ecosystems to adapt. The world is now more interconnected, more densely populated and impacted by people, has less diverse and abundant biodiversity, and is, therefore, less flexible than in the past, making this a time of unprecedented vulnerability. The Intergovernmental Panel on Climate Change (IPPC) 5th Assessment Report reminded us that climate change and associated human health impacts are inevitable. Climate change impacts on wild and domestic animal health are similarly inevitable, even under the most optimistic scenarios.

CLIMATE CHANGE AND THE DETERMINANTS OF HEALTH

Climate change directly and indirectly affects the determinants of health in multiple, interacting ways, across a range of scales (19). Climate change effects on health can be compound, indirect, and sometimes ambiguous, making study, predictions, and preparation for the total dynamics often problematic (20). Most attention in the animal health literature has focused on how climate change will impact infectious diseases, especially vector-borne disease (12). Experience in human health to date is suggesting that noninfectious diseases, such as heat stress, malnutrition, and impacts on respiratory and cardiovascular diseases, are creating the greatest health impacts (21). This situation gives rise to recommendations for animal health professionals to explore broad areas of inquiry when investigating animal health impacts from climate change. Maintaining a focus on infectious diseases at the exclusion of other determinants of health will severely restrict the resources and attention needed to protect animal health and safeguard the planet in the face of a rapidly changing climate.

Direct effects of climate change will result from shifting patterns and extremes of temperature and precipitation that result in dangerous weather (ex. hurricanes, droughts, heat waves). Indirect effects will arise when ecosystems are damaged or disrupted to the point that they can no longer provide or sustain life supporting functions such as the provision of clean air and safe, abundant food and water. Ecosystem effects and changes in weather can also affect the presence, distribution, and amounts of pathogens or pollutants in an environment. Societies' responses to climate change can indirectly affect animal health through changes to animals' access to their determinants of health, such as competition for land uses, changing food systems, human settlement, and resettlements, or income shifts that reduce financial resources to spend on animal care. Climate-associated impacts on animal health, in turn, affect human health by changing exposures to zoonotic pathogens and foodborne contaminants, reducing food security, limiting income opportunities from animals or animal products, or affecting mental and cultural wellness.

The IPPC proposed that for the foreseeable future, climate change will mainly exacerbate pre-existing health problems (22). High on their list of anticipated health effects were injury, disease, and death due to more intense heat waves and fires, diminished food production, and increased risks of food-, water-, and vector-borne diseases. Altered global food production and consumption patterns, diminished availability and quality of ecological services, and unanticipated threats must also be accounted for when anticipating the animal health implications of climate change.

WHAT IS EXPECTED FOR ANIMAL HEALTH IMPACTS?

Fewer climate change studies have been done on animal health compared to human health, creating uncertainties about anticipated animal health effects (23). Details on the types, mechanism, and magnitudes of harms anticipated to impact animal health are provided in subsequent chapters. What follows is a brief introduction to some of the most anticipated animal health effects of climate change.

1. Infectious diseases
   a. Vector-borne, water-borne, wind-borne, and enteric infections are expected to increase due to changes in food webs, timing of life-cycles, transmission pathways, and food habits (24). There is a rapidly growing literature on climate impacts on host-pathogen relationships, and on the emergence and dynamics of infectious diseases. Predicting the patterns of change in infectious disease epidemiology will be complicated by the complex relationships between ecological change, social change, microbial ecology, host susceptibility, and exposure pathways, each of which will be modified by climate change.

2. Heat stress
   a. Rising temperatures will cause suffering and premature death, reduced productivity and fertility (24), metabolic alterations, oxidative stress (25), and altered immune and endocrine systems, thereby enhancing disease susceptibility (26). Increased temperatures in marine and aquatic systems threaten fisheries and aquaculture. Climate change can decrease water quality, make harmful algae blooms more frequent, decrease water oxygen levels, and reduce feeding and growth, all of which can affect disease susceptibility (27).

3. Extreme weather
   a. The number of natural disasters worldwide has more than quadrupled in the last 50 years, creating an urgency to ensure preparedness, mitigation, response, and resilience to them (28). Typhoons, hurricanes, floods, fires, and drought have animal health and welfare implications that in turn will have public health and economic repercussions. Extreme events can alter and destabilize critical wildlife habitat or domestic animal housing, directly kill animals, and make food and water unsafe or unavailable. Citizens' desires to rescue or protect animals in the face of extreme weather events can risk public safety and impact mental health.

4. Contaminants
   a. Climate change alterations to food webs, lipid dynamics, ice and snow melt, and organic carbon cycling will affect pollutant levels in water, soil, air, plants, and animals. Flooding and melting events will remobilize contaminants and redistribute them onto grazing lands, thus contaminating animals and animal products (29). Changing patterns of insect pests are expected to change when and how much pesticide enters the environment (29). Increased production, frequency, and distribution of mycotoxins (30) and toxic algae (31) have implications for both animal health and food safety in terrestrial and aquatic systems. There is compelling evidence that increasing temperatures may alter the biotransformation of contaminants to more bioactive metabolites (32), and there may be synergistic effects of contaminants with parasitic or infectious diseases.

5. Effects on other determinants of health
   a. Climate changes will influence how wild and domestic animals access their needs for daily living and how management decisions will need to change to adapt to these changing determinants of animal health. Oceans are acidifying. Desertification is occurring in some regions, while ice and snow loss are affecting others. Droughts are leading to crop failures and fires. There will be a reduction in water resources in most dry subtropical regions, and the frequency of droughts will likely increase, intensifying competition for water. The annual cycles of plants and animals are changing, affecting food webs and disease cycles.

ANIMAL HEALTH AND SOCIETAL CLIMATE CHANGE RESILIENCE

Communities become more vulnerable to the effects of climate change when they are highly connected to or in contact with a climate induced threat, are more sensitive or prone to being harmed by that threat, or have little capacity to adapt to the new situation (20). A community's climate change resilience can be thought of as the product of economic, social, and environmental (or natural) capital that facilitate its coping capacity. Resilience is greater when all three capitals are well developed, while it is weaker when one or two of these capitals are under-developed or absent (20). Animals are integral parts of human society and of coupled human and natural systems, and thus are important sources of economic, social, and environmental capital (32).

Data on the role of animals in maintaining and restoring human resilience is scant despite animals being a consistent part of the human experience for millennia. Services that humanity derives from animals include provisioning (e.g. production of food and income), regulating (e.g. control of disease and pests), supporting (e.g. nutrient cycles), and cultural (e.g. spiritual and recreational benefits). Most discussion on the roles of animals in community resilience have historically focused on their contributions to economic capital (through animal related jobs and animal production) and environmental capital (mostly as food sources). For example, through their pest control activities, bats save agriculture and forestry economies billions of dollars per year as well as help reduce pesticide resistance and unanticipated occupational safety risks due to pesticide use (33). Similarly, the economic dependence of the USA industrial sectors on animal-mediated pollination service was estimated at US$10.3–21.1 billion in 2015 (34). Hunting, trapping, and fishing not only contribute to culture, food security, and the quality of life for many communities around the globe, but they also generate billions in community revenue or savings annually. Wild animal meat represents an important source of protein for many people, particularly in rural, indigenous, and vulnerable communities (35). The overall benefit:cost ratio of an effective global conservation program has been estimated at least 100 :1 (36). The United Nations Food and Agricultural Organization (FAO) highlighted the reliance of food systems on biodiversity and the continual decline of biodiversity within the agricultural sector. This lack of diversity compromises agroecosystems' resilience against climate change, pests, and pathogens (37). Climate-induced shocks can deplete livestock assets which, for many poor people, means collapsing into chronic poverty with long-term effects on their livelihoods or ability to climb up the poverty ladder (38).

Animals benefit people through more than just their contribution to material welfare and livelihoods. For example, many Indigenous communities' sense of identity and belonging is linked to their historic and ongoing use of wildlife for work, recreation, and spiritual purposes. Empowering communities to protect and promote animal welfare and health improves livelihoods while building resilience to emerging threats (39). People's ability to face and manage adverse conditions, emergencies, or disasters is promoted by nature (40). Failure to direct attention toward animal determinants of community resilience is a risky strategy, especially, but not exclusively, for animal-dependent communities.

The health and resilience benefits from animals do not fall on all people equally because the circumstances of our lives create inequalities in our access to the sources of health found in nature and animals. Their relative contribution to individual and community resilience can vary over a life course and across geographic locations. Because there is no universal definition of community resilience and because animal contributions to resilience can be modified by social determinants, causal attributions of specific animal contributions to resilience outcomes can be difficult to identify. The relationship between environmental hazards (i.e. a zoonotic pathogen) or environmental deficits (i.e. fishery declines) seem more straightforward to track and have more traditionally been components of nonhuman indicators for human well-being (41). Critics of typical indices that look at animal-associated harms or deficits suggest that community resilience is better served by viewing the environment and animals as a source of positive determinants of health and enabling factors for well-being rather than as a constraining variable (42).

Community climate change resilience is built by protecting community health and wellness, ensuring community members can access and use their economic, social, and natural capital and be self-sufficient because they have options to transform in times of change (43). Examples and evidence about the influences of human-animal-environmental health interdependence on climate change vulnerability and resilience are found throughout this book.

THE CHALLENGE OF MEASURING HEALTH VULNERABILITY

Vulnerability, put simply, is the result of the interplay between factors that put individuals, populations, and/or places at risk and factors that reduce the ability to respond to threats. Vulnerability to the climate change health impacts varies widely by species, location, and management system but is ultimately determined by exposure to the effects of climate change and capacity to adapt to or cope with those effects (44). Populations most vulnerable to climate change's health risks will be those facing the largest pre-existing health burden and those with the least capacity to adapt.

Departing from this simple conception of vulnerability brings a diversity of definitions and perspectives that have evolved over time and vary by discipline and interests. There is a rich and growing literature on climate change vulnerability. Although that literature is anthropocentrically biased, there is growing interest and investment in assessing the climate change vulnerability of wild and domestic animals. Most climate change and health vulnerability assessments focus on four main questions; (i) is the population going to be exposed to the threat with sufficient intensity and frequency to result in a harm (exposure assessment); (ii) is the population prone to be affected by the threat due to its susceptibility and the influences of concurrent stressors (sensitivity assessment); (iii) what is the nature and significance of effects if adequate exposure and sensitivity exists (impact assessment), and (iv) can the population respond to or adjust to the impact to lessen its harm(s) (adaptive capacity assessment)? The details of how each of these assessments are conducted and combined is the subject of ongoing research and debate.

There is a rapidly growing number of climate change vulnerability indices. They are used to communicate about climate change, select species or places to be prioritized for management, inform management decisions, identify monitoring needs, and inform adaptation decisions. There is not, however, a universally accepted way to assess climate change vulnerability consistent across circumstances and species. The relative contribution of the different vulnerability determining factors vary over a life course, across geographic locations, between species and with varying environmental and social circumstances. Our ability to understand and predict the outcome of climate changes is hampered if we consider change and stressors disconnected from dynamic interactions with local social and environmental factors (45). This situation creates significant challenge in coming to agreement on how best to measure and assess vulnerability.

There are four question to ask before adopting a vulnerability index for your use (46). First, are the objectives of the index stated, and do they align with your objectives? Second, the vulnerability of what (ex. is it a local population, a species, a farming system, and/or an animal use) and to what (ex. is it to extreme weather, infectious diseases, or multiple hazards) is being considered? Third, can the proposed measurement criteria be assessed? Are there the methods and reliable data available, and are the impacts of measured factors on vulnerability known? Fourth, who is involved in the vulnerability assessment? The socioecological model of health discussed above highlighted the influence of human judgments on assessing health. Values, expectations, and other social constructs can affect how available health and vulnerability data and information are assessed; therefore, who is "at the table" to make these judgments will influence the outcomes of vulnerability assessments.

Identifying thresholds for unacceptable change or securing consensus on what constitutes vulnerability are context dependent undertakings informed by local priorities, perspectives, and capacities. While there are accepted thresholds for a wide variety of environmental harms (ex. water quality standards, air pollution guidelines, food safety standards), there remains considerable debate and uncertainty on thresholds that indicate important changes in health vulnerabilities. "Analysis of thresholds is complicated by nonlinear dynamics and by multiple factor controls that operate at diverse spatial and temporal scales" (47). This necessitates intensive and expensive site-specific investigations to determine meaningful thresholds, precluding the selection of generic thresholds that can be applied across various landscapes, species, and times. Because the links between health and climate can be both direct and indirect, displaced in space and time, and influenced by a range of moderating forces, many of the standard health methods for establishing cause-effect relationships cannot be applied. Animal health workers cannot isolate their climate change response from their responses to other threats. Climate change is part of a larger syndrome of systemic environmental and social changes. It must be put into the broader context of other large-scale changes, such as land-use changes, habitat loss, and loss of biodiversity (48). Deciding if something is vulnerable to climate change is a complicated and context-rich task.

SUMMARY

Health is a multidimensional capacity that is a cumulative effect of interacting assets, deficits, and threats that affect an organism's or population's ability to thrive in a way that conforms with opportunities, capacities, and social and scientific expectations. This multidimensionality creates innumerable ways that climate change can influence health outcomes through its impacts on interacting determinants of health.

Although the ultimate goals of preventing disease, maintaining productivity, and sustaining systems that promote health will remain unchanged, the form and scope of animal health research and services need to change. The bulk of information on animal health impacts deal with infectious diseases and to a lesser extent impacts of extreme weather and fires, but growing experience and expectations require a wider interest in noncommunicable diseases and the relationship of animal health with social and ecological resilience.

The components of effective adaptation to the future will be unpredictable and emergent rather than predictable and planned (49) because of the unprecedent rate of social and environmental change and the complexity of interactions between co-occurring global threats and anthropogenic climate change. Animal health programs will need to shift from an almost exclusive focus on diseases and harms to an animal health focus if they are to mitigate as well as adapt to climate change. This shift will require strategic partnerships with a breadth of disciplines that can influence animal determinants of health as well as the social systems that influence them. This will necessitate thinking of animal health and climate change not only from risk management to but also capacity-building for healthy, resilient animal populations and animal health systems.

REFERENCES

1. Stephen, C. and Wade, J., 2019. Testing the waters of an aquaculture index of well-being. *Challenges*, *10*(1), p. 30.
2. Gunnarsson, S., 2006. The conceptualisation of health and disease in veterinary medicine. *Acta Veterinaria Scandinavica*, *48*(1), pp. 1–6.
3. Kirkwood, J.K. and Sainsbury, A.W., 1996. Ethics of interventions for the welfare of free-living wild animals. *Animal Welfare* Societies in *Potters Bar*, *5*, pp. 235–244.
4. Nicks, B. and Vandenheede, M., 2014. Animal health and welfare: Equivalent or complementary. *Revue scientifique et technique (International Office of Epizootics)*, *33*, pp. 97–101.
5. Stephen, C. and Wade, J., 2018. Wildlife population welfare as coherence between adapted capacities and environmental realities: A case study of threatened lamprey on Vancouver Island. *Frontiers in Veterinary Science*, *5*, p. 227.
6. Fraser, D., Weary, D.M., Pajor, E.A. and Milligan, B.N., 1997. A scientific conception of animal welfare that reflects ethical concerns. *Animal Welfare*, *6*(3), pp. 187–205.
7. Huber, M., Knottnerus, J.A., Green, L., van der Horst, H., Jadad, A.R., Kromhout, D., Leonard, B., Lorig, K., Loureiro, M.I., van der Meer, J.W. and Schnabel, P., 2011. How should we define health?. *BMJ*, *343*, p. d4163. doi: 10.1136/bmj.d4163.
8. World Health Organization. "The Ottawa Charter for Health Promotion." *World Health Organization*. WHO, 1986. https://www.who.int/healthpromotion/conferences/previous/ottawa/en/ Last accessed Feb3, 2022.
9. Eriksson, M. and Lindström, B., 2006. Antonovsky's sense of coherence scale and the relation with health: a systematic review. *The Journal of Epidemiology and Community Health*, *60*(5), pp. 376–381.
10. World Health Organization, 2020. The Social Determinants of Health. https://www.who.int/health-topics/social-determinants-of-health#tab=tab_1. Last accessed Feb 3, 2022
11. Wittrock, J., Duncan, C. and Stephen, C., 2019. A determinants of health conceptual model for fish and wildlife health. *Journal of Wildlife Diseases*, *55*(2), pp. 285–297.
12. Stephen, C. and Wade, J., 2020. Missing in action: Sustainable climate change adaptation evidence for animal health. *Canadian Veterinary Journal*, *61*(9), p. 966.
13. Rock, M.J., Adams, C.L., Degeling, C., Massolo, A., & McCormack, G.R., 2015. Policies on pets for healthy cities: A conceptual framework. *Health Promotion International*, *30*(4), pp. 976–986. 10.1093/heapro/dau017
14. Dhakal, S., Stephen, C., Ale, A. and Joshi, D.D., 2012. Knowledge and practices of pig farmers regarding Japanese encephalitis in Kathmandu, Nepal. *Zoonoses and Public Health*, *59*(8), pp. 568–574.
15. Dhakal, S., Joshi, D.D., Ale, A., Sharma, M., Dahal, M., Shah, Y., Pant, D.K. and Stephen, C., 2014. Regional variation in pig farmer awareness and actions regarding Japanese encephalitis in Nepal: Implications for public health education. PloS One, *9*(1), p. e85399.
16. Robertson, C., Pant, D.K., Joshi, D.D., Sharma, M., Dahal, M. and Stephen, C., 2013. Comparative spatial dynamics of Japanese encephalitis and acute encephalitis syndrome in Nepal. (7), p. e66168.
17. Bell, M. and Greenberg, M.R., 2018. Climate change and human health: links between history, policy, and science. *The American Journal of Public Health*, *105*, pp. S54–S55. 10.2105/AJPH.2018.304437
18. McMichael, A.J., 2012. Insights from past millennia into climatic impacts on human health and survival. *Proceedings of the National Academy of Sciences of the United States of America*, *109*(13), pp. 4730–4737.

19. Black, P.F. and Butler, C.D., 2014. One Health in a world with climate change. *Revue scientifique et technique*, *33*, pp. 465–473.
20. Kais, S.M. and Islam, M.S., 2016. Community capitals as community resilience to climate change: Conceptual connections. *The International Journal of Environmental Research and Public Health*, *13*(12), p. 1211.
21. Confalonieri, U., Menne, B., Akhtar, R., Ebi, K.L., Hauengue, M., Kovats, R.S., Revich, B. and Woodward, A., 2007. Human health. Climate Change 2007: Impacts, Adaptation and Vulnerability. In: Parry, M.L., Canziani, O.F., Palutikof, J.P., van der Linden, P.J. and Hanson, C.E. eds., *Contribution of Working Group II to the Fourth Assessment Report of the Intergovernmental Panel on Climate Change*. Cambridge, UK: Cambridge University Press, pp. 391–431.
22. Smith, K., Woodward, A., Campbell-Lendrum, D., Chadee, D., Honda, Y., Liu, Q., Olwoch, J., Revich, B., Sauerborn, R., Aranda, C. and Berry, H., 2014. Human Health: Impacts, Adaptation, and Co-Benefits. In Climate Change 2014: Impacts, Adaptation, and Vulnerability. Part A: Global and Sectoral Aspects. *Contribution of Working Group II to the fifth assessment report of the Intergovernmental Panel on Climate Change*. Cambridge, UK: Cambridge University Press, pp. 709–754.
23. Lubroth, 2012. Climate Change and Animal Health in 2012. *Building Resilience for Adaptation to Climate Change in the Agriculture Sector: Proceedings of a Joint FAO/OECD Workshop 23–24 April*. Meybeck et al eds. FAO Inter-Departmental Working Group.
24. Forman, S., Hungerford, N., Yamakawa, M., Yanase, T., Tsai, H.J., Joo, Y.S., Yang, D.K. and Nha, J.J., 2008. Climate change impacts and risks for animal health in Asia. *Revue scientifique et technique (International Office of Epizootics)*, *27*, pp. 581–597.
25. Lacetera, N., 2019. Impact of climate change on animal health and welfare. *Animal Frontiers*, *9*(1), pp. 26–31.
26. Das, R., Sailo, L., Verma, N., Bharti, P. and Saikia, J., 2016. Impact of heat stress on health and performance of dairy animals: A review. *Vet World*, *9*(3), pp. 260.
27. Handisyde, N.T., Ross, L.G., Badjeck, M.C. and Allison, E.H., 2006. The effects of climate change on world aquaculture: a global perspective. Aquaculture and Fish Genetics Research Programme, Stirling Institute of Aquaculture. *Final Technical Report, DFID, Stirling*. p. 151.
28. Gallagher, C., Jone B. and Tickel J., 2020. Towards Resilience: The One Health Approach in Disasters. In: Zinsstag, J., Schelling, E., Crump, L., Whittaker, M., Tanner, M. and Stephen, C. eds., *One Health: The Theory and Practice of Integrated Health Approaches*. Wallingford, UK: CAB International.
29. Tirado, M.C., Clarke, R., Jaykus, L.A., McQuatters-Gollop, A. and Frank, J.M., 2010. Climate Change and Food Safety: A Review. *Food Research Intl*, *43*(7), pp. 1745–1765.
30. Van der Fels-Klerx, H.J., Liu, C. and Battilani, P., 2016. Modelling climate change impacts on mycotoxin contamination. *World Mycotoxin J*, *9*(5), pp. 717–726.
31. Griffith, A.W. and Gobler, C.J., 2020. Harmful algal blooms: A climate change co-stressor in marine and freshwater ecosystems. *Harmful Algae*, *91*, p. 101590.
32. Noyes, P.D., McElwee, M.K., Miller, H.D., Clark, B.W., Van Tiem, L.A., Walcott, K.C., Erwin, K.N. and Levin, E.D., 2009. The toxicology of climate change: Environmental contaminants in a warming world. *Environment International*, *35*(6), pp. 971–986.
33. Dietz, T. and York, R., 2015. Animals, capital and sustainability. *Human Ecology Review*, *22*(1), pp. 35–54.
34. Boyles, J.G., Cryan, P.M., McCracken, G.F. and Kunz, T.H., 2011. Economic importance of bats in agriculture. *Science*, *332*(6025), pp. 41–42.

35. Chopra, S.S., Bakshi, B.R. and Khanna, V., 2015. Economic dependence of US industrial sectors on animal-mediated pollination service. *Environmental Science Teachers*, *49*(24), pp. 14441–14451.

36. Wyatt T., 2013. The security implications of the illegal wildlife trade. *J Social Criminology*. pp. 130–158.

37. Balmford, A., Bruner, A., Cooper, P., Costanza, R., Farber, S., Green, R.E., Jenkins, M., Jefferiss, P., Jessamy, V., Madden, J. and Munro, K., 2002. Economic reasons for conserving wild nature. *Science*, *297*(5583), pp. 950–953.

38. IPBES, 2019. Summary for policymakers of the global assessment report on biodiversity and ecosystem services of the Intergovernmental Science-Policy Platform on Biodiversity and Ecosystem Services. In: Díaz, S., Settele, J., Brondízio, E.S., Ngo, H.T., Guèze, M., Agard, J., Arneth, A., Balvanera, P., Brauman, K.A., Butchart, S.H.M., Chan, K.M.A., Garibaldi, L.A., Ichii, K., Liu, J., Subramanian, S.M., Midgley, G.F., Miloslavich, P., Molnár, Z., Obura, D., Pfaff, A., Polasky, S., Purvis, A., Razzaque, J., Reyers, B., Roy Chowdhury, R., Shin, Y.J., Visseren-Hamakers, I.J., Willis, K.J., and Zayas, C.N. eds.. *IPBES secretariat*, Bonn, Germany. p. 56.

39. Thornton, P.K., Herrero, M.T., Freeman, H.A., Okeyo Mwai, A., Rege, J.E.O., Jones, P.G. and McDermott, J.J., 2007. Vulnerability, climate change and livestock-opportunities and challenges for the poor. *Journal of Semi-Arid Tropical Agricultural Research*. *41*. 0000-0002-1854-0182

40. Njisane, Y.Z., Mukumbo, F.E. and Muchenje, V., 2020. An outlook on livestock welfare conditions in African communities—A review. *Asian-Australasian J Animal Sci*, *33*(6), p. 867.

41. Keim, M.E., 2008. Building human resilience: The role of public health preparedness and response as an adaptation to climate change. *Am J Prev Med*, *35*(5), pp. 508–516.

42. Kjellstrom, T., Friel, S., Dixon, J., Corvalan, C., et al. 2007. Urban environmental health hazards and health equity. *Journal of Urban Health*, *84*(1), pp. 86–97.

43. Stephen, C. and Duncan, C., 2017. Can wildlife surveillance contribute to public health preparedness for climate change? A Canadian perspective. *Climatic Change*, *141*(2), pp. 259–271.

44. Wulff, K., Donato, D. and Lurie, N., 2015. What is health resilience and how can we build it? *The Annual Review of Public Health*, *36*, pp. 361–374.

45. Yohe, G.W., 2001. Mitigative capacity-The mirror image of adaptive capacity on the emissions side. *Climatic Change*, *49*(3), p. 247.

46. Bozelli, R.L., Caliman, A., Guariento, R.D., Carneiro, L.S., Santangelo, J.M., Figueiredo-Barros, M.P., Leal, J.J., Rocha, A.M., Quesado, L.B., Lopes, P.M. and Farjalla, V.F., 2009. Interactive effects of environmental variability and human impacts on the long-term dynamics of an Amazonian floodplain lake and a South Atlantic coastal lagoon. *Limnologica*, *39*(4), pp. 306–313.

47. Preston, B.L., Yuen, E.J. and Westaway, R.M., 2011. Putting vulnerability to climate change on the map: A review of approaches, benefits, and risks. *Sustainability Science*, *6*(2), pp. 177–202.

48. Groffman, P.M., Baron, J.S., Blett, T., Gold, A.J., Goodman, I., Gunderson, L.H., Levinson, B.M., Palmer, M.A., Paerl, H.W., Peterson, G.D. and Poff, N.L., 2006. Ecological thresholds: The key to successful environmental management or an important concept with no practical application? *Ecosystems*, *9*(1), pp. 1–13.

49. Stephen, C. and Soos, C., 2021. The implications of climate change for veterinary services. *Revue scientifique et technique*, *40*(2), pp. 421–430.

3 Climate Change Action: An Overview

Craig Stephen

CONTENTS

DOI: 10.1201/9781003149774-3

KEY LEARNING OBJECTIVES

- Be able to differentiate mitigation and adaptation and recognize how strategies to achieve those climate change management outcomes need to be tailored to the situation, context, and animal health goals for specific circumstances.
- Describe the four core areas of animal health action (protecting animal health, protecting the contributions of animals to community resilience, mitigating climate change contributions from animals and animals as sentinels).
- Describe the links between animal determinants of health and climate change harm reduction.

IMPLICATIONS FOR ACTION

- Program evaluation and implementation science are urgently needed to expand the quality and quantity of evidence for use in selecting adaptation interventions that will be effective, acceptable, feasible, and sustainable.
- Investment and attention to the determinants of health through harm reduction processes is a current strategy that will protect the capacities and resources animals need to retain health in a changing environment.
- Climate change actions for animal health need to be sustainable and therefore must be developed with social and ecological justice in mind to avoid unintended consequences.

OVERVIEW OF CLIMATE CHANGE ACTION

At its core, this book is about action. Our driving goal is to help people responsible for, interested in, or benefiting from healthy animals gain knowledge and perspectives to influence circumstances that protect and promote animal health in the face of climate change. Animal health stewardship means promoting a continuum of care that will prevent anticipated impacts, resist unanticipated impacts, and ensure recovery without persistent and irreversible harms (1). Animal health-sector climate change actions have primarily been concerned with: (i) protecting animals from the adverse effects of climate change, (ii) mitigating the effects of animals and their care on climate change, (iii) protecting animal assets that help people cope with climate change, and (iv) using animals as sentinels for emerging climate change harms. The way one

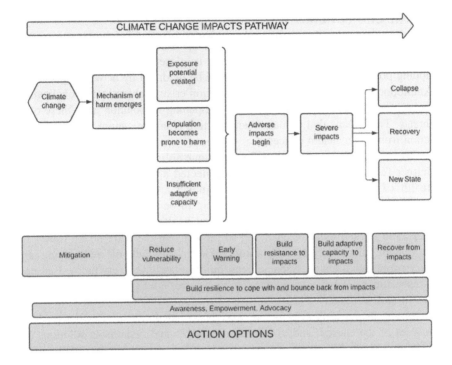

FIGURE 3.1 Climate change actions for animal health (green) will vary along the pathway from climate change to health outcomes (blue).

prioritizes and achieves these different goals depends on where one enters the chain of events from the emergence of a climate change mechanism of harm to health impacts (Figure 3.1).

Without human interventions, animals have limited options to cope with climate change. They can move to locations with more suitable climate conditions. They can exploit their inherent adaptability to match the changed conditions. They can die. Sensile species, like shellfish, are excluded from the first option as are animals whose migratory pathways have been obliterated or are so damaged as to preclude movement to more suitable habitats. Climate change may happen too quickly for some species to outrun its effect by emigration to new settings. Specialist species may not have the flexibility to adapt to changed conditions. Species with greater phenotypic plasticity may be better suited to cope with rapid environmental changes (2).

Human actions can help expand options for animals to adapt to or resist the effects of climate change. We can modify their living conditions and habitats to retain climate variables within tolerable limits. We can selectively breed animals or import new species better suited to changed conditions. We can enable animal movements through migration corridors. We can provide veterinary assistance to help them avoid or recover from climate-related diseases.

CLIMATE CHANGE ACTION IS COMPLEX

The world is "interprobleminary." It is made up of multiple, simultaneous assets, deficits, and problems that interact to pull us closer to or further from critical tipping points. Climate change actions will be modified by or will impact other Anthropocene stressors, such as the extinction crisis, habitat loss, globalization, and urbanization. Climate change is not happening in isolation from these other threats. This means that climate change action planning cannot be done in isolation from other health concerns.

In a world of concurrent problems, unique solutions for each problem are neither feasible nor effective (3). Unconnected approaches to managing the health of people, animals, and environments are failing to meet today's complex health challenges and are proving to be unsustainable (4). Climate change and other major impacts on animal health, such land use, food systems, and livelihoods, are changing so rapidly that there are no set routes for successful adaptation for the future (5). In the face of multiplying and overlapping challenges, agencies can quickly become overwhelmed as new threats add to the pressure of existing hazards and persistent obstacles to health and welfare.

Strong interdisciplinary partnerships are needed to better understand the "causes of the causes" and to identify acceptable and sustainable solutions (6). Climate change has made it clear that population health needs will rapidly outstrip resources if the focus remains on the status quo (7). However, evidence suggests that the status quo prevails. For example, a systematic review found that most peer-reviewed publications on livestock and climate change adaptation focused on incremental adaptation through technical and management improvements rather than on systemic adaptation options, such as altering farming system components, making institutional and policies changes, or making transformative changes in farming systems and farmer livelihoods (8).

CLIMATE CHANGE ACTION IS ABOUT HUMAN DECISIONS

Regardless of the target for climate change action, implementation of those actions will require people to decide to act in a certain way. Society, not scientists, decides what we want to protect, how much we want to invest, and what we are willing to lose. Helping people find and understand information, become comfortable with the value, feasibility, and acceptability of a proposed change, and believe they can make the change are critical first steps in helping people decide to act on climate change (9). Creating or finding opportunities to act, showing the value or benefits of change, and developing the social support to motivate maintaining the change helps people turn their decisions into action. Climate change actions can happen when people know what they need to do, believe that they can do it in their circumstances, are equipped with ideas, resources, and support to sustain their actions, and believe they must do it because of social or legal obligations (Figure 3.2).

FIGURE 3.2 Four building blocks of a behavior-change approach to climate change action.

It is a different task to accumulate evidence on the presence of a hazard, its spread, and the harms it causes rather than on effective ways to prevent, mitigate, or adapt to those harms under real-world conditions. The effectiveness, acceptability, and sustainability of climate change actions are sensitive to the circumstances in which the actions are conceived, developed, implemented, and evaluated. Sustainable and effective actions cannot happen until we can discern what works for whom, in what circumstances, in what respects, and how these actions can be done effectively and efficiently to successfully translate aspirations into action. Given that interventions to reduce animal health impacts from climate change will principally be targeting human decisions and behaviors, it is important to account for how characteristics of people's social groups and context affect their behaviors (10).

A GENERAL APPROACH TO KNOWN THREATS

Healthy animals must be able to respond to, cope with, or adapt to multiple challenges, changes, or hazards (e.g. stressors) brought about by climate change in their dynamic social or physical environment, whether those challenges are predictable or not (11). Climate change will amplify pre-existing problems that are known or expected to occur and create unanticipated threats. Managing unanticipated threats is the subject of Chapter 12.

Four general actions can be used to deal with known threats: (i) adaptable surveillance and monitoring of expected health outcomes or hazards for early warning; (ii) health intelligence to detect and respond to circumstances that are altering population exposure, sensitivity, and/or capacity to adapt to known climate change hazards and risk (i.e. tracking population vulnerability); (iii) strengthening of core animal health capacities and building of sustainable policies and infrastructures to manage increased effects and changing distributions of existing or reasonably expected threats; and (vi) reducing socio-economic in-equities that affect access to animal health services and resources (11).

CLIMATE CHANGE MITIGATION

What Is Mitigation?

Discussions of climate change action are replete with the words *mitigation* and *adaptation*. Mitigation refers to the things people can do to reduce greenhouse gases either by reducing their emission or finding ways to accumulate and store carbon-containing compounds. Effective mitigation will reduce climate change related effects (both negative and positive) and will have a long-term impact because greenhouse gases reside in the atmosphere for several decades. Mitigation, whether done at a local, national, or global scale, will have local, national, and global impacts on the rate of climate change. While it may be too late to avoid any effects of anthropogenic climate change, it is not too late to alter its future course. Attempts to subdue the acceleration of climate change and dampen the extend of change through mitigation must be part of climate change action.

Mitigating the Climate Impacts of Animals

The farm animal sector is the single largest anthropogenic user of land and a significant contributor to climate change (12). Animal agriculture generates 14–18% of the global greenhouse gas emissions via the fertilizer used for feed crops, on-farm energy use, feed transport, enteric fermentation in ruminants, manure, animal product processing and transport, and land-use change (13,14). It has been estimated that in the early 2000s, livestock contributed 44% of an-thropogenic methane, 53% of anthropogenic nitrous oxide, and 5% of anthro-pogenic carbon dioxide emissions (14). With the anticipated growth in demand for livestock products due to the rising global middle class, if left unchecked, the contributions of livestock to climate change will be even more profound.

Pet food production has been estimated to contribute 1.1–2.9% of global agriculture greenhouse gas emissions (15). The global pet care and supplies market is expected to increase by almost 60% by 2027 (16). Pet ownership will compound the environmental impacts of human dietary choices as ownership increases and consumers demand pet food with a higher content and quality of meat (17).

Mitigation strategies have centered around six key themes (16,18). First is to increase carbon sequestration through agricultural land-use practices that do not eliminate forests or other natural areas; using crop and pasture management methods and improved herd efficiencies that decrease the need for more land conversion for agriculture. The second theme is animal nutrition innovations to decrease diet components that are carbon intensive in production, changing feed composition to reduce reliance on meats for pets, reducing methanogenic components in ruminant diets, using feed additives to decrease methane production, and combating pet obesity to reduce food waste. The third theme is better manure management by using more efficient diets and new means to process and dispose of wastes. The fourth focuses on fertilizer use for animal feed crop production. Shifting trends in human and pet dietary preferences is the fifth theme, particularly focused on decreasing meat demands or increasing the use of human waste food or by-products for animals. Lastly, animal climate change contributions could be mitigated by selecting animals that are less climate sensitive and therefore need fewer inputs to support their needs in unsuitable climate conditions; examples are smaller pets that eat less or no meat, or shifting livestock species for those with lower carbon footprints.

MITIGATING THE CLIMATE IMPACTS OF ANIMAL CARE

Despite most literature focusing on the carbon footprint of animal production, veterinary health care practices can also be targets of mitigation strategies. One estimate set the climate footprint of the human healthcare sector to be equivalent to the fifth biggest carbon-emitting country in 2019 (19). Similar estimates for veterinary practice are not available; however, lessons from the human healthcare sector are transposable to the veterinary sector. A healthcare center's carbon footprint can be reduced by direct action to reduce waste and energy use in the practice setting, as well as by reducing the need for patients to come to healthcare facilities (or, in the case of food animal veterinary practice, the need for veterinarians to travel to their clients) and using water conservation (20). Experience in the COVID-19 pandemic opened many opportunities for telemedicine that ensured appropriate health care but reduced travel by patients, animal owners, or healthcare providers. A variety of national veterinary associations and organizations promote the idea of "green" practices that help design and deliver veterinary services that reduce not just the carbon footprint of these services but also other environmental impacts. Although past surveys have found veterinary students and practicing veterinarians are concerned about the impacts of climate change on animal health, a 2021 survey of veterinary teaching hospitals in the United States found little evidence that sustainable behaviors were being practiced (21). A lack of knowledge on effective and efficient actions and costs were barriers to incorporating more environmentally sustainable actions into these veterinary practices.

ADAPTATION AND RESILIENCE

WHAT IS ADAPTATION?

Adaptation refers to ways that people, societies, animals, or ecosystems adjust to current or impending climate change to reduce harms or exploit benefits that might arise. Adaptation involves reducing risk and predisposition to be adversely affected by climate change by building capacity to cope with climate impacts and mobilizing that capacity by implementing decisions and actions. Adaptation does not shift the course of climate change. It deals with the steps taken by natural, social, and economic systems, as well as individuals, to live with the effects of climate change. Adaptation action is more local because it is concerned with affecting peoples' behavior, which is a locally contextual issue.

As climate change forecasts worsen, there has been a growing sense of urgency to invest in adaptation. The 2018 special report of the Intergovernmental Panel on Climate Change (IPCC) emphasized the necessity of limiting global warming to 1.5°C to reduce health risk (22). The world was not on target to meet that goal in 2022, nor did it seem there was consensus on the path to do so. Therefore, greater emphasis is being directed toward climate change adaptation.

The climate will be slow to respond to any new initiatives, and the trajectory back to climate conditions to which species, ecosystems, and societies have evolved will take untold years, even if we could eliminate all drivers of adverse climate change today. It has been estimated that anthropogenic climate change will be irreversible on centennial to millennial timescales even after a complete cessation of CO_2 emissions (23). Adaptation action will, therefore, be a pressing issue for several generations of animal health professionals.

WHAT IS RESILIENCE?

Resilience is another commonly heard word in climate change discourse. Resilience is a complex idea that may be defined differently in different contexts. The IPCC defined resilience as "the ability of a system and its component parts to anticipate, absorb, accommodate, or recover from the effects of a hazardous event in a timely and efficient manner, including through ensuring the preservation, restoration, or improvement of its essential basic structures and functions" (24). In terms of human resilience to life-threatening experiences, resilience definitions have included "a stable trajectory of healthy functioning after a highly adverse event; a conscious effort to move forward in an insightful and integrated positive manner as a result of lessons learned from an adverse experience; the capacity of a dynamic system to adapt successfully to disturbances that threaten the viability, function, and development of that system; or a process to harness resources in order to sustain well-being" (25). Ecological resilience has been defined by some as the ability of an ecosystem to "regain fundamental structures, processes and functions following disturbance" (26). Part of the problem in finding a single definition of resilience is that people look at different scales and for different reasons, ranging from mental health in the present to planetary boundaries in the future. Despite these differences, common across many conceptions of resilience are the time needed to return to equilibrium following disturbance (recovery), the ability to deflect disturbance and avoid impact (resistance), and the range of disturbance that can be coped with or tolerated without needing to change (robustness) (27). Resilience reflects the capability of bouncing back to a pre-event state, as well as the capacity to cope with emerging post-event situations and changes (28). Resilience is something that can be supported, promoted, and sustained at all stages of climate change impact pathways.

SUSTAINABLE ADAPTATION

The concept of sustainable adaption draws our attention to the need to consider the wider or longer-term impacts and implications of adaptation actions (29). Not every climate change action is universally good. There will be winners and losers. Actions that benefit one group could harm another, and approaches to address climate risk could affect other environmental concerns. When planning adaptation action, one must recognize (i) that there are multiple concurrent stressors; (ii) actions against one harm can affect adaptation for another; (iii) differing values and interests will affect adaptation outcomes; (iv) local knowledge is needed for effective adaptation action; (v) adaptation actions should not add to climate change, limit the ability to respond, or negatively impact other parts of society, the economy, or the natural environment, and (vi) adaptation actions should be effective, efficient, equitable, and evidence based (30,31). Social justice and ecological integrity are, therefore, important conceptual filters when planning or selecting actions (30).

Animal health is being increasingly constrained by changing production methods, competition for food, water, and resources, endemic and emerging diseases, and rapidly changing environments (32). Climate change actions need to support health early rather than only control disease later. A multifactorial approach to health promotion can help make explicit some of the external drivers of health, which could in turn help to identify a wider suite of stakeholders, interventions, and policy options to prevent harms before they arise (33). Multimodal approaches to animal health protection explore: (i) how the systems need to change to support integrated and comprehensive health programs; (ii) the education and training needs of people who can create conditions favorable to animal health; (iii) monitoring and feedback to show what works for who, and (iv) demonstrable support to shift to a culture of health protection rather than disease detection and response alone. Attention is needed on the incentives of people who work with animals to change practices to cooperatively identify means of improving animal health, protect the determinants of animal health, and facilitate their own innovation and problem solving. Better evidence is needed to identify and operationalize strategies to move us from a disease-focused agenda to a health agenda that monitors and manages the individual, ecological, and social determinants of animal health.

The State of Animal Health Adaptation Knowledge
Animal health workers have awakened to the reality of climate change and its implications. However, the general discourse is too often limited to identifying and managing emerging risks, rather than building resilience in advance of

climate harms (1). Climate change is part of the animal health dialog but with comparatively little attention on what actions animal health professionals can undertake now to better protect wild and owned animals from current and future climate change harm. When climate change and animal health are discussed, it is most often in relation to infectious diseases that will have public health implications or threaten food systems. The voices of the stewards of healthy animal populations are relatively infrequently heard, especially in comparison to public health and economic concerns.

A 2020 scoping literature review concluded that there was little peer-reviewed evidence related to sustainable animal health climate change adaptation planning or action (34). Few papers or websites were explicitly concerned with adaptation action. Those few that were discussed adaptation in very general terms rather than evaluating the effectiveness, efficiency, equity, or evidence upon which to base actions. The bulk of the literature was predominantly interested in hypothesizing what hazards might occur, describing the spread of putative climate-sensitive hazards, or proposing or documenting disease impacts from unusual or extreme weather conditions that could be expected with climate change. Attention on climate change impacts on animal health continues to rapidly grow; but, at government, international, and scientific levels, attention seems more focused on speculating the course of known climate-sensitive diseases rather than on building resilient wild and domestic animal populations adaptable to a wide suite of known and unknown harms that will come from climate change. Limited awareness of best health protection practices is a barrier to climate change adaptation (35).

National Climate Adaptation Plans

All countries are being encouraged and supported to create National Adaptation Plans (NAPs). NAPs are intended to help countries undertake comprehensive medium- and long-term climate adaptation planning that builds on each country's existing adaptation activities and helps integrate climate change into national decision making across sectors. Their development is a continuous and iterative process that follows a country-driven, participatory, and transparent approach. Countries make both general and sector-specific NAPs. Ensuring animal health is accounted for in the NAP process is critical to protecting animals, as well as the values and service they produce for community climate resilience and adaptability.

NAPs for agriculture are particularly important because (i) agricultural sectors are among the most sensitive sectors to changing climate conditions and the most highly exposed to the impacts of climate change; (ii) livestock production and fisheries are critical to food security, nutrition, livelihoods, and incomes, thus contributing to community resilience; and (iii) agriculture influences natural resources, including land, water, biodiversity, and genetic resources, and so can influence ecosystem adaptation to climate change (36). But the evidence base for selecting adaptation actions that are effective across different social and ecological contexts, produce systemic and lasting change, and consider unintended consequences of actions is extremely limited (8,34).

NAPs should consider how to protect biodiversity from the impacts of climate change and to conserve and manage ecosystems so that they increase the adaptive capacity of people and wildlife. While general advice on how to plan for climate change is plentiful, evaluations of the actual practice of adapting to climate change through conservation planning are scarce (37). General approaches to action plans for conservation focus on conserving geophysical diversity, identifying and protecting climate refugia, and promoting cross environment connectivity (38).

ANIMAL HEALTH CONTRIBUTIONS TO HUMAN COMMUNITY RESILIENCE

Human vulnerability to climate change is a complex mix of socio-economic, health, and environmental dimensions. Resilience to anticipated health risks is strongly affected by nonclimatic factors, such as socioeconomic status, social capital, access to the needs for daily living and environmental quality. Emerging evidence suggests that climate change responses for society require a broad public health approach that works with health-determining sectors and functions (such as the food sectors) and agencies influencing the social determinants of health (such as income, social capital, and mental health) (39). Much of the dialog on animals' role in human vulnerability emphasizes animals as sources of harms, such as foodborne or other zoonotic infections. But animals also contribute to the positive determinants of climate change adaptation and resilience by giving people options and resources for accessing their needs for living and having resources with which to cope with climate change (Table 3.1). The nature, amount, and variability in biodiversity determines the sustainability and flow of ecosystem services, and therefore, is a key determinant of a community's capacity to adapt to future health challenges (40). Through their ongoing contributions to community health, animals contribute to climate change resilience.

Animals can be a direct source of income, a means to store wealth; provide food and other resources that otherwise would need to be purchased; and help reduce gender income disparities (42). People who rely on livestock, nature, or fishing for their food and income – over one-third of the world's population – are often the most vulnerable (43). Livestock can produce pathways toward health equity by securing current and future assets, facilitating greater participation of the poor in livestock-related markets, and sustaining and improving the productivity of agricultural systems that produce human health benefits (44). Sustainable food production can improve health through increased self-esteem and empowerment in formerly marginalized groups; increase the status of women; support better maternal and child health and nutrition; and improve local employment and reversed migration (45). Animal protein fuels the diets of the growing middle class and wealthy. Animals play a major role in aboriginal cultures, provide income in rural and remote communities, support our mental health, and contribute to scientific innovations. Wildlife impact many environmental and

TABLE 3.1

Animal Contributions to Community Health. (Adapted from 41)

Determinants of human population health	Examples of influence of animals on determinants of human health
Income and Social Status	• Economic resource (consumptive and nonconsumptive use) • Social status through animal use and ownership communities
Employment/Working Conditions	• Sense of identity and purpose for people working with or for animals • Healthy working conditions (ex. safe non-built environments; zoonoses)
Social Environment	• Promotion of healthy lifestyle choices (ex. recreational fishing, bird watching, dog walking, and other nature uses) • Group membership (wildlife watching, hunting, etc.)
Physical Environments	• Source of commercial, community, and subsistence food • Biodiversity and disease regulation • Source of pathogen and pollution exposure in natural settings • Trauma • Ecosystem integrity
Personal Health Practices and Coping Skills	• Mental health value of natural spaces and human-animal bonds • Confidence in safe use of food and other natural resources • Autonomy over food choices
Healthy Child Development	• Connectivity with nature and animals
Culture	• Spiritual connection with animals • Cultural identity and values

social determinants of human health and resilience through both direct and indirect mechanisms, several of which are strongly interconnected. Protecting these animal-based assets must therefore be part of plans to build community resilience to climate change.

Empowering communities to protect and promote animal welfare and health improves livelihoods while building resilience to emerging threats (46). Animal health can contribute to food security while decreasing the likelihood and impact of infectious diseases (46). Animal health and welfare is about more than animal husbandry and disease control. They are also about ensuring the compatibility of the animal's environments with its innate adaptations (47) and protecting the determinants of health (see Chapter 5). Healthy animals are more resilient to the effects of endemic disease, and they carry and transmit fewer pathogens affecting people and animals. This results in fewer opportunities for pathogen spillover into people or wildlife and less need for farmers to use antimicrobials, thus preserving treatment options for the future. Keeping livestock healthy reduces pathogen and contaminant loads in foods of animal origin and into the environment.

VULNERABILITY ASSESSMENT

The large combination of species, spaces, and harms to consider when planning an animal health action agenda is overwhelming. Resources to deal with them are limited. Climate change vulnerability assessment can be an initial step in the adaptation planning process to help identify which species or systems are most likely to be affected and why they are particularly vulnerable, including the effects of interactions between climate change and other stressors. There are many frameworks and methods for vulnerability assessments. What best describes and measures vulnerability can vary between different disciplines and perspectives. The selection of the best method should be informed by stakeholder and decision makers' information needs, the purpose of the assessment, information availability and reliability, and resources available.

Generally, vulnerability is the product of the likelihood of sufficient exposure to a threat, sensitivity of a population to harms from that threat, and the capacity of the population to resist, cope, and recover from those harms (Figure 3.3). Knowledge gaps plus the reality that local circumstances will affect the most influential contributors to vulnerability preclude prescription of a specific, standard set of recommendations on the best variables to monitor for assessing animal health vulnerability. The most useful variables to be assessed and tracked will be context specific and will require a combination of data, experience, collaboration, and judgment. Vulnerability assessments of animal health are rare in climate change planning, but they are increasingly being used in wildlife conservation and agriculture-planning purposes. Examples of guiding documents to undertake vulnerability assessment can be found in references 48, 49, and 50.

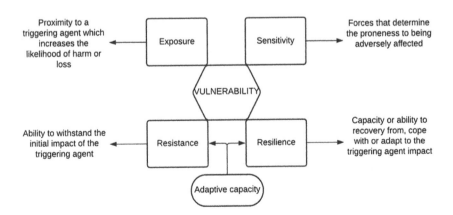

FIGURE 3.3   General framework of the components of vulnerability.

HARM REDUCTION AS A PHILOSOPHY AND OPTION FOR ACTION

THE RELEVANCE OF HARM REDUCTION TO CLIMATE CHANGE ACTION

Harm reduction is a set of perspectives and practices to reduce negative consequences of addictions, without requiring elimination of the addictive substance or behavior. It acknowledges the desirable goal of ending the use of a harmful substance but recognizes that people will want to keep engaging in those uses. It is a relevant approach to consider when planning climate change action because climate change is the result of humanity's addiction to fossil fuels and consumptive lifestyles.

The global economy and human development agendas depend on consumption and production that is enabled by an addiction to fossil fuels. Metaphorically like drug or alcohol addiction, the use of fossil fuels provides us with immediately desirable outcomes (like ease of travel, industrial development for jobs, fertilizers for agriculture) but causes future detrimental effects.

Like other addictions, dependence on fossil fuels or lifestyles that promote greenhouse gas production may cause people to sincerely want to do something to address climate change but accomplish little to achieve that outcome (51). When we aim to protect animal health in the face of climate change, divergent values and uncertain science lead to contentious situations where short-term interests conflict with long-term benefits, and when problems are construed in very different ways. In the quest for scientific certainty before we act, society, species, and ecosystems continue to suffer harm. Harm reduction does not argue against eliminating addictive things or for accepting them, but rather, it aims to decrease the negative consequences in the face of uncertainty and conflicting opinions about the "best" response and actions.

ADDICTION MANAGEMENT AS A CLIMATE CHANGE ACTION ANALOGY

Our climate change–causing addictions are intractable societal problems for which the impacts and solutions exceed the reach of any one individual or organization. Such problems are so conflict-prone and seemingly intractable that they defy a straightforward solution and quick fixes. Societies have often reacted to addictions by regulating who can access the substance (ranging from being freely available to prohibition) or via market forces (from subsidies to high tax burdens) (52). These approaches often require fundamental social and legal changes, changes that are slow to come. It is often wrongly assumed that sufficient research will produce effective policies. Tactics like regulation, litigation, and research may be necessary but are often insufficient to reduce harms in the present reality and are not necessarily the best way to mobilize locally acceptable, collaborative actions. Scientific advances or changes in technology or regulations needed to eliminate climate change hazards or harms can take considerable time to implement. Many hazards cannot be quickly eliminated because of the often-slow pace of scientific, social, and political change. The pervasive uncertainties and simplifying assumptions about the relationship of animal health and climate change create a gap between what science provides and what society demands. Different standards and expectations for how much and what types of harms should be attacked can discourage actions, especially when the science remains inconclusive. Harm reduction approaches encourage us to act now, on goals and targets on which we agree, to make incremental improvements in health despite debate, conflict, and uncertainty on what constitutes the right thing to do.

Harm reduction is a needed response to climate change impacts on animal health because transitions away from fossil fuel use through prohibitions or pricing are too slow and globally inconsistent. Harms are, therefore, accelerating. Harm reduction thinking and acting can help find ways to protect animal health despite ongoing consumerism and fossil fuels uses that magnify and sustain negative climate change effects. Its origins lie in public health actions to protect entire communities, rather than only addressing effects of addiction on individuals.

HARM REDUCTION IN PRACTICE

Harm reduction is both a goal and a process. It is a goal shared across many health and environmental sciences. Most people interested in animal health ultimately want to reduce harm to individuals, species, or environments, whether through clinical care, preventive actions, or health promotion. Harm reduction, as a process, focuses on developing local relationships for collective actions leading to incremental improvements in health in the face of addictions. The harm reduction process is a collaborative effort to understand and modulate the complex of underlying social and environmental factors that promote harmful outcomes by working at individual, community, and society levels (53).

The harm reduction process recognizes that harms from persistent hazards linked to an addiction cannot solely be addressed by targeting one part of the network (e.g. the user) but instead require a broad societal response. Harm reduction binds together various interests and evidence to adapt actions to a local context, making action more likely, acceptable, and sustainable. It promotes collaborative policy and action for horizontal, cooperative approaches to protecting health. Harm reduction works to minimize harms through nonjudgmental strategies by enhancing skills and knowledge of those helping animals live safer, more sustainable, healthier lives. Harm reduction processes that inspire action on shared priorities can make incremental gains in animal health, preserving capacity to adapt to future harms.

Many climate change issues involve multiple concurrent harms and dimensions (Figure 3.4). Harm reduction processes for animal health and climate change operate in a landscape of interacting influences and constantly changing conditions. Reciprocal care harm reduction asks us to consider the interconnections of harms and how actions to reduce one harm in one situation influence harms in other species, circumstances, or generations (54). It must be attentive to the inter-species and inter-generational implications of interventions

Dimensions of Climate Change Animal Health Harms

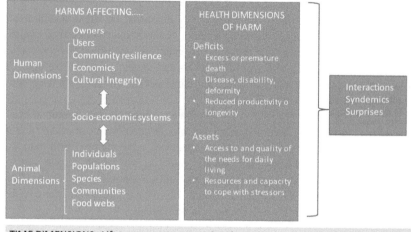

TIME DIMENSIONS - Lifecourse, transgenerational, intergenerational

FIGURE 3.4   Animal health harms from climate change have implications that cross species, generations, dimensions, and times. Systems approaches that looks across species and generations underlie the reciprocal care approach to harm reduction.

to prevent or reduce the adverse consequences to all members of the One Health community rather than only targeting individual health outcomes or hazards in one target group. Cooperative approaches that provide insight into cross-sectoral and socio-ecological systems context are critical to working in inter-species health (55). Taking a systems approach to harm reduction can provide a wider perspective to discover alternative means of solving problems and avoid unintended consequences.

HARM REDUCTION GOALS

There are four general harm reduction goals: (1) prevent persistent, irreversible, or severe harms after a harm has emerged; (2) prevent or contain harmful consequence by early intervention; (3) reduce the likelihood that a harmful situation arises, and (4) maintain conditions that are not harmful or hazardous. Total harm faced by a population can, in general, be reduced by reducing the total amount of harm (which is achieved by reducing exposure or sensitivity), or by reducing the total impact of harm (achieved by increasing capacity to cope or reducing cumulative effects) (Figure 3.5).

HARM REDUCTION AS A PROCESS

The harm reduction process inspires and empowers collective action to make incremental improvements toward a healthy situation in the presence of

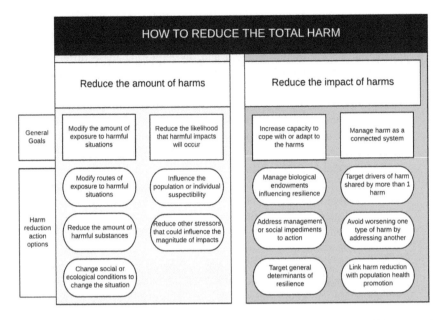

FIGURE 3.5 A taxonomy of harm reduction targets. This classification links general actions to harm reduction goals. Harm reduction programs require multilevel actions targeting more than one goal. (From 54).

uncertainty and conflict. It is a perspective and a set of strategies that applies to all the determinants of health and not merely problematic risks. It involves pragmatic, multidisciplinary approaches to remove barriers to the implementation of knowledge to protect health in situations where the hazard or harmful situation cannot soon be eliminated.

Harm reduction in animal health can provide new tactics to overcome entrenched perspectives and inaction to ensure progress on shared goals. It works by reducing the more immediately harmful consequences of an activity through pragmatic, realistic programs feasible under current conditions. This goal is achieved by promoting relationships, structures, and processes to make gains toward safer situations by incrementally reducing the negative health, social, and ecological consequences to individuals, communities, and ecosystems, without relying on elimination of the hazard. Stephen (54) summarized five defining characteristics of a harm reduction process.

1. *It looks for entry points for actions around which all can agree.* Climate change action involves a diversity of individuals who must find ways to act together toward a shared vision that results in mutually beneficial interactions. Harm reduction processes build collaboration on shared goals and reduce conflict to find solutions throughout the chain of causation. Social and organizational factors that influence actions and

opportunities to prevent, mitigate, or cope with harms are targeted. The process does not blame or judge the participants. It helps people see different aspects of the problem and, by exploring these differences, find solutions that go beyond their own perspective of the problem. Rather than seeing the issue as someone else's problem, successful collaborators can see their role and responsibilities to help reduce harms extending beyond their interests. Incorporating the context in which environmental, organizational, and personal factors interact increases the likelihood of finding shared priorities. Some degree of negotiation will be needed to create a shared vision that will help collaborators see how working toward collective interests will meet the interests of themselves or their organization. Harm reduction processes build new forms of strategic collaboration and governance that allow actions while debate remains on the scope and mechanisms of harm.

2. *It creates an enabling environment for collective actions.* Comprehensive and collective approaches are needed to overcome the consequences of climate change. Harm reduction processes should facilitate conditions for collaboration and cooperation to help collaborators develop a vision of what they can practically achieve and a hierarchy of achievable steps that, taken one at a time, can lead to a healthier situation. The process emphasizes actions that can benefit multiple parties and lead to progress on shared goals. It helps participants understand and endorse the process for making decisions and for moving what they know into action. The pathway from knowledge to harm reduction action requires trust. Trust can be built by being honest in negotiations, communicating purposefully and regularly, behaving in accordance with agreements, and not taking advantage of others or events when the opportunity is available. Trust, commitment, and a deeper understanding of the value of collective action is gained by focusing on series of incremental small wins toward the long-term goal. People need to trust both the information provided as well as the information providers. To be successful in communicating and building trust, one must recognize, value, and respect the diverse and complex value systems operating at the animal, health, and society interface.

3. *It is tries to find pragmatic solutions.* Gains that are feasible within the current circumstances and state of knowledge are sought rather than relying on the hope of a preferred future before acting. Harm reduction teams work on finding strengths, possibilities, and opportunities to reduce negative consequences rather than emphasizing discovery of the proximate cause of harms, inventorying deficits and obstacles, or attributing blame to others. It is about working with what we have and who we have today to make incremental improvements. Inspiring people to act in a way that addresses a suite of harms needs to do more than present the facts. The mere acquisition of new information often does little to affect risk perceptions and willingness to act. Information

sharing needs to be tailored to the individual characteristics of targeted audiences, including an assessment of their readiness for change. Blending objective information with an understanding of peoples' emotions, values, and personal experiences is essential when promoting reciprocal harm reduction actions.

4. *It is inclusive and local.* Harm reduction helps individuals and communities own their responses to complicated health challenges. It emphasizes bottom-up, locally developed planning. This reflects its focus on working with the knowledge, resources, and relationships that are currently affecting the problems of concern in the context in which they are occurring. Harm reduction actions focus on the current situation to initiate pragmatic actions to the greatest needs. It works on achieving the most immediate and realistic goals first. No one approach works for everyone in all situations. Generic recommendations need to be adapted to local circumstances to produce gains that can be built on over time to lessen present harms while preparing for tomorrow's risks. Those affected by or affecting harms and risks need to be involved in cocreating harm reduction strategies tailored to a specific situation. Harm reduction is fluid and dynamic, allowing flexibility as people and problems fluctuate. This process can be divergent, with some approaches that get stagnated by the need to gather an irrefutable body of evidence before interventions are authorized to proceed. Harm reduction recognizes that behaviors, their associated harms, and their proposed solutions are highly dependent on belief systems and culture within a setting and, therefore, are highly contextual in nature (56).

5. *It is integrative.* Animal health-related climate change harms are interrelated (Figure 3.4). The usual approach of examining one type of harm in isolation from another reduces chances of finding common pathways or opportunities to reduce or eliminate risks and harms. Examining them together helps build consensus on goals and find actions that may have benefits across domains. Seeking consensus on biological harms without accounting for social harms can increase conflict and delay actions. Harm reduction shifts the focus for change from technical and biological matters alone to include social innovations and opportunities.

While harm reduction policies, programs, and practices are well known and effective for substance abuse harms, they lack a broader adoption by animal health professionals. Harm reduction concepts have been used in managing some risks to endangered species, wild-farmed fish interactions, and invasive species concerns (57), but evidence of their use in One Health settings is sparse (56). The addiction analogy used in this chapter suggests that a harm reduction process could enable collaborations that address social, animal, and ecological situations that will affect animals' ability to cope with current climate change threats and resist future challenges.

ADVOCACY AND AWARENESS AS ESSENTIAL CLIMATE CHANGE ACTIONS

Advocacy is a critical tool to influence people and policies to bring about change and act more fairly. The issue of whether animal health researchers and practitioners should be advocates is fraught with debate. On the one hand, some people worry that if the evidence is not moved into action by those who generate the evidence, the evidence could be misinterpreted, misused, or not used. On the other hand, there are fears that if one takes an advocacy position, one's evidence might be taken as biased and skewed toward a personal perspective. The role of advocates is not to push forward their personal agenda but to strategically plan how to mobilize knowledge to action. An advocate is tasked with structuring an argument in favor of animal health but within the reality of the social context in which actions will be taken and from a solid basis in evidence.

An effective advocate must be able to tailor advocacy strategies and messages that meet people at their stage of willingness to change. People go through five steps before adopting a new idea: (i) they need to become aware of the new idea; (ii) they need to become motivated and able to find out more; (iii) they need to see how the change applies to their own needs and circumstances; (iv) they decide to try (or reject) the new idea; and (v) they need to confirm that their decision helped them meet their goals to continue its application (58). A change is more likely to be adopted if its advantages can be demonstrated to those who are being asked to change and can be feasibly applied. The advocate and their targets need to be able to hear from, respond to, and influence each other.

PROMOTING HOPE

It has long been understood that the way an issue is framed has important consequences for people's willingness to engage and act (59). Climate change hope and concern are positively related to climate change action, while despair is negatively related to action (60). Feelings that nothing can be done discourage people from taking steps to mitigate or adapt to climate change. It sometimes all seems too big, too far progressed, and too far beyond one's capabilities that people can become frustrated and frozen in inaction. One of the benefits of harm reduction thinking in climate change is that it instills hope. Harm reduction can help us shift from focusing on the risk environments in which a variety of factors interact to increase the chances of climate change related harms to focusing on the enabling resources, relations, and places that will help animals now and reduce demand on animal health services down the road.

Achieving and celebrating small successes leads to consistent engagement in harm-reduction-based programs that can give rise to efforts to change what we do to meet future goals and plans (61). By finding lowest barriers to actions with a high potential for outcomes that are valued by all involved stakeholders, harm reduction can create a sense of optimism. By creating circumstances to maintain health and welfare today and mitigating existing risks and vulnerabilities, collaborators can see the value of actions and therefore develop some hope that sustained collaboration can create benefits in the future. The hope is to entrench support for sufficient and timely behavioral change consistent with the scale of the problem by facilitating and sustaining ongoing interventions that meet people where they are in terms of needs, capacities, and expectations, and by addressing goals that are important to the very people we wish to influence.

SUMMARY

The need for climate change action is growing for animal health at a much faster pace than the evidence available to select effective, efficient, acceptable, and sustainable actions. Actions need to consider not only the harms that may come from animals or afflict animals from climate change, like new diseases. They also need to understand how to avoid worsening climate conditions due to the ways we manage and care for animals and how to best preserve the values animals contribute to human climate change resilience. Emerging evidence and experience in public health suggests that to be fully effective, climate change and health management efforts must not only tend to the preventive and curative functions of the health sector, but also must protect and manage the determinants of health that fall outside their usual scope of practice (62). The same can be expected for animal health.

A multi-sectoral approach is needed to mount timely and effective responses to changing climate-related risks because many of the factors influencing animals' determinants of resilience and adaptability fall outside the usual purview of the animal health sector. Active coordination with clear roles and responsibilities defined across sectors are needed. Recommendations in the report "Advancing the Science of Climate Change" (63) can be modified to formulate an animal health and climate change agenda that encourages multi-sectoral collaborations (1). This agenda includes:

- Promoting animal health, linked with ecological and human health, as part of society-wide climate change plans and actions.
- Investigating and communicating the implications of climate change for animal health affecting conservation, sustainable food production systems, food security, public health, and community resilience.
- Bridging the knowing-to-doing gap by facilitating collaborative action in the face of disagreements or institutional inertia.
- Strengthening links between animal health practice, policy, and research, with social, ecological, economic, and health sectors in concert with communities to build locally relevant actions.
- Making animal health knowledge accessible to climate change planning decision makers.
- Bringing climate change into animal health undergraduate and graduate curricula to develop highly trained individuals who can be knowledge brokers and facilitators of local, place-based actions to address climate change action as appropriate in different places/contexts.

- Expanding our thinking of animal health and climate change from only risk management to also capacity-building for healthy animal populations and animal health systems.

Current strategies must concurrently focus on protecting options for change (such as genetic diversity, habitat connectivity or environmental conditions), seeing new threats fast enough to limit their effects, ensuring access to animal health services that can adapt and surge with changing climate conditions, and protecting the determinants of health upon which climate change resilience rests. The need to match the evolved needs of the animals to their changing environmental realties can be assisted through new ways to care for animals, ensuring animals can move to more suitable conditions and matching the animals we keep to the conditions in which they are expected to live. While we await new investments in program evaluation and implementation science to pick the best interventions that attend to climate change as an interprobleminary and complex phenomenon, we can continue to advocate to protect the raw materials for adaptation that animal health provides and continue to find ways to reduce the contribution of animals to the global burden of greenhouse gases.

What does success look like in the realm of animal health and climate change? It depends, in part of who you are and where you are in terms of your beliefs and expectations, but it is not a single thing (Figure 3.6). Many may not be able to implement ideal interventions due to financial limits, lack of technology or knowledge, lack of political or social support, or other reasons. Harm reduction meets people where they are and helps to create enabling circumstances so that they can move toward actions that are effective, acceptable, and sustainable within their circumstances. Providing choices that can fit the needs of people who need to act is an essential part of the harm reduction process. While significant effort is and should be directed at reducing greenhouse gas emissions,

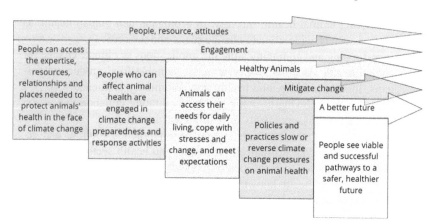

FIGURE 3.6 A continuum of success in combating climate change impacts on animal health.

there is room for people to create the conditions that incrementally make the world healthier by focusing on conditions that reduce the total amount and total impact of harms to animals through locally developed and collective harm reduction processes.

REFERENCES

1. Stephen, C., Carron, M. and Stemshorn, B., 2019. Climate change and veterinary medicine: Action is needed to retain social relevance. *The Canadian Veterinary Journal*, 60(12), p. 1356.
2. Bonamour, S., Chevin, L.M., Charmantier, A. and Teplitsky, C., 2019. Phenotypic plasticity in response to climate change: The importance of cue variation. *Philosophical Transactions of the Royal Society B*, 374(1768), p. 20180178.
3. Fried, L.P., Piot, P., Frenk, J.J., Flahault, A. and Parker, R., 2012. Global public health leadership for the twenty-first century: Towards improved health of all populations. *Global Public Health*, 7(sup1), pp. S5–S15.
4. Kock, R., Queenan, K., Garnier, J., Nielsen, L.R., Buttigieg, S., De Meneghi, D., Holmberg, M., Zinsstag, J., Rüegg, S. and Häsler, B., 2018. Health Solutions: Theoretical Foundations of the Shift From Sectoral to Integrated Systems. In Rüegg, S.R. Häsler, B. Zinsstag, J., *Integrated Approaches to Health: A Handbook for the Evaluation of One Health*. Wageningen Academic Publishers, 2018 Sep 10(2018), pp. 22–37.
5. Jackson, L., van Noordwijk, M., Bengtsson, J., Foster, W., Lipper, L., Pulleman, M., Said, M., Snaddon, J. and Vodouhe, R., 2010. Biodiversity and agricultural sustainagility: From assessment to adaptive management. *Current Opinion in Environmental Sustainability*, 2(1–2), pp. 80–87.
6. Mabry, P.L., Olster, D.H., Morgan, G.D. and Abrams, D.B., 2008. Interdisciplinarity and systems science to improve population health: A view from the NIH Office of Behavioral and Social Sciences Research. *American Journal of Preventive Medicine*, 35(2), pp. S211–S224.
7. Smith, J.A., 2018. Public Health, Governance, and the Anthropocene. *The Houston Journal of Health Law & Policy*, 18, p. 169.
8. Escarcha, J.F., Lassa, J.A. and Zander, K.K., 2018. Livestock under climate change: A systematic review of impacts and adaptation. *Climate*, 6(3), p. 54.
9. Stephen, C., 2020. Helping People Make Healthy Decisions for Themselves, Animals, and Nature. In *Animals, Health, and Society*. CRC Press, pp. 153–166.
10. Diez-Roux, A.V., 1998 Feb. Bringing context back into epidemiology: Variables and fallacies in multilevel analysis. *American Journal of Public Health*, 88(2), pp. 216–222.
11. Stephen, C. and Soos, C., 2021. The implications of climate change for veterinary services. *Revue scientifique et technique (International Office of Epizootics)*, 40(2), pp. 421–430.
12. Koneswaran, G. and Nierenberg, D., 2008. Global farm animal production and global warming: Impacting and mitigating climate change. *Environmental Health Perspectives*, 116(5), pp. 578–582.
13. Steinfeld, H., Gerber, P., Wassenaar, T.D., Castel, V., Rosales, M., Rosales, M. and de Haan, C., 2006. *Livestock's Long Shadow: Environmental Issues and Options*. Food & Agriculture Org.

14. Gerber, P.J., Steinfeld, H., Henderson, B., Mottet, A., Opio, C., Dijkman, J., Falcucci, A. and Tempio, G., 2013. *Tackling Climate Change Through Livestock: A Global Assessment of Emissions and Mitigation Opportunities.* Food and Agriculture Organization of the United Nations (FAO).
15. Alexander, P., Berri, A., Moran, D., Reay, D. and Rounsevell, M.D., 2020. The global environmental paw print of pet food. *Global Environmental Change, 65,* p. 102153.
16. Protopopova, A., Ly, L.H., Eagan, B.H. and Brown, K.M., 2021. Climate change and companion animals: Identifying links and opportunities for mitigation and adaptation strategies. *Integrative and Comparative Biology.*
17. Okin, G.S., 2017. Environmental impacts of food consumption by dogs and cats. *PLoS One, 12*(8), p. e0181301.
18. Rojas-Downing, M.M., Nejadhashemi, A.P., Harrigan, T. and Woznicki, S.A., 2017. Climate change and livestock: Impacts, adaptation, and mitigation. *Climate Risk Management, 16,* pp. 145–163.
19. Fathy, R., Nelson, C.A. and Barbieri, J.S., 2021. Combating climate change in the clinic: Cost-effective strategies to decrease the carbon footprint of outpatient dermatologic practice. *International Journal of Women's Dermatology, 7*(1), p. 107.
20. Tomson, C., 2015. Reducing the carbon footprint of hospital-based care. *Future hospital journal, 2*(1), p. 57.
21. Schiavone, S.C., Smith, S.M., Mazariegos, I., Salomon, M., Webb, T.L., Carpenter, M.J., Baumgarn, S. and Duncan, C.G., 2021. Environmental sustainability in veterinary medicine: An opportunity for teaching hospitals. *Journal of Veterinary Medical Education,* 2022 Mar 1; *49*(2), pp. 260–266.
22. Allen, M., Antwi-Agyei, P., Aragon-Durand, F., Babiker, M., Bertoldi, P., Bind, M., Brown, S., Buckeridge, M., Camilloni, I., Cartwright, A. and Cramer, W., 2019 . Technical Summary: Global warming of 1.5°C. An IPCC Special Report on the impacts of global warming of 1.5°C above pre-industrial levels and related global greenhouse gas emission pathways, in the context of strengthening the global response to the threat of climate change, sustainable development, and efforts to eradicate poverty. available at: https://www.ipcc.ch/sr15/. Accessed July 13, 2022.
23. Thorne, P.W., Willett, K.M., Allan, R.J., Bojinski, S., Christy, J.R., Fox, N., Gilbert, S., Jolliffe, I., Kennedy, J.J., Kent, E. and Tank, A.K., 2011. Guiding the creation of a comprehensive surface temperature resource for twenty-first-century climate science. *Bulletin of the American Meteorological Society, 92*(11), pp. ES40–ES47.
24. IPCC, 2012. Glossary of terms. In: Managing the Risks of Extreme Events and Disasters to Advance Climate Change Adaptation. In: Field, C.B., Barros, V., Stocker, T.F. and Dahe, Q. eds., 2012. *Managing the Risks of Extreme Events and Disasters to Advance Climate Change Adaptation: Special Report of the Intergovernmental Panel on Climate Change.* Cambridge University Press, pp. 555–564.
25. Southwick, S.M., Bonanno, G.A., Masten, A.S., Panter-Brick, C. and Yehuda, R., 2014. Resilience definitions, theory, and challenges: Interdisciplinary perspectives. *European Journal of Psychotraumatology, 5*(1), p. 25338.
26. Chambers, J.C., Allen, C.R. and Cushman, S.A., 2019. Operationalizing ecological resilience concepts for managing species and ecosystems at risk. *Frontiers in Ecology and Evolution, 7,* p. 241.
27. Allen, C.R., Angeler, D.G., Chaffin, B.C., Twidwell, D. and Garmestani, A., 2019. Resilience reconciled. *Nature Sustainability, 2*(10), pp. 898–900.

28. Kais, S.M. and Islam, M.S., 2016. Community capitals as community resilience to climate change: Conceptual connections. *International Journal of Environmental Research and Public Health*, *13*(12), p. 1211.
29. Eriksen, S. and Brown, K., 2011. Sustainable adaptation to climate change. *Climate and Development*, 3(1), pp. 3–6.
30. Eriksen, S., Aldunce, P., Bahinipati, C.S., Martins, R.D.A., Molefe, J.I., Nhemachena, C., O'brien, K., Olorunfemi, F., Park, J., Sygna, L. and Ulsrud, K., 2011. When not every response to climate change is a good one: Identifying principles for sustainable adaptation. *Climate and Development*, *3*(1), pp. 7–20.
31. Prutsch A., Grothmann T., Schauser I., Otto S. and McCallum S., 2010. Guiding principles for adaptation to climate change in Europe. *ETC/ACC Technical Paper*, pp. 1–32.
32. Thornton, P.K., 2010. Livestock production: Recent trends, future prospects. *Philosophical Transactions of the Royal Society B: Biological Sciences*, *365*(1554), pp. 2853–2867.
33. Wittrock, J., Duncan, C. and Stephen, C., 2019. A determinants of health conceptual model for fish and wildlife health. *Journal of Wildlife Diseases*, *55*(2), pp. 285–297.
34. Stephen, C. and Wade, J., 2020. Missing in action: Sustainable climate change adaptation evidence for animal health. *The Canadian Veterinary Journal*, *61*(9), p. 966.
35. Séguin, J., Berry, P., Bouchet, V., Clarke, K.L., Furgal, C., Environmental, I. and MacIver, D., 2008. Human health in a changing climate: A Canadian assessment of vulnerabilities and adaptive capacity. *Human Health in a Changing Climate*, *1*.
36. Karttunen, K., Wolf, J., Garcia, C. and Meybeck, A., 2017. *Addressing Agriculture, Forestry and Fisheries in National Adaptation Plans*. FAO.
37. Fischman, R.L., Meretsky, V.J., Babko, A., Kennedy, M., Liu, L., Robinson, M. and Wambugu, S., 2014. Planning for adaptation to climate change: Lessons from the US National Wildlife Refuge System. *BioScience*, *64*(11), pp. 993–1005.
38. Game, E.T., Lipsett-Moore, G., Saxon, E., Peterson, N. and Sheppard, S., 2011. Incorporating climate change adaptation into national conservation assessments. *Global Change Biology*, *17*(10), pp. 3150–3160.
39. World Health Organization. 2014. *Strengthening Health Resilience to Climate Change*. Geneva: World Health Organization, pp. 24–28.
40. Keune, H., Kretsch, C., De Blust, G., Gilbert, M., Flandroy, L., Van den Berge, K., Versteirt, V., Hartig, T., De Keersmaecker, L., Eggermont, H. and Brosens, D., 2013. Science–policy challenges for biodiversity, public health and urbanization: Examples from Belgium. *Environmental Research Letters*, *8*(2), p. 025015.
41. Stephen, C. and Duncan, C., 2017. Can wildlife surveillance contribute to public health preparedness for climate change? A Canadian perspective. *Climatic Change*, *141*(2), pp. 259–271.
42. FAO, 2009. Livestock, Food Security and Poverty Reduction. In *The State of Food and Agriculture 2009*. Food and Agriculture Organization. Available from http://www.fao.org/3/a-i0680e.pdf. Last accessed Feb 2, 2022.
43. Halwart, M., Funge-Smith, S., and Moehl, J., 2003. The role of aquaculture in rural development. *Review of the State of World Aquaculture, FAO, Rome*, pp. 47–58.
44. Kristjanson, P., Waters-Bayer, A., Johnson, N., Tipilda, A., Njuki, J., Baltenweck, I. et al., 2010. *Livestock and Women's Livelihoods: A Review of the Recent Evidence*. ILRI Discussion Paper No. 20. Nairobi: ILRI.
45. Pretty, J.N. and Hine, R., 2001. *Reducing Food Poverty with Sustainable Agriculture: A Summary of New Evidence*. Colchester: University of Essex. Center for Environment and Society, https://www.iatp.org/sites/default/files/Reducing_Food_Poverty_with_Sustainable_Agricul.pdf. Last accessed Feb 2, 2002.

46. Njisane, Y.Z., Mukumbo, F.E. and Muchenje, V., 2020. An outlook on livestock welfare conditions in African communities—A review. *Asian-Australasian Journal of Animal Sciences, 33*(6), p. 867.

47. Fraser, D., Duncan, I.J., Edwards, S.A., Grandin, T., Gregory, N.G., Guyonnet, V., Hemsworth, P.H., Huertas, S.M., Huzzey, J.M., Mellor, D.J. and Mench, J.A., 2013. General principles for the welfare of animals in production systems: The underlying science and its application. *The Veterinary Journal, 198*(1), pp. 19–27.

48. Glick, P., Stein, B.A. and Edelson, N.A., 2011. *Scanning the Conservation Horizon: A Guide to Climate Change Vulnerability Assessment.* Washington, DC: National Wildlife Federation, p. 168.

49. Fellmann, T., 2012. The assessment of climate change-related vulnerability in the agricultural sector: Reviewing conceptual frameworks. *Building Resilience for Adaptation to Climate Change in the Agriculture Sector, 23*, p. 37.

50. Barsley, W., De Young, C. and Brugère, C., 2013. Vulnerability assessment methodologies: An annotated bibliography for climate change and the fisheries and aquaculture sector. *FAO Fisheries and Aquaculture Circular, 1083*, p. I.

51. Suranovic S., 2013. Fossil fuel addiction and the implications for climate change policy. *Global Environmental Change, 23*(3), pp. 598–608.

52. Courtwright, D.T. 2012. A short history of drug policy or why we make war on some drugs but not on others. History Faculty Publications. Available at: https://digitalcommons.unf.edu/ahis_facpub/23. Accessed July 13, 2022.

53. Ezard, N., 2001. Public health, human rights and the harm reduction paradigm: From risk reduction to vulnerability reduction. *International Journal of Drug Policy, 12*(3), pp. 207–219.

54. Stephen, C., 2020. Harm Reduction for Reciprocal Care. In *Animals, Health, and Society.* CRC Press, pp. 95–112.

55. Head, B.W. and Alford, J., 2015. Wicked problems: Implications for public policy and management. *Administration & Society, 47*(6), pp. 711–739.

56. Gallagher, C.A., Keehner, J.R., Hervé-Claude, L.P. and Stephen, C., 2021. Health promotion and harm reduction attributes in One Health literature: A scoping review. *One Health*, p. 100284.

57. Stephen, C., Wittrock, J. and Wade, J., 2018. Using a harm reduction approach in an environmental case study of fish and wildlife health. *EcoHealth, 15*(1), pp. 4–7.

58. Kaminski, J., 2011. Diffusion of innovation theory. *Canadian Journal of Nursing Informatics, 6*(2), pp. 1–6.

59. Myers, T.A., Nisbet, M.C., Maibach, E.W. and Leiserowitz, A.A., 2012. A public health frame arouses hopeful emotions about climate change. *Climatic Change, 113*(3), pp. 1105–1112.

60. Stevenson, K. and Peterson, N., 2016. Motivating action through fostering climate change hope and concern and avoiding despair among adolescents. *Sustainability, 8*(1), p. 6.

61. Lee, H.S. and Zerai, A., 2010. "Everyone deserves services no matter what": Defining success in harm-reduction-based substance user treatment. *Substance Use & Misuse, 45*(14), pp. 2411–2427.

62. World Health Organization (WHO). 2015. Strengthening health resilience to climate change. *Technical Briefing.* Available at: https://www.who.int/phe/climate/conference_briefing_1_healthresilience_27aug.pdf Last accessed Feb 2, 2002.

63. National Research Council. 2011. *Advancing the Science of Climate Change.* National Academies Press. Available from: https://www.nap.edu/read/12782/chapter/1. Last accessed Feb 2, 2022.

4 The Study and Classification of Climate-Associated Diseases in Animals

Colleen Duncan, Michelle Dennis, and Cheryl Sangster

CONTENTS

KEY LEARNING OBJECTIVES

- Important tools for investigating climate-associated illness in animals are pathology, the study of disease within individuals, and epidemiology, the study of disease within populations.
- Climate associated illness in animals, as with people, can be classified by the pathway of exposure, such as temperature, air quality, extreme events, vector-borne disease, safe and nutritious food, water-related illness, and mental health and well-being.

DOI: 10.1201/9781003149774-4

- Owing to unique habitats and dependency on natural habitat climate change is causing diseases in wild animals that do not have a human corollary, such as those associated with changes in water quality.
- Linking climate change and animal disease can be challenging and necessitates a team approach and long-term commitments.

IMPLICATIONS FOR ACTION

- Collaboration between different diagnostic systems is necessary to ensure sufficient samples are available to identify novel conditions, explore risk factors, and detect changes in disease frequency and distribution.
- Linking environmental exposures to animal health outcomes necessitates better collaborations across disciplines.
- Financial and personnel resources are needed to both better investigate climate associated disease in animals and share the information with policy makers and the public to inform action.

INTRODUCTION

The study of animal health can be conducted on several levels and is informed by people with expertize in many disciplines. In the case of climate-associated illness, impacts are described and studied at the individual or population level. This work is heavily informed by two health disciplines, pathology and epidemiology. In this chapter, we will explore ways in which climate change impacts on animal health are investigated at the individual and group level, common pathways of climate-associated illness and resources that would support a better system for the study of climate associated illness in domestic and wild animals.

CONVENTIONAL STUDY OF DISEASE IN ANIMALS

THE STUDY OF DISEASE IN INDIVIDUAL ANIMALS

Pathology is a branch of medical science focused on disease. Pathologists are often involved in determining the cause of morbidity (illness) or mortality (death) in animals through detailed examinations of the whole body (gross examination) and at the tissue and cellular level (microscopic examination). The discipline of pathology is particularly focused on the mechanisms or ways in which disease develops (pathogenesis). The way cells and the tissue they comprise respond to an injury, such as with inflammation or scarring, is one of the main ways that diseases are classified. In addition to mechanisms of response, disease can be classified in other ways. Systemic pathology refers to pathology

TABLE 4.1

Examples of How Climate-Associated Disease Can Be Classified by Mechanism (Pathogenesis), Body System and Cause (Etiology)

Climate change associated disease	Pathogenesis	Body system	Etiology
Bushfire Smoke Inhalation the Smoky Mouse (*Pseudomys fumeus*) (1)	Inflammation	Respiratory (lungs)	Toxins produced by unprecedented climate change induced bush fires
Avian Malaria in Hawaiian Forest Birds (2)	Ischemic necrosis from failed blood supply	Cardiovascular	Infectious organism increasing its range as a result of global warming
Sub-lethal temperatures that sterilize *Drosophila* species (3)	Cell death	Reproductive	Increasingly high ambient temperatures occurring more frequently as a result of global warming

within a particular organ system, such as cardiovascular pathology or reproductive pathology. Finally, diseases can also be classified according to their primary cause or etiology, such as infectious, genetic, or nutritional (Table 4.1).

Climate change is a unique health threat since it spans all three of the disease classifications described above. It can alter an animal's ability to respond to an injury, for example by suppressing the immune system and causing an impaired inflammatory response; it can affect all systems in the body, with some, such as respiratory, cardiovascular, and reproductive being particularly susceptible; and finally, it can change the frequency, geography, and severity of etiologies that cause disease in animals, such as increased range of disease carrying insects. Additionally, some bias may occur in that available animals or tissues do not completely represent all of the factors that contribute to disease in the animal. For example, a bear killed by animal control officers for entering areas of human habitation to look for food may be hungry and malnourished because of reduced available food due to climatic environmental impacts. To better understand the numerous risk factors that contribute to disease states occurring with climate change, we need to aggregate information from many animals.

THE STUDY OF DISEASE IN GROUPS OF ANIMALS

Epidemiology is the study of the frequency, distribution, patterns, and determinants of disease in populations. Information gleaned from population studies is essential to the development of disease control and prevention strategies. Epidemiological studies are either descriptive or analytical. Descriptive epidemiology focuses on characteristics of those affected by the disease ("who"), which disease they have ("what"), time scale of cases ("when"), and geographic distribution of cases ("where"). Analytical epidemiology seeks to address the "why" and the "how" by comparing groups of individuals to test hypotheses about causal relationships.

The concept of causation is complex. Very rarely does a single factor, such as an infectious organism, a toxin, or an elevated temperature, cause disease in 100% of the animals exposed to it. Whether disease results from exposure is influenced in part by a variety of characteristics specific to the animal (host), characteristics of the factor (agent) to which the animal is exposed, and conditions in which the animal resides (environment). This host-agent-environment relationship is a concept referred to as the epidemiological triad or disease triangle. Using this framework, a change in climate represents a shift in numerous environmental drivers that may influence the occurrence of disease (Figure 4.1).

FIGURE 4.1   The epidemiological triad highlights the relationship of the host, agent, and environment as it pertains to disease. Historically represented as a triangle with three points (b), the triad is increasingly reconfigured as (a) to reflect the importance of the environment as a factor that can influence the host-agent relationship and ultimately the disease outcomes.

In reality, several different host, agent, and environmental factors often act together to cause disease in an individual. These are often referred to as "component causes" and visualized as wedges of a circle (Figure 4.2), the "causal pie" (4). When the constellation of component causes forms a complete circle, that collection of factors is "sufficient cause" for disease to occur.

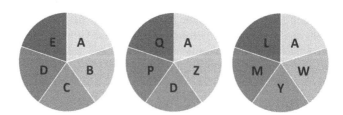

FIGURE 4.2   Schematic example of Rothman's causal pies (4). The single-component causes (wedges of the pie) are typically not enough to cause disease on their own; however, when present together, a collection of component causes becomes enough (sufficient) to cause disease. Each of the three complete pies represents sufficient cause for a hypothetical disease. Some component causes (ex. 'D') are present in multiple causal pies. When a component cause is present in all cases (ex. 'A'), it is referred to as a necessary cause.

The collection of component causes may differ between individuals, but it can still be sufficient to cause the disease. If a particular component is present in every case of disease, it is considered a "necessary cause" because without it, there is no disease. The concept of causal pies is helpful in the study of climate-associated illness since it provides a way of identifying factors upon which action can be taken.

STUDYING ANIMAL DISEASE IN OUR CHANGING CLIMATE

Although pathology and epidemiology are highly effective in describing the cause and effect of traditional disease, the disciplines are not sophisticated enough to incorporate the added complexity of climate change. Therefore, a new model of classifying climate associated disease is needed.

CLASSIFYING CLIMATE-ASSOCIATED DISEASE

Human health professionals have broadly classified climate-associated illness according to a series of exposure pathways (Figure 4.3, (5,6)). This approach has been adapted from use in chemical risk assessments to link climate change associated exposures to health outcomes. Because there is significant variability in the physical, temporal, and social context in which any exposure may cause harm, factors influencing an individual or group's vulnerability and resilience must also be considered. This aligns with the causal pie concept described above, where the particular exposure pathway is a necessary cause and a variety of physical and social conditions specific to the animals make up the other component causes.

A benefit of classifying climate-associated disease by exposures as opposed to mechanisms of disease or organ systems is that information gleaned through the

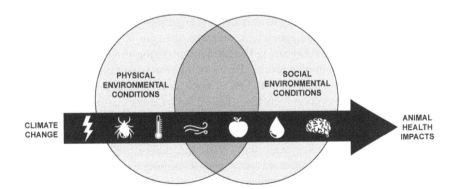

FIGURE 4.3   The seven exposure pathways for the classification of climate-associated disease include (from left to right) extreme weather, vector borne diseases, temperature, air quality, safe and nutritious food, water quality, and mental health (modified from (6,7)).

summation of related case data can be used to inform mitigation efforts specific to the exposure. While such a system has not been widely adopted in veterinary medicine, examples of domestic and wild animal diseases are recognized in the same exposure routes as are commonly studied in humans. By using a shared classification convention, human and animal health professionals may be better positioned to share data, which in turn deepens our understanding of how climate change impacts the health of all species. The following seven exposure pathways transcend species and provide a helpful framework for studying climate-associated disease in all species.

1. Temperature:

Heat-associated illness involves disruption of the body's ability to thermo-regulate, leading to blood vessel damage and inflammation throughout the body (8). The severity of heat-associated illness is variable, ranging from mild heat stress to severe disease in multiple organ systems and even death. In animals, the impact of temperature has been best studied in production animals where heat has long been recognized as a stressor impacting production (9,10), but now with climate change, the problem is intensifying (10–12). For example, some species, like dairy cattle, appear particularly vulnerable to heat stress, leading to de-creased milk production (13–15). Substantial research, effort, and cost is in-volved in efforts to mitigate the impact of heat stress in the dairy industry now and in the future (16–18).

Temperature associated disease is not just restricted to livestock. Heatstroke in dogs is increasingly tied to anomalous heat events with particular breeds (snub-nosed) and size (large) of dogs at particular risk (19). Although more challenging to quantify, heat associated illness has been identified in a variety of terrestrial wildlife (20), and even pollinators (21). As with domestic animals, these are projected to increase (22) and ecosystem impacts can persist for years after a heat event (23).

The stress caused by rising temperatures is arguably more apparent in aquatic animals than in terrestrial animals; sea temperatures are now the warmest they have been in over 400,000 years (24), and the impacts for some marine taxa have already reached catastrophic levels. One of the most striking examples is thermal stress in corals, which can initiate loss of pigmented symbiotic dinoflagellates, visually detectable as coral bleaching (Figure 4.4). Corals that remain bleached will essentially starve or succumb to disease without the photosynthetic con-tributions to metabolism from their lost symbionts. Around half of the world's coral reefs have been lost since the 1950s, and during this time, rising tem-peratures have triggered three circumtropical bleaching events and perpetuated coral diseases (25). Large derangements in water temperature are seen with marine heat waves, which occur all over the world, with greater frequency and duration in recent decades (26). Beyond impacting corals, marine heat waves can trigger mass mortality events involving numerous species at once, from sea stars to mussels to fish (27,28). Even mild derangements compromise the immune

FIGURE 4.4   Bleaching endangered pillar coral (*Dendrogyra cylindrus*) on a degraded Caribbean reef in October 2016, during the last global bleaching event. The white areas (arrows) of the coral reflect loss of symbiotic pigmented dinoflagellates, which are required for coral metabolism. Photo credit M. Dennis.

system of a wide range of aquatic species, while at the same time facilitating growth of certain infectious agents, leading to greater occurrence of infectious disease.

Mitigating the impacts of increasing temperatures is particularly challenging for certain wildlife species, and the reproduction impacts at the population level may be serious enough to threaten extinction. In Europe, egg-laying in a variety of songbirds is timed to correspond with spring emergence of oak leaf-eating caterpillars, such that chicks hatch when the caterpillar population is reaching its peak, providing ample food to feed the young birds. In an early spring, birds can respond to the same temperature cues as the oak trees and caterpillars, laying their eggs earlier to remain in synch with their food source (29). However, climate change is resulting in unprecedentedly warmer and more erratic spring temperatures, resulting in markedly earlier oak leaf emergence and faster caterpillar development (30). The normal adjustment of song bird egg-laying has not managed to keep up, resulting in mismatch between chick hatch and peak caterpillar production, and ultimately, insufficient food to successfully rear chicks. Although the birds have some capacity to adjust reproductive factors, such as initiating earlier incubation, predictive modeling suggests these strategies will eventually be insufficient to account for the timing mismatch and could lead to species extinction (30,31). The reproductive outlook is also dismal for long-lived reptile species with temperature-dependent sex determination, including sea turtles, crocodilians, and some freshwater turtles. In these species, rising nest-incubation temperatures will cause increasing proportions of female hatchlings, eventually culminating in feminized populations incapable of reproducing.

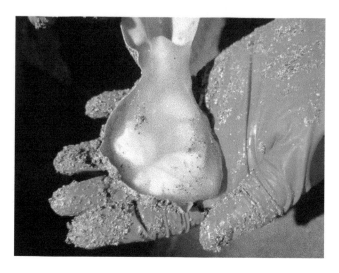

FIGURE 4.5 Leatherback sea turtle (*Dermochelys coriacea*) egg from the Eastern Caribbean with coagulated yolk and albumin, possibly indicating lethal incubation temperature. Photo credit M. Dennis.

Recent studies in the warmer northern regions of the Great Barrier Reef indicate that green sea turtles are now hatching 99% females (32). Moreover, rising nest-incubation temperatures can result in embryo mortality, essentially through cooking eggs during embryonal development (Figure 4.5), threatening to thwart reproduction entirely for some species (33).

2. Air quality:

Rising temperatures, and greater occurrence of wildfires and drought, result in increasing air pollutants, including particulate matter, smoke, allergens (especially pollen), and ozone. The development of lung disease secondary to smoke and air pollution is relatively consistent across species. Contaminants like ground-level ozone and particulate matter can cause direct injury to the respiratory tract or impair the tissue's natural defense mechanisms, leaving it vulnerable to infection. In response to this damage, the lung produces an increased amount of connective tissue (scar tissue) that blocks normal gas exchange and stiffens the lung, resulting in chronic respiratory conditions. But air pollution doesn't just damage the respiratory tract. Fine particulate matter (often referred to as $PM_{2.5}$) can cross from the lung into the bloodstream where it induces inflammation-causing cardiovascular disorders like stroke and myocardial infarctions. Animal models suggest that air pollution can cause primary lung and heart disease, and can also worsen underlying problems, similar to that seen in humans (34–36).

In animals, the effects of air pollution can be seen at both the individual and population level. Studies of dairy cattle have found a significant association between ambient air pollution and cattle mortality (37,38) as well as milk pro-duction and inflammation (39). Studies in companion animals have demonstrated deleterious effects of air pollution on cardiopulmonary function (40,41), Wildfire smoke is known to harm both domestic and wild animals (42,43). In a recent study of 12 polo horses naturally exposed to wildfire smoke and elevated $PM_{2.5}$ the authors reported that it was improved environmental conditions, rather than therapy, that reduced the horses' clinical symptoms and increased their athletic performance, as measured by treadmill speed and VO_2 peak (44). Air pollution, including wildfire smoke, also has a significant effect on captive and free-ranging wildlife populations (1,45). Air quality even has flow-on effects in the aquatic environment. For example, desert dust can increase the abundance of the marine pathogenic bacterium, *Vibrio* (46).

3. Extreme events:

One of the most dramatic manifestations of climate change are extreme events. Heat waves, wildfires, storms, drought, and floods are all climate-associated events that are increasing in frequency, intensity, and duration (47). Impacts from these events are influenced by the type of disaster, location, species in-volved, and coping capacity, and, in some instances, particularly heat waves (48), risks are generally underappreciated. While immediate injuries and deaths can be substantial, so too are the long-term impacts secondary to displacement, infrastructure damage, food and water contamination, or infectious disease outbreaks (49). Disaster-planning and preparedness efforts increasingly involve domestic animals, but more work is needed (Chapter 9).

Extreme events also have significant impacts for wild animals and the systems that sustain them. In a review of more than 500 observational studies of ecological responses to extreme events (cyclones, drought, flood, cold waves and heat waves), negative ecological responses were most commonly documented, including large population declines, local extirpation, and failure of species to recover to pre-disturbance levels (50). In aquatic ecosystems, marine heat waves, intense rainfall events, and tropical storms have been well documented in causing severe mortality to corals, kelps, seagrasses, and mangroves, all fundamental habitat-forming or-ganisms (46,51). On the other hand, exceptionally cold winters accompanying climate change similarly can result in diverse and complex illnesses, as seen with cold-stress syndrome in manatees (*Trichechus* sp) (52). Affected manatees have a variety of nonspecific health problems, including emaciation, lymphoid depletion, skin wounds, enterocolitis, and opportunistic infections. The complex presentation reflects compromise to metabolism, nutrition, and immunological function together, all of which must be addressed by treatment strategy.

4. Vector-borne diseases:

Many infectious diseases are transmitted by arthropod vectors like mosquitos, ticks, fleas, and flies. While the specific diseases and animal hosts vary by geographic location, climate change is a significant driver of vector-borne diseases (53). Longer warm season conditions appropriate for survival of the cold-blooded vectors result in longer infectious-disease seasons; warming has expanded the geographic distribution of many vectors and the diseases they carry. This increasing burden of vector-borne disease has major impacts for people and animals around the world. The climate-induced northward expansion of blue-tongue virus serotype 8 has resulted in hundreds of millions of dollars lost to livestock producers in Europe (54–57). In North America, Lyme disease is the most common vector-borne disease of humans, and it also impacts animals. Its occurrence and geographical distribution have increased, along with the expanded range and density of its vector, the black-legged tick (58).

5. Water related illness:

Water is a critical determinant of animal health (Chapter 5), and the mechanisms through which climate change increases and expands water-related illness in people is similar in animals. Both infectious agents and toxins are highly influenced by changes in climate, particularly temperature and precipitation. Warming waters facilitate the growth, survival, and toxicity of many disease agents, particularly certain bacteria, fungi, and parasites (59,60). Rising temperature and precipitation often occur in parallel with eutrophication of water sources that can bring on harmful algal blooms, which have been associated with both domestic and wildlife mortality events (61–63). Extreme weather events involving massive rainfall can facilitate water-related illnesses. During high precipitation events, terrestrial pathogens can even be effectively transmitted to aquatic animals (64). Conversely, drought can concentrate disease agents within water sources, limit availability of water for animals dependent on natural sources, congregate animals at water sources, allowing for easier pathogen transmission, and impact the availability of food. Examples of drought impacting a range of animal species can be found around the world (65,66).

6. Safe and nutritious food:

Our ability to produce and distribute safe, secure, and nutritious food is clearly linked to climate (67). Food safety and security is well described in Chapter 5 as a key determinant in animal health. Briefly, climate change compounds malnutrition, which is already a leading cause of poor human health globally (68). The increasing occurrence of total crop failure associated with extreme weather events threatens our ability to feed domestic animals in addition to humans. In addition, warming temperatures increase the survival and expand the geographic range of production, limiting agricultural pests such as insects, pathogens, and weeds. Food contaminants, like toxin-producing fungi, are more common in drought-stressed crops and proliferate more rapidly in warm conditions, causing significant disease in a variety of species (69). Similarly, temperature and moisture changes alter microbial growth and persistence, resulting in higher pathogen loads on food and more rapid proliferation of organisms in storage, causing spoilage (70). Climate change can even alter the composition of plants, particularly protein, zinc, and iron, making them less nutritious (71).

7. Mental health and well-being:

In people, direct and indirect mental health impacts associated with climate change are vast and increasing (72,73). There is no parallel study of mental health in animals; however, animal welfare is a topic of vast research and policy. One of the most commonly recognized structures is the Five Freedoms of animal welfare. Developed initially for the UK livestock industry and formalized by the UK Farm Animal Welfare Council, these guiding principals have been widely adopted by animal health and governance groups around the world (74). The freedoms include:

1. freedom from hunger, malnutrition, and thirst;
2. freedom from fear and distress;
3. freedom from heat stress or physical discomfort;
4. freedom from pain, injury, and disease; and
5. freedom to express normal patterns of behavior.

Through pathways and mechanisms described previously, climate change violates all of these, often concurrently. Animal health is also impacted as an indirect consequence of human mental health and well-being since people may be unable to provide appropriate care or protection to domestic and wild animals when impacted by climate change. For example, veterinarians already struggle with significant mental health issues (75,76), and the additional stress and burden from climate change is a growing concern for some in the profession (77).

UNIQUE ATTRIBUTES IN THE STUDY OF ANIMAL DISEASE

The exposure pathways approach used in human health is readily applicable to animals. That is, all categories used in the human framework remain relevant in animals, whether domestic or wild, terrestrial or aquatic, and several examples are provided for each category throughout this book. It would be counter-productive to reinvent something specific to animals while the existing system is sufficient and amendable. However, it is important to recognize that the health of free-ranging animals is even more closely dependent on the environment than for humans and is managed differently from humans and domestic animal health. This means that we can't expect exposure-pathway specific mitigations to work for all animals and humans alike. This idea is especially apparent when considering the broad impacts of temperature on wild animal reproduction, discussed above.

One shortcoming of the present human-centric exposure pathways is that they do not capture the complexity of freshwater or marine ecosystems, and not surprisingly, since humans do not live in an aquatic habitat. The environmental consequences of climate change are different in water than air, and many of these have dramatic and diverse impacts on aquatic animal health. Considering that over 70% of the Earth's surface consists of aquatic habitat, this is an important gap to address. Freshwater and marine ecosystems are priceless resources to which we are culturally connected and dependent for food, goods, water, agriculture, carbon cycling, and even oxygen. It is important that our approaches to studying health during climate change accounts for these ecosystems, which contain some of the greatest biodiversity of the planet and are at great risk. UNESCO predicts that by 2100, more than half of the world's marine species will be on the brink of extinction (78), and the IUCN estimates that one-third of freshwater biodiversity faces extinction (79). Aspects of the proposed framework above need to be expanded to account for the following attributes of water quality, including: changes to stratification of the water column, water flow, salinity, nutrient levels, acidity, or dissolved oxygen. These factors are closely intertwined and may be directly fatal. For example, higher sea temperatures and nutrient loading from runoff associated with extreme weather events will together increase the occurrence and negative impact of low-oxygen dead zones, areas where dissolved oxygen is so low that life cannot survive (80). These dead zones presently account for more than 245,000 km^2 of ocean surface area (roughly the same size as the United Kingdom, (78)). Elevated dissolved organic matter can also fuel harmful algal blooms, or even influence mixed pathogenic microbial growth, as seen on sea star body surfaces in sea star wasting disease (81). Salinity influences pathogenicity of many aquatic micro-organisms, so much so that adjusting salinity is often used as a treatment approach in controlled environments. In climate change, severe rainfall events may reduce estuarine salinity in affected regions, and diseases caused by pathogens that favor low salinity would have greater impacts (82). Periods of drought or rising sea levels can lead to increasing salinity in other habitats, favoring pathogens that thrive in high salinity (83). Fluctuations in salinity can also be directly injurious to aquatic animals (84). The rising concentration of atmospheric carbon dioxide leads to greater carbon dioxide dissolved in seawater, which lowers pH. This ocean acidification reduces availability of calcium carbonate, impeding coral and mollusk growth and predisposing to erosion, dissolution, or other negative health effects (24,85).

ADDRESSING CHALLENGES

The challenges posed by climate change perhaps present the best argument for an integrated ("one health") approach (86) compared to any disease scenario in the history of human and veterinary medicine. Human and animal health professionals need to come together with combined resources to tackle shared threats. In doing so, classification schemes for climate-associated illness can be expanded to allow for unique attributes specific to animals, for example, including challenges to reproduction for wild animals and water quality for aquatic animals. In this way, exposure pathways can consider environmental quality more broadly.

Despite the clear utility of conventional tools like pathology and epidemiology, there are several ways to enhance existing diagnostic systems so that information gleaned from these disease investigations can be better used to study climate associated animal disease. Most critical is the need to be able to better link environmental and animal disease data to inform action. This issue has also been raised by those studying the human health impacts of climate change (87,88). While there is extensive study in both domains, much of the resulting health and environmental information is effectively siloed, and innovation is needed to work across disciplines. There are abundant animal health and disease data aggregated within veterinary clinics and diagnostic laboratories; however, this data is not well curated for use in the study of known and anticipated harms associated with climate change. Development and support of innovative and integrated animal surveillance systems, as discussed in Chapter 13, are needed. Such efforts require not only technological advancements, but also long-term financial and personnel support to ensure they remain relevant and appropriate to address current and future concerns of both the scientific community and the general public.

In addition to building data relationships, there is a need to develop opportunities for collaboration. Support of diverse teams will help to advance our understanding of climate-associated disease. There is a long history of animals serving as sentinels and models of disease in humans, which can help to inform and advance public health efforts around climate change. Similarly, animal health professionals would benefit from some of the resources and infrastructure that are more common in human health as opposed to animal health. There is also a need to better link animal health professionals working in different geographic areas and animal health sectors. Veterinary medicine lags behind human health systems with respect to medical record synthesis, informatics, and data sharing. These systems could help to increase the power of veterinary research in this space. Tools and support for information sharing both within the profession and across health disciplines could also help to elucidate information on the pathogenesis of climate associated illness in all species.

Educational opportunities for animal health professionals at all stages of their careers could be enhanced (see also Chapter 16). Trainees in areas like pathology and epidemiology must be encouraged to think about the impacts of environmental exposures and the role of their work as part of broader initiatives and synthesize multiple streams of information. Developing skills like natural disease investigation and interdisciplinary scholarship will prepare students for responding to future threats and filling future rolls. Trainees with interest in the intersection of climate and animal health should be supported to work in this less conventional area while ensuring that there is sufficient skill development to prepare for the workforce.

There is great value in the dissemination of findings from these investigations to the public and policymakers to help build awareness of current and emerging health issues, which in turn can inspire and support action. Communication of science to the public, or for policy, is a skill that requires training and practice. Supporting diagnosticians to develop these skills, or work with teams who can assist in communication efforts, will result in more effective messaging and be more likely to inspire action to mitigate the animal health impacts of climate change.

Finally, resources for the support of diagnostic and epidemiologic study of climate associated illness in animals is needed. Veterinary diagnostic laboratories provide a range of critical services and are of great value to the animal health community, both locally and nationally (89,90). Such laboratories are largely user financed. As mentioned previously, surveillance and diagnostic systems are often designed to study single agents, with diseases of economic importance prioritized. Focusing instead on causal pathways that transcend species necessitates a broader scope and more interdisciplinary teams. Such an approach may require alternative funding strategies; however, this need may be addressed, at least in part, by providing information that other groups can use to address the shared health concerns arising from climate change. For example, wildlife or livestock health programs can support public health initiatives around climate change (91,92) in a number of ways. Resources, both financial and personnel time, must be available for research on naturally occurring disease associated with climate change. These synthesis products are needed to build awareness of the topic within the public and support policy actions that mitigate climate harms.

CONCLUSION

Animal disease associated with climate change can be seen in any species, any body system, and can be caused by any of the known disease pathways. Classifying these diseases according to exposure pathways conventionally used in human health, and studying disease at the population level, will help to better understand these conditions and, hopefully, build support for intervention strategies that can protect the health of all species.

REFERENCES

1. Peters, A., Hume, S., Raidal, S., Crawley, L. and Gowland, D., 2020. Mortality associated with bushfire smoke inhalation in a captive population of the smoky mouse (Pseudomys fumeus), a threatened Australian rodent. *Journal of Wildlife Diseases*, 57(1), pp. 199–204.
2. Liao, W., Atkinson, C.T., LaPointe, D.A. and Samuel, M.D., 2017. Mitigating future avian malaria threats to Hawaiian forest birds from climate change. *PLOS ONE*, 12(1), p. e0168880.
3. Parratt, S.R., Walsh, B.S., Metelmann, S., White, N., Manser, A., Bretman, A.J., et al., 2021. Temperatures that sterilize males better match global species distributions than lethal temperatures. *Nature Climate Change*, 11(6), pp. 481–484.
4. Rothman, K.J., 1976. Causes. *Am J Epidemiol*, 104(6), pp. 587–592.
5. Smith, K.R., Woodward, A., Campbell-Lendrum, D., Chadee, D.D., Honda, Y., Liu, Q., Olwoch, J.M., Revich, B. and Sauerborn, R., 2014. *Human Health: Impacts, Adaptation, and Co-benefits*. In: Cambridge, United Kingdom and New York, NY, USA: Cambridge University Press.
6. USGCRP, 2016. *The Impacts of Climate Change on Human Health in the United States: A Scientific Assessment*. Washington, DC.

7. Smith, K.R., Woodward, A., Campbell-Lendrum, D., Chadee, D.D., Honda, Y., Liu, Q., et al., 2014. Human health: impacts, adaptation, and co-benefits. In: Field, C.B., Barros, V.R., Dokken, D.J., Mach, K.J., Mastrandrea, M.D., Bilir, T.E., Chatterjee, M., Ebi, K.L., Estrada, Y.O., Genova, R.C., Girma, B., Kissel, E.S., Levy, A.N., MacCracken, S., Mastrandrea, P.R., and White, L.L. eds. *Climate Change 2014: Impacts, Adaptation, and Vulnerability Part A: Global and Sectoral Aspects Contribution of Working Group II to the Fifth Assessment Report of the Intergovernmental Panel on Climate Change.* Cambridge, UK and NY: Cambridge University Press, pp. 709–754.
8. Bouchama, A. and Knochel, J.P., 2002. Heat stroke. *New England Journal of Medicine, 346*(25), pp. 1978–1988.
9. Nienaber, J.A., Hahn, G.L. and Eigenberg, R.A., 1999. Quantifying livestock responses for heat stress management: A review. *International Journal of Biometeorology, 42*(4), pp. 183–188.
10. Summer, A., Lora, I., Formaggioni, P. and Gottardo, F., 2018. Impact of heat stress on milk and meat production. *Animal Frontiers, 9*(1), pp. 39–46.
11. Gunn, K.M., Holly, M.A., Veith, T.L., Buda, A.R., Prasad, R., Rotz, C.A., et al., 2019. Projected heat stress challenges and abatement opportunities for U.S. milk production. *PLOS ONE, 14*(3), p. e0214665.
12. Lacetera, N., 2018. Impact of climate change on animal health and welfare. *Animal Frontiers, 9*(1), pp. 26–31.
13. Key, N., Sneeringer, S. and Marquardt, D., 2014. Climate change, heat stress, and U.S. dairy production. *SRPN: Food Production (Topic).*
14. Polsky, L. and von Keyserlingk, M.A.G., 2017. Invited review: Effects of heat stress on dairy cattle welfare. *Journal of Dairy Science, 100*(11), pp. 8645–8657.
15. West, J.W., 2003. Effects of heat-stress on production in dairy cattle. *Journal of Dairy Science, 86*(6), pp. 2131–2144.
16. Ji, B., Banhazi, T., Perano, K., Ghahramani, A., Bowtell, L., Wang, C., et al., 2020. A review of measuring, assessing and mitigating heat stress in dairy cattle. *Biosystems Engineering, 199*, pp. 4–26.
17. Ferreira, F.C., Gennari, R.S., Dahl, G.E. and De Vries, A., 2016. Economic feasibility of cooling dry cows across the United States. *Journal of Dairy Science, 99*(12), pp. 9931–9941.
18. Fournel, S., Ouellet, V. and Charbonneau, É., 2017. Practices for alleviating heat stress of dairy cows in humid continental climates: A literature review. *Animals, 7*(5), p. 37.
19. Hall, E.J., Carter, A.J., and O'Neill, D.G., 2020. Incidence and risk factors for heat-related illness (heatstroke) in UK dogs under primary veterinary care in 2016. *Scientific Reports, 10*(1), p. 9128.
20. Ratnayake, H.U., Kearney, M.R., Govekar, P., Karoly, D., and Welbergen, J.A., 2019. Forecasting wildlife die-offs from extreme heat events. *Animal Conservation, 22*(4), pp. 386–395.
21. Potts, S.G., Biesmeijer, J.C., Kremen, C., Neumann, P., Schweiger, O., and Kunin, W.E., 2010. Global pollinator declines: Trends, impacts and drivers. *Trends in Ecology & Evolution, 25*(6), pp. 345–353.
22. Soroye, P., Newbold, T., and Kerr, J., 2020. Climate change contributes to widespread declines among bumble bees across continents. *Science (American Association for the Advancement of Science), 367*(6478), pp. 685–688.
23. Suryan, R.M., Arimitsu, M.L., Coletti, H.A., Hopcroft, R.R., Lindeberg, M.R., Barbeaux, S.J., et al., 2021. Ecosystem response persists after a prolonged marine heatwave. *Scientific Reports, 11*(1), p. 6235.

060

24. Hoegh-Guldberg, O., Mumby, P.J., Hooten, A.J., Steneck, R.S., Greenfield, P., Gomez, E., et al., 2007. Coral reefs under rapid climate change and ocean acidification. *Science*, *318*(5857), pp. 1737–1742.
25. Hughes, T.P., Barnes, M.L., Bellwood, D.R., Cinner, J.E., Cumming, G.S., Jackson, J.B.C., et al., 2017. Coral reefs in the anthropocene. *Nature*, *546*(7656), pp. 82–90.
26. Oliver, E.C.J., Donat, M.G., Burrows, M.T., Moore, P.J., Smale, D.A., Alexander, L.V., et al., 2018. Longer and more frequent marine heatwaves over the past century. *Nature Communications*, *9*(1), p. 1324.
27. Garrabou, J., Gómez-Gras, D., Ledoux, J-B, Linares, C., Bensoussan, N., López-Sendino, P., et al., 2019. Collaborative database to track mass mortality events in the Mediterranean sea. *Frontiers in Marine Science*, *6*(707).
28. Genin, A., Levy, L., Sharon, G., Raitsos, D.E. and Diamant, A., 2020. Rapid onsets of warming events trigger mass mortality of coral reef fish. *Proc Natl Acad Sci U S A*, *117*(41), pp. 25378–25385.
29. Burgess, M.D., Smith, K.W., Evans, K.L., Leech, D., Pearce-Higgins, J.W., Branston, C.J., et al., 2018. Tritrophic phenological match–mismatch in space and time. *Nature Ecology & Evolution*, *2*(6), pp. 970–975.
30. Visser, M.E., Noordwijk, A.J.V., Tinbergen, J.M. and Lessells, C.M., 1998. Warmer springs lead to mistimed reproduction in great tits (Parus major). *Proceedings of the Royal Society of London Series B: Biological Sciences*, *265*(1408), pp. 1867–1870.
31. Simmonds, E.G., Cole, E.F., Sheldon, B.C. and Coulson, T., 2020. Phenological asynchrony: A ticking time-bomb for seemingly stable populations? *Ecology Letters*, *23*(12), pp. 1766–1775.
32. Jensen, M.P., Allen, C.D., Eguchi, T., Bell, I.P., LaCasella, E.L., Hilton, W.A., et al., 2018. Environmental Warming and feminization of one of the largest sea turtle populations in the world. *Current Biology*, *28*(1), pp. 154–159.e4.
33. Choi, E., Charles, K.E., Charles, K.L., Stewart, K.M., Morrall, C.E. and Dennis, M.M., 2020. Leatherback sea turtle (Dermochelys coriacea) embryo and hatchling pathology in Grenada, with comparison to St. Kitts. *Chelonian Conservation and Biology*, *19*(1), pp. 111–123, 13.
34. Bartoli, C.R., Wellenius, G.A., Coull, B.A., Akiyama, I., Diaz, E.A., Lawrence, J., et al., 2009. Concentrated ambient particles alter myocardial blood flow during acute ischemia in conscious canines. *Environmental Health Perspectives*, *117*(3), pp. 333–337.
35. Wellenius, G.A., Coull, B.A., Godleski, J.J., Koutrakis, P., Okabe, K., Savage, S.T., et al., 2003. Inhalation of concentrated ambient air particles exacerbates myocardial ischemia in conscious dogs. *Environmental Health Perspectives*, *111*(4), pp. 402–408.
36. Takata, S., Aizawa, H., Inoue, H., Koto, H. and Hara, N., 1995. Ozone exposure suppresses epithelium-dependent relaxation in feline airway. *Lung*, *173*(1), pp. 47–56.
37. Cox, B., Gasparrini, A., Catry, B., Fierens, F., Vangronsveld, J. and Nawrot, T.S., 2016. Ambient air pollution-related mortality in dairy cattle: Does it corroborate human findings? *Epidemiology*, *27*(6), pp. 779–786.
38. Cox, B., Gasparrini, A., Catry, B., Fierens, F., Vangronsveld, J. and Nawrot, T., 2015. Cattle mortality as a sentinel for the effects of ambient air pollution on human health. *Arch Public Health*, *73*(Suppl 1), p. P22-P.
39. Beaupied, B.L., Martinez, H., Martenies, S., McConnel, C.S., Pollack, I.B., Giardina, D., et al., 2021. Cows as canaries: The effects of ambient air pollution exposure on milk production and somatic cell count in dairy cows. *Environmental Research*, p. 112197.

40. Lin, C-H., Lo, P-Y, and Wu, H-D. An observational study of the role of indoor air pollution in pets with naturally acquired bronchial/lung disease. *Veterinary Medicine and Science. n/a(n/a)*.
41. Lin, C-H, Lo, P-Y, Wu, H-D, Chang, C. and Wang, L-C., 2018. Association between indoor air pollution and respiratory disease in companion dogs and cats. *Journal of veterinary internal medicine, 32*(3), pp. 1259–1267.
42. Erb, W.M., Barrow, E.J., Hofner, A.N., Utami-Atmoko, S.S. and Vogel, E.R., 2018. Wildfire smoke impacts activity and energetics of wild Bornean orangutans. *Scientific Reports, 8*(1), p. 7606.
43. Marsh, P.S., 2007. Fire and smoke inhalation injury in horses. *Veterinary Clinics of North America: Equine Practice, 23*(1), pp. 19–30.
44. Bond, S.L., Greco-Otto, P., MacLeod, J., Galezowski, A., Bayly, W. and Léguillette, R., 2020. Efficacy of dexamethasone, salbutamol, and reduced respirable particulate concentration on aerobic capacity in horses with smoke-induced mild asthma. *Journal of veterinary internal medicine, 34*(2), pp. 979–985.
45. Liang, Y., Rudik, I., Zou, E.Y., Johnston, A., Rodewald, A.D. and Kling, C.L., 2020. Conservation cobenefits from air pollution regulation: Evidence from birds. *Proceedings of the National Academy of Sciences, 117*(49), pp. 30900–30906.
46. Westrich, J.R., Ebling, A.M., Landing, W.M., Joyner, J.L., Kemp, K.M., Griffin, D.W., et al., 2016. Saharan dust nutrients promote Vibrio bloom formation in marine surface waters. *Proc Natl Acad Sci U S A, 113*(21), pp. 5964–5969.
47. IPCC, 2021. Summary for policymakers. In: *Climate Change 2021: The Physical Science Basis. Contribution of Working Group I to the Sixth Assessment Report of the Intergovernmental Panel on Climate Change*. In Press.
48. Campbell, S., Remenyi, T.A., White, C.J. and Johnston, F.H., 2018. Heatwave and health impact research: A global review. *Health & Place, 53*, pp. 210–218.
49. Crist, S., Mori, J. and Smith, R.L., 2020. Flooding on beef and swine farms: A scoping review of effects in the Midwestern United States. *Preventive Veterinary Medicine, 184*, p. 105158.
50. Maxwell, S.L., Butt, N., Maron, M., McAlpine, C.A., Chapman, S., Ullmann, A., et al., 2019. Conservation implications of ecological responses to extreme weather and climate events. *Diversity and Distributions, 25*(4), pp. 613–625.
51. Babcock, R.C., Bustamante, R.H., Fulton, E.A., Fulton, D.J., Haywood, M.D.E., Hobday, A.J., et al., 2019. Severe continental-scale impacts of climate change are happening now: Extreme climate events impact marine habitat forming communities along 45% of Australia's coast. *Frontiers in Marine Science, 6*.
52. GD B, RA M, SA R, SJ G, AB J, 2003. Pathological features of the Florida manatee cold stress syndrome. *Aquatic Mammals, 29*(1), pp. 9–17.
53. Rocklöv, J. and Dubrow, R., 2020. Climate change: An enduring challenge for vector-borne disease prevention and control. *Nature Immunology, 21*(5), pp. 479–483.
54. Purse, B.V., Mellor, P.S., Rogers, D.J., Samuel, A.R., Mertens, P.P.C. and Baylis, M., 2005. Climate change and the recent emergence of bluetongue in Europe. *Nature Reviews Microbiology, 3*(2), pp. 171–181.
55. Wilson, A. and Mellor, P., 2008. Bluetongue in Europe: Vectors, epidemiology and climate change. *Parasitology Research, 103*(1), pp. 69–77.
56. Velthuis, A.G., Saatkamp, H.W., Mourits, M.C., de Koeijer, A.A. and Elbers, A.R., 2010. Financial consequences of the Dutch bluetongue serotype 8 epidemics of 2006 and 2007. *Prev Vet Med, 93*(4), pp. 294–304.
57. Wilson, A.J. and Mellor, P.S., 2009. Bluetongue in Europe: Past, present and future. *Philosophical Transactions of the Royal Society B: Biological Sciences, 364*(1530), pp. 2669–2681.

58. Ogden, N.H., Mechai, S. and Margos, G., 2013. Changing geographic ranges of ticks and tick-borne pathogens: drivers, mechanisms and consequences for pathogen diversity. *Front Cell Infect Microbiol, 3*, p. 46.

59. Semenza, J.C., Herbst, S., Rechenburg, A., Suk, J.E., Höser, C., Schreiber, C., et al., 2012. Climate change impact assessment of food- and waterborne diseases. *Critical Reviews in Environmental Science and Technology, 42*(8), pp. 857–890.

60. Charron, D.F., Thomas, M.K., Waltner-Toews, D., Aramini, J.J., Edge, T., Kent, R.A., et al., 2004. Vulnerability of waterborne diseases to climate change in Canada: A review. *Journal of Toxicology and Environmental Health, Part A, 67*(20-22), pp. 1667–1677.

61. O'Neil, J.M., Davis, T.W., Burford, M.A. and Gobler, C.J., 2012. The rise of harmful cyanobacteria blooms: The potential roles of eutrophication and climate change. *Harmful Algae, 14*, p. 313–334.

62. Oberholster, P.J., Botha, A.M. and Myburgh, J.G., 2009. Linking climate change and progressive eutrophication to incidents of clustered animal mortalities in different geographical regions of South Africa. *African Journal of Biotechnology, 8*(21), pp. 5825–5832.

63. Trevino-Garrison, I., DeMent, J., Ahmed, F.S., Haines-Lieber, P., Langer, T., Ménager, H., et al., 2015. Human illnesses and animal deaths associated with freshwater harmful algal blooms—Kansas. *Toxins, 7*(2), pp. 353–366.

64. Hicks, C.L., Kinoshita, R. and Ladds, P.W., 2000. Pathology of melioidosis in captive marine mammals. *Aust Vet J, 78*(3), pp. 193–195.

65. Knight, M.H., 1995. Drought-related mortality of wildlife in the southern Kalahari and the role of man. *African Journal of Ecology, 33*(4), pp. 377–394.

66. Robertson, G., 1986. The mortality of kangaroos in drought. *Wildlife Research, 13*(3), pp. 349–354.

67. Chakraborty, S. and Newton, A.C., 2011. Climate change, plant diseases and food security: an overview. *Plant Pathology, 60*(1), pp. 2–14.

68. Swinburn, B.A., Kraak, V.I., Allender, S., Atkins, V.J., Baker, P.I., Bogard, J.R., et al., 2019. The global syndemic of obesity, undernutrition, and climate change: The Lancet Commission report. *The Lancet, 393*(10173), pp. 791–846.

69. Malekinejad, H. and Fink-Gremmels, J., 2020. Mycotoxicoses in veterinary medicine: Aspergillosis and penicilliosis. *Vet Res Forum, 11*(2), pp. 97–103.

70. Misiou, O. and Koutsoumanis, K., 2021. Climate change and its implications for food safety and spoilage. *Trends in Food Science & Technology*.

71. Myers, S.S., Zanobetti, A., Kloog, I., Huybers, P., Leakey, A.D.B., Bloom, A.J., et al., 2014. Increasing CO2 threatens human nutrition. *Nature, 510*(7503), pp. 139–142.

72. Bourque, F. and Cunsolo Willox, A., 2014. Climate change: The next challenge for public mental health? *International Review of Psychiatry, 26*(4), pp. 415–422.

73. Hayes, K., Blashki, G., Wiseman, J., Burke, S. and Reifels, L., 2018. Climate change and mental health: risks, impacts and priority actions. *International Journal of Mental Health Systems, 12*(1), pp. 28.

74. Health WOfA. Animal Welfare [Available from: https://www.oie.int/en/what-we-do/animal-health-and-welfare/animal-welfare/.

75. Nett, R.J., Witte, T.K., Holzbauer, S.M., Elchos, B.L., Campagnolo, E.R., Musgrave, K.J., et al., 2015. Risk factors for suicide, attitudes toward mental illness, and practice-related stressors among US veterinarians. *J Am Vet Med Assoc, 247*(8), pp. 945–955.

76. Best, C.O., Perret, J.L., Hewson, J., Khosa, D.K., Conlon, P.D. and Jones-Bitton, A., 2020. A survey of veterinarian mental health and resilience in Ontario, Canada. *Can Vet J, 61*(2), pp. 166–172.

77. Kramer, C.G., McCaw, K.A., Zarestky, J. and Duncan, C.G., 2020. Veterinarians in a changing global climate: Educational disconnect and a path forward. *Frontiers in Veterinary Science*, 7(1029).
78. UNESCO. Facts and figures on marine biodiversity [Available from: http://www.unesco.org/new/en/natural-sciences/ioc-oceans/focus-areas/rio-20-ocean/blueprint-for-the-future-we-want/marine-biodiversity/facts-and-figures-on-marine-biodiversity/.
79. IUCN. Freshwater biodiversity [Available from: https://www.iucn.org/theme/species/our-work/freshwater-biodiversity.
80. Altieri, A.H. and Gedan, K.B., 2015. Climate change and dead zones. *Global Change Biology*, 21(4), pp. 1395–1406.
81. Hewson, I., 2021. Microbial respiration in the asteroid diffusive boundary layer influenced sea star wasting disease during the 2013-2014 northeast Pacific Ocean mass mortality event. *Marine Ecology Progress Series*, 668:pp. 231–237.
82. Yasunari, K., Jeffrey, D.S., Wolfgang, K.V., Howard, K. and Vicki, S.B., 2003. Infectivity and pathogenicity of the oomycete Aphanomyces invadans in Atlantic menhaden Brevoortia tyrannus. *Diseases of Aquatic Organisms*, 54(2), pp. 135–146.
83. Burreson, E. and Ragone Calvo, L., 1996. Epizootiology Of Perkinsus Marinus disease of oysters in Chesapeake Bay, with emphasis on data since 1985. *Journal Of Shellfish Research*, 15(1), pp. 17–34.
84. Duignan, P.J., Stephens, N.S. and Robb, K., 2020. Fresh water skin disease in dolphins: A case definition based on pathology and environmental factors in Australia. *Scientific Reports*, 10(1), p. 21979.
85. Gazeau, F., Parker, L.M., Comeau, S., Gattuso, J-P, O'Connor, W.A., Martin, S., et al., 2013. Impacts of ocean acidification on marine shelled molluscs. *Marine Biology*, 160(8), pp. 2207–2245.
86. Zinsstag, J., Schelling, E., Waltner-Toews, D. and Tanner, M., 2011. From "one medicine" to "one health" and systemic approaches to health and well-being. *Preventive Veterinary Medicine*, 101(3), pp. 148–156.
87. Tong, S. and Ebi, K., 2019. Preventing and mitigating health risks of climate change. *Environmental Research*, 174, pp. 9–13.
88. Ebi, K.L., Ogden, N.H., Semenza, J.C. and Woodward, A., 2017. Detecting and attributing health burdens to climate change. *Environmental Health Perspectives*, 125(8), p. 085004.
89. Schulz, L.L., Hayes, D.J., Holtkamp, D.J. and Swenson, D.A., 2018. Economic impact of university veterinary diagnostic laboratories: A case study. *Prev Vet Med*, 151, pp. 5–12.
90. Dunne, G. and Gurfield, N., 2009. Local veterinary diagnostic laboratory, a model for the one health initiative. *Veterinary Clinics of North America: Small Animal Practice*, 39(2), pp. 373–384.
91. Stephen, C. and Duncan, C., 2017. Can wildlife surveillance contribute to public health preparedness for climate change? A Canadian perspective. *Climatic Change*, 141(2), pp. 259–271.
92. De Garine-Wichatitsky, M., Binot, A., Ward, J., Caron, A., Perrotton, A., Ross, H., et al., 2021. "Health in" and "health of" social-ecological systems: A practical framework for the management of healthy and resilient agricultural and natural ecosystems. *Frontiers in Public Health*, 8.

5 Climate Change and the Determinants of Animal Health

Carrie McMullen, Jane Parmley, and Craig Stephen

CONTENTS

DOI: 10.1201/9781003149774-5

KEY LEARNING OBJECTIVES

- The health impacts of climate change extend beyond direct effects of pollutants, pathogens, parasites, and extreme weather. Climate change will also impact the foundational capacities and resources animals need to be adaptive and resilient in a changing environment.
- Animal health needs to be recognized as the shared responsibility of multiple sectors that influence the determinants of health to ensure animals can meet their needs for daily living, cope with the stressors expected from climate change, and continue to meet our economic, cultural, and social expectations of them.
- Understanding health as a cumulative effect of interacting individual, environmental, and social determinants can reveal opportunities for action by the animal health sector and partners outside of that sector.

IMPLICATIONS FOR ACTION

- Protecting and ensuring access to the determinants of health is multi-solving and can help animals adapt to a suite of climate change threats today and in the future.
- Interactions among determinants of health and variability in quality and access to them will make intervention impacts, and responses to climate change, difficult to predict. Health managers will need to be alert for unanticipated effects on other determinants of health or other species when trying to protect an animal health determinant.
- Protecting animal health in the face of climate change will need cross-sectoral collaborations since control of many determinants of health fall outside of the typical animal health sector.

INTRODUCTION

CLIMATE CHANGE AND THE DETERMINANTS OF HEALTH

Chapter 2 introduced the idea of health as a cumulative effect of interacting individual, social, and environmental factors that affect animals' capacity to cope with the challenges of daily living. These interacting factors are known as the determinants of health (see Figure 2.1). The determinants of health are those individual and collective factors and conditions that enable an individual or population to be healthy. The determinants of health approach considers not only adverse health outcomes like disease, but also the positive dimensions of health. Fully understanding what affects health status requires understanding the physiological, environmental, community, social, and political processes that shape how animals interact with the world around them to meet their needs for daily living, cope with stressors and challenges, and meet our expectations of them. Health, therefore, is not just about the state of an individual but also about the state of our environmental, social, and economic conditions. This view of health demands that we similarly look at climate-health interactions across the suite of health determinants and not limit our attention only to diseases.

Climate change health impacts cannot be compartmentalized (e.g., considering precipitation without considering the effects of heat; examining climate effects without accounting for urbanization). Climate has multiple direct and indirect animal health impacts because of its multitude of influences on the determinants of health. Climate change occurs simultaneously with disease threats, habitat loss, social changes, and behavioral responses that result in cumulative impacts on animal determinants of health. Climate change, in some respects, could be conceived as a source of interconnected and cumulative effects of health and social challenges that affect populations (e.g., syndemics) (1).

The implications of intersecting climate impacts on the determinants of health are threefold. First, animal health management recommendations for climate change adaptation need to be attentive to the effects of decisions not only on the targeted determinant of concern (e.g., increasing vector-borne disease prevalence), but also on the implications for other determinants of health (e.g., food security impacts if vector control hits nontarget pollinator species). Second, actions to protect one species or population from climate change will need to be wary of the impacts on other species given that animal populations can cohabit the same places and their use of various habitats can overlap or compete. Third, these multiple and overlapping health-climate-species interactions will make it very difficult to predict how climate change will affect animal health. While we might achieve sufficient predictive power to anticipate a specific disease impact (e.g., heat stress), no animal population experiences just one threat or accesses just one health asset at a time. The complex, cumulative, and socially informed nature of health makes it a messy issue wherein predictions are elusive (Figure 5.1).

While much of the literature emphasizes the prediction, detection, and monitoring of potential harms, with much less attention on development and evaluation of management options (2), throughout this chapter, we offer some illustrations of possible responses to climate change threats to the determinants of health in "Response Example" text boxes.

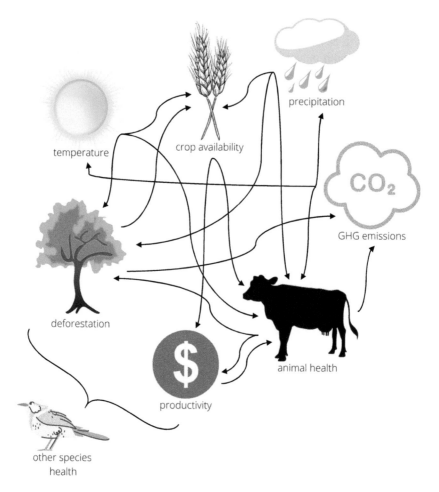

FIGURE 5.1 Depiction of the intersection of climate change impacts on the determinants of animal health: (a) decisions affect more than the targeted determinant of health or species, (b) actions to protect a single species (e.g., cattle) will have impacts and potential unintended consequences against another species or population (e.g., birds), and (c) these multiple and overlapping factors make it difficult to predict climate impacts on the determinants of health. The complex and interwoven connections (solid arrows), combined with the uncertainty of climate change, creates the inability to predict the direction and magnitude of each factor on another, and most interactions are bidirectional and cyclic (i.e., temperature affects animal health, which lessens productivity, and, therefore, lessens the income a farmer is able to put toward veterinary services and feed).

SCOPE OF THE CHAPTER

Scholarly work on the relationship of climate change and animal determinants of health is widely dispersed across veterinary, agricultural, ecological, and biological sources. We used a rapid narrative review of the literature to develop an overview of how the animal health community is discussing health-climate

TABLE 5.1

Search Strategy to Identify Review Articles of Climate Change Impacts on Animal Health, along with Mitigation or Adaptation Solutions[*]

#1	(climate change)
#2	AND (wildlife OR livestock OR animal OR dog OR cat OR feline OR canine OR aquaculture)
#3	AND (mitigation OR adaptation OR resilience OR "determinants of health")
#4a	AND Document type: "Review"
#4b	AND Document type: "case study" OR "case studies"

Note
[*] This search was limited to titles, abstracts, and keywords.

interactions to frame key messages and find illustrative examples for this chapter. We used Web of Science and CAB Direct (via CAB Interface) to identify English language review papers with no date limit. The search term "climate change" was combined with animal types (e.g., wildlife), and with the terms "mitigation," "adaptation," "resilience," and "determinants of health." Boolean operators "AND" or "OR" were used to combine relevant search terms (Table 5.1). The searches were conducted from July to September 2021.

The searches were limited to review articles to identify a broad range of literature already synthesized by others investigating this topic. To find additional illustrative examples, case studies were also searched for by limiting document type to case studies; Google Scholar was also used to identify case studies. Relevancy screening was conducted by one reviewer to identify relevant pivotal papers across a broad range of animal species. Key messages were extracted from papers of interest by the same reviewer throughout the screening process. These key messages were grouped based on *a priori* categories developed by Wittrock et al. (3), identified in Chapter 2. Many of the papers explored multiple determinants of animal health, and, therefore, an article could be included in multiple thematic categories. Of 250 articles screened, 61 articles were included based on applicability of the topic (i.e., climate change and the determinants of animal health), study design (i.e., review articles and case studies), variety (i.e., if saturation was reached in livestock determinants, other animal types were focused on), and illustrative impact (i.e., innovative or impactful response examples). The literature included in this chapter was not exhaustive; however, the findings from this review were supplemented by earlier reviews undertaken to write Stephen and Wade (2) and Stephen and Soos (4).

Very early in our search, it became evident that much of the literature focused on ways that terrestrial livestock or aquaculture facilities can adapt to climate change to protect human food security. Few reviews identified animal health for animals' sake as their primary focus. For example, it was challenging to find studies on mitigation or adaptative action to protect animal food security. This gap required us to draw some implicit connections between disparate articles. For example, some articles described how climate change could affect crop health, while others discussed how food insecurity via crop failures could affect animal health, thereby decreasing productivity. Articles that investigated both direct and indirect connections between climate change and animal health outcomes were rare. Many articles focused on reducing production of greenhouse gas (GHG) emissions within a livestock production system to mitigate climate change. Although lessening GHG emissions in all areas of society is important for protecting animal health long term, these connections were not commonly recognized in the literature we found.

The effects of climate change on animal health are context- and species-specific, and the intertwining cumulative effects of several forces create additional uncertainty in just how severe animal health impacts will be. This chapter does not, and indeed cannot, discuss all possible interactions between climate change and health. This limitation, in part, is due to gaps in the literature and in part due to the complexity of these interactions. Instead, we provide an overview of common themes that emerged in our literature reviews to give the reader a sense of the variety of health impacts that animals will experience due to climate change. Table 5.2 provides definitions of the determinants of health we used to thematically categorize the literature, along with some illustrative examples. In Table 5.2, the determinants are also grouped into three foundations of health (meeting the needs for daily living, having capacity to cope with stressors and change, and meeting societal expectations), which were used to frame the following discussion of the literature.

MEETING THE NEEDS FOR DAILY LIVING

The needs for daily living are the basic requirements for health. In the absence of secure and safe food, water, and places to live, animals will lack the capacity to meet their needs, avoid threats, and cope with the challenges of daily living. Meeting the needs for daily living is one of the six main themes in the determinants of health model for wild animal species (3). Within needs for daily living are food security, water security, and physical security, including habitat use and availability (3). Requirements for meeting the needs for daily living vary across animal species and uses, which is evidenced through their varying dependence on specific food sources and habitats in each of livestock, aquaculture, wildlife, and companion animal populations.

FOOD SECURITY

Accessing Appropriate Food
The climate of a particular area is the largest determinant to the life found there (24). Long-term changes in weather patterns affect food supplies, thereby changing the assemblages of species present. An accessible and sustainable food

TABLE 5.2
Overview of Animal Determinants of Health (ADH) with Illustrative Examples of Potential or Realized Climate Impacts on the Determinants

ADH	Description	Examples of climate change impacts on ADH
Meeting the needs for daily living		
Food security	Animals, or the people providing animals with food, can feasibly and consistently access sufficient, safe, appropriate, and nutritious food that meets the animals' evolved nutritional needs.	• Food security is threatened by increased dispersal of existing and emerging pests, affecting crop production and yield in previously untouched areas (5). • Animals will require larger quantities of food to compensate for higher energy requirements under heat stress or cold stress (6,7). • Extreme weather can mobilize contaminants that leach into pasture lands and flood waters (8).
Water security	Safeguarding sustainable access to adequate quantities of safe water will sustain lives, support ecosystems, and enable socio-economic development.	• High water levels have contributed to terrestrial mammal population declines in the Peruvian Amazon (9). • Increased runoff into coastal systems contaminates shellfish, increases harmful algal blooms, and reduces dissolved oxygen in the water (10). • Rising water temperatures and dams that prevent salmon from returning to spawning sites co-contribute to population decline (11).
Physical security	The physical environment can support the needs of animals and is safe from natural, social, and environmental hazards (e.g., extreme weather or overexploitation).	• Hunting pressures on elephants increase as they migrate into human-dominated regions to feed (12). • Climate change affects the estuarine habitat and its associated nursery quality for juvenile fish (13).
Capacity to cope with stressors and change		
Innate capacity to cope	Animals have inherent capabilities that allow them to respond to climate pressures (e.g., phenotypic plasticity or population genetic diversity).	• Genetic modification of livestock has been explored to increase thermotolerance (14,15). • Many animal species exhibit behaviors to increase heat loss (16).
Bolstering capacity to cope	Access to capacities that allow coping (e.g., migration paths to alternative habitats, or animal husbandry modifications).	• Risk-management and preparedness activities can reduce the impact of climate change (17). • Farm-level modifications can increase adaptive capacity to climate change, such as planting vegetation along structure walls to reduce solar radiation (18).

(Continued)

TABLE 5.2 (Continued)
Overview of Animal Determinants of Health (ADH) with Illustrative Examples of Potential or Realized Climate Impacts on the Determinants

ADH	Description	Examples of climate change impacts on ADH
Multi-solving	Social interventions that ensure access to animals' needs for living can be maintained under new conditions and provide additional benefits.	• Animal welfare education and investment can concurrently protect animal health and sustain animal production (19).
Meeting Expectations		
Productivity	Productivity in domestic animals depends on the quantity of goods and services produced and the quantity of resources used to obtain them (i.e., labor, animals, land, feed, veterinarians, etc.).	• Optimal milk and meat production is affected by many aspects of poor animal welfare (20,21).
Social pressures	Social pressures arise when addressing the cultural, regulatory, or scientific expectations for animal populations through human decisions and actions.	• When producers are uncertain about impacts and risks of climate change, the adoption of adaptation strategies will be affected (22). • Agricultural subsidy and trade policies can reduce incentives for producers to respond to economic and environmental changes (23).

supply is a primary factor in determining animal population growth and productivity (25). Climate establishes which plants can grow, which grazers can eat those plants, which predators can eat those grazers, and which animals will be able to produce expected values for people. Predation and social interactions influence access to food and, thus, food security (26).

Climate change is already driving significant changes in species' geographic ranges with subsequent impacts on food webs. For example, the number of feeding links between fish species in Mediterranean Sea food webs is projected to decrease by 73.4% off the continental shelf (27). Such large-scale impacts on marine food web structure will have deep consequences on ecosystem functioning. Others foretell of climate change weakening marine food webs through reduced energy flow to higher trophic levels and a shift toward more detritus-based systems, leading to food web simplification and altered producer–consumer dynamics (28).

People are responsible for providing confined livestock with access to sufficient, safe, and nutritious foods, especially where animals are unable to disperse and food supply is controlled (26). Climate change impacts on forage quality and availability and on the capacity of farmers to access adequate safe food of suitable nutritional quality will be impacted by climate change, especially in regions already nearing climate extremes and/or experiencing reduced human capacity to adapt because of poverty, insecurity, or lack of mobility.

A well-noted indirect consequence of climate change on animal food security results from decreased crop quality. Uncertainty in the quantity and nutritional quality of crops for feed due to climate change (6) limits a producer's ability to meet the food needs for their livestock. Higher temperatures paired with lower precipitation decrease crude protein, digestible organic matter, and the forage quality of potential feed stuffs for cattle (6,29–31). In areas of the United States where ranching is common, heat waves result in low-quality forage, increased abundance of noxious plants, and invasion of woody plant species (32), which are rarely consumed by most livestock. The impacts of climate change on food security for animals are not always negative, especially at higher latitudes. One review noted that climate change has resulted in longer growing seasons in some areas of the world, thereby increasing crop production (33).

Food Safety

Climate change will affect human food safety through increased mycotoxins in plant products, pesticide residues affected by changes in pest pressure, marine biotoxins, the presence of pathogenic bacteria in foods following extreme weather conditions, such as flooding and heat waves, and more (5). For species with food chains overlapping human food chains, such as pets like dogs and cats, the concerns for human food safety directly apply. For other species, climate drivers will likely change pathogen and pollutant distribution within nonhuman food chains as well. Alterations to food webs, lipid dynamics, ice and snow melt, and organic carbon cycling will affect pollutant levels in water, soil, air, plants, and animals (34). Flooding and melting events will remobilize contaminants and redistribute them onto grazing lands, thus contaminating animals and animal products (8).

Changing Food Needs

Changing climate systems will affect the energetic needs of some species. For example, climate change exacerbates heat stress through increased temperatures. Heat stress supresses appetite but increases energy requirements; these increased energy requirements affect growth rate, rumen function, and grazing behaviors (6,29). The physiology of ectothermic animals, such as fish, is strongly affected by environmental temperatures and thus too are their energetic needs. For example, in lakes high in the Rocky Mountains, unless lake productivity improves with climatic changes, temperature-responsive increases in energetic demands may result in lower trout growth (7). Feed intake and feed efficiency, meaning conversion of feed to energy for life and production parameters, is also affected by increased temperature and worsened due to reduced feed intake under heat stress (35,36).

Phenology is the study of relationships between climate and periodic biological phenomena, such as migration, insect eruptions, or flowers blooming. The food requirements for raising young are often linked to the seasonal peak in food availability for wild species. For example, nesting of songbirds in the boreal forest coincides with the spring emergence of insects, ensuring an abundant food supply for their nestlings. Some bird species have not yet changed their timing of breeding sufficiently to coincide with changes in peaks of food availability, a response that appears to be contributing to declines in their population (37).

WATER SECURITY

Water Availability

Variability of temperature and rainfall is expected to increase around the world leading to more drought in some areas and floods in others (38). Tropical areas may see more rainfall, and already dry areas mid-latitude and in the interior of continents will likely see less rainfall (38). As climate change increases the frequency, intensity, and duration of extreme weather events, we may see areas alternating between periods of flood and drought (39). This alternation will require animal health managers to adapt to new "normals." Dairy farmers in the United States, for example, are being counseled to develop emergency plans for increased frequency of flash floods and the associated increases in environmental mastitis (40). Major shifts in Amazon wildlife populations have been correlated with intensification of floods and drought cycles (9), which affect the availability and safety of harvested wildlife and fish.

Drought has long been a primary driver of domestic and wild species survival, persistence, and health, but climate change has accelerated hydrological processes to make drought more frequent and more intense (41). Water insecurity from drought brings food insecurity due to impeded plant growth, increased fire risk, increased opportunity for pathogen and parasite transmission (due to animals concentrating around limited water supplies), dehydration, and death.

Excess water, from extreme rainfall and floods, can have immediate and long-term impacts on the determinants of health. Floods not only decrease water safety via changed exposure to environmental hazards such as pathogens or pollutants, but they can also cause physical damage to animal infrastructure, cause drowning and hypothermia, and alter ecosystems by displacing people and animals.

Water Safety

Runoff is the flow of surface and groundwater into rivers, lakes, and oceans. It occurs due to precipitation, melting snow, or irrigation, and can carry hazards such as fertilizer, bacteria, pesticides, and sediment that reduce water quality (42). These threats are magnified during extreme weather events (42). Higher precipitation can increase the spread of pathogens through contaminated feed and water in terrestrial livestock, aquatic species, and wildlife populations, especially via faecal contamination (33). Further, runoff of fertilizer or sewage into coastal systems contaminates shellfish, increases harmful algal blooms, and reduces dissolved oxygen in the water (10). In some regions of the world, industrial and agricultural wastes are not safely disposed of, which increases the risk of mixing of floodwaters and drinking water during monsoon seasons (43), threatening both animal and human health.

Water Cycle and Watershed Effects

Variability of the hydrological cycle makes many of the effects of climate change on water security difficult to predict (38). Bangladesh, for example, experiences droughts, floods, and drastic temperature swings within a single season (44). The variety of effects that can be experienced is illustrated by the impacts on aquaculture in this country. Temperature swings affect the growth rate and metabolism of farmed tilapia and can lead to deformities in juvenile fish (43). Sea-level rise brings a host of water-security problems, such as salt intrusion into coastal groundwater and seasonal or episodic flooding of land-based aquaculture facilities due to storm surges (10). The prevention of shrimp escapes from aquaculture sites due to flooding requires new adaptive responses, such as the installment of netting around tanks (10). Shrimp and prawn production further is affected by disease outbreaks and parasite introductions due to sea-level rise and loss of habitat availability in inland systems due to drought (45). Adaptive interventions such as Integrated Multi-Trophic Aquaculture (IMTA) systems have been proposed as a response to these water-security challenges in Bangladesh (see Response Example 5.1).

RESPONSE EXAMPLE 5.1   NEW ANIMAL PRODUCTION SYSTEMS FOR CLIMATE ADAPTATION

Researchers in Bangladesh assessed the possibility of integrated multi-trophic aquaculture (IMTA) systems to adapt to climate change (45). IMTA systems combine fish, crustaceans, and seaweeds to create a mini-ecosystem. Fed fish (e.g., shrimp and finfish) exist in middle to upper trophic levels with organic extractive species (e.g., crustaceans such as mussels) beneath. Seaweeds then produce dissolved oxygen, consume carbon dioxide, and provide cleaner water for the entire ecosystem.

Key message: new agricultural production practices, such as IMTA systems, can help mitigate the impacts of climate change on domestic livestock and the ecosystems on which they depend.

Human modifications of water sources can further exacerbate water insecurity. For example, dams across rivers affect both the quality (temperature) and availability of water. A case study from the Sacramento River showed that construction of dams prevented chinook salmon (*Oncorhynchus tshawytscha*) from returning to habitual spawning sites (11). This salmon population decline was worsened by rising water temperatures (11). Response opportunities have been developed to mitigate the effects of the dams on populations already stressed by climate change (see Response Example 5.2).

RESPONSE EXAMPLE 5.2   HUMAN INTERVENTIONS CAN
HELP PROTECT DETERMINANTS OF ANIMAL HEALTH

To combat the impacts of climate change on Sacramento River salmon, managers schedule release of cold water from dams to support temperature needs for earlier springtime spawning and increased survival during drier years (11). Recognizing the benefits of ecosystem-based management over single-species management is important to ensure that changes made in support of salmon populations do not harm other native river species. Maintaining habitat heterogeneity helps distribute the risk of climate change across the Sacramento River landscape (11). Thus, population redundancy is prioritized to decrease risk of extinction in addition to restoring physical and biological river processes and reconnecting migratory corridors to historical spawning sites.

Key message: humans have made a lot of changes to most ecosystems, but we can also adjust these structures to better help animals cope with climate change.

Physical Security

Physical security exists when a place can support the needs of animals and is safe from natural and environmental hazards. Here we specifically use the word *place* instead of *space*. The notion of *place* encompasses not only spatial boundaries but also the relationships that occur within those boundaries. Climate change is not only changing the nature of what exists in a space, but also our relationships with those spaces.

Heat

Climate change will directly reduce physical safety through excess heat. The experience in the human health sector has clearly demonstrated the importance of heat-related threats (46). Extreme heat has already interacted with drought and habitat alterations to kill animals in wildfires in South America, Europe, North America, and Australia. Impacts of extreme heat events are expected to become more frequent, severe, and widespread in the future as witnessed by flying-fox (*Pteropus* spp.) deaths during heat waves in Australia (47). Production, health, welfare, and economic impacts of heat stress have been well characterized for a variety of animal agriculture systems and are anticipated to increase with climate change (48). Heat stress not only causes suffering and premature death, but also can reduce productivity and fertility in both aquatic and terrestrial species (4). In some cases, climate change may increase productivity or reproduction. For example, domestic cat overpopulation is worsened when climate change results in longer summer seasons, and thus, increased kitten survival and earlier sexual maturity (49). People, too, experience the effects of hot weather conditions, resulting in changes in how they care for their animals. Obesity and behavioral problems can result when we ignore our duty to exercise our pets when temperatures are not favorable (49).

Although climate-driven changes in temperature, precipitation, and other factors do not occur in isolation, laboratory-based studies can provide concrete evidence of the potential effects of climate change on animal behavior, such as memory, spatial, and associative learning. Under heat stress, zebra finches did not exhibit preference for the mating call of a zebra finch over a different species (50), providing evidence that heat stress increases the chance of a biological mismatch that reduces their reproductive success. Larger cognitive decline and lower brain weights were seen in rodents born to chronically heat-stressed mothers, and in trout maternal heat stress can reduce the ability of offspring to locate food (50).

Acidification

Ocean acidification will negatively impact many marine organisms and cause indirect effects on marine ecosystems. When CO_2 dissolves in seawater, it generates carbonic acid (H_2CO_3), which acidifies the water. Ocean acidification is expected to affect marine life in three ways: (1) lower carbonate concentration impedes the calcification process for shellfish and corals; (2) lower pH changes acid-base regulation and other physiological processes; and (3) increased dissolved CO_2 alters the ability of primary producers to photosynthesize (51). Some view ocean acidification as the most pressing and critical issue facing the oceans due to the anticipated irreversible ecosystem changes. Food system pathways are already being altered when acidification threatens coral reefs, which influences nursery areas for wild fish that form the basis for coastal fisheries – a critical threat to ecosystem integrity and conservation as well as food security and income in poor coastal communities (52).

Conflict

Warming temperatures, changes in precipitation, and more extreme weather, such as droughts, wildfires, and monsoons, render historically suitable habitats unviable, causing people and animals to change where they live and their interactions therein. For example, climate change is increasing range expansion of African apes as traditional protected lands become inhabitable (53). Food security and water security are threatened within these protected lands, and so the apes are forced outside the protected areas to find food and water. By leaving the protected lands, hunting pressure and other human threats (e.g., vehicle collisions) have increased for the apes, necessitating expansion of anti-poaching efforts beyond the limits of protected areas (53).

Climate change also changes animal-to-animal competition for resources and spaces (12) and predator-prey interactions. For example, chemical alarm cues important for predator avoidance in ambon damselfish (*Pomacentrus amboinensis*) are expected to degrade much faster in more acidic water (54). Coral reef damselfishes reared under high CO_2 environments lose ability to distinguish between olfactory cues that are important for locating reefs, which affects habitat selection and predator avoidances (54).

Climate change is already having a measurable impact on species distributions, reproduction, and behavior. This impact includes continual range-shifts in some species, which will cause new invasive species concerns. Biological invasions cause species loss, changes in distribution, and habitat degradation; challenges that climate change worsens. In some species of farmland birds, seasonal harvest of crops now overlaps with nesting periods, creating a need for nest protection (55). Climate change also leads to shifting range distribution of bird species to the north, leading to hybridization and lower reproductive success, or loss of a species entirely – a current concern for the black-capped chickadee (*Poecile atricapillus*) in North America (56).

Scarcity of water, food, and livelihoods will lead to increasingly desperate human populations and augment the risk of human–human conflict (57). Human conflict can decrease access to veterinary services, thus increasing endemic animal disease exposure and impacts and reducing agricultural productivity (58). Violent civil strife has been linked to degradation of environmental resources necessary for animal health, as well as over-exploitation of wildlife (59). When people migrate to more favorable climates or flee areas stricken by natural disasters or conflict, long-distance transportation can stress livestock and companion animals and increase risk of exotic diseases (49).

Natural Disaster

Typhoons, hurricanes, fires, floods, and extreme heat are all anticipated to increase in severity and frequency under conditions of global warming (60). The impacts of these disasters can be catastrophic (61). From September 2019 to March 2020, approximately three billion wild amphibians, reptiles, birds, and terrestrial mammals were killed or displaced due to over 15,000 fires that destroyed extensive regions of Australian habitat (62). Disasters can expose food animals to contaminated water, feed, and air in addition to direct physical harm. Pet abandonment during natural disasters affects the welfare of animals and their owners. If enough animals are killed in a disaster, it may threaten a community's or a country's food security. Natural disasters have negative animal health implications that in turn will have public health and economic implications, yet many countries omit animals from their national and regional contingency planning (63).

Habitat and Housing

Habitat loss is one of the greatest threats to biodiversity, and its effects are compounded by climate change (64). A meta-analysis found that negative impacts of habitat loss and fragmentation have been disproportionately severe in areas with very high temperatures and declining rainfall, although impacts varied (65). For domestic species, human alteration of housing or rearing conditions will be needed to protect animals from excess heat, extreme weather, and other climate effects. Housing and husbandry modifications have been suggested to help domestic animals adjust to climate. Such modifications are needed in both major livestock species and poultry systems, as well as minor livestock species and companion animals, and must consider inexpensive adaptation strategies that work for lower input systems (66). For wild animals, it is important to manage the landscape post-extreme weather event. Watershed management and access to high ground is important post-flood, and where plants or downed trees have been removed, artificial housing structures may be needed (16).

As the world changes due to climate change, habitat fragmentation is also increasing. As climates become intolerable, species exhibit range expansion to areas that are more suitable and where resources are more plentiful. These new locations are populated through movement of animals along migration corridors. Urbanization and roadway structures present barriers to migration, and motor vehicle accidents are important causes of wildlife mortality (67). Ensuring connectivity between existing protected areas is an important adaptation activity (68) (see Response Example 5.3). Climate change not only affects plant abundance, but also the plant species that survive amid increasing climate pressures. This is a concern for the giant panda in China, where climate change has reduced the diversity of bamboo species (69). Such effects will impact reforestation and conservation efforts that must concurrently consider protecting habitat for species already present, as well as anticipating which species will be present under predicted climate change scenarios.

RESPONSE EXAMPLE 5.3   REDESIGNING THE HUMAN-ANIMAL INTERFACE

Increasing climate pressures are driving wildlife to new regions with better food, water, and physical security. Wildlife movement is hampered by transportation infrastructure (road and railways) and fragmented habitat. Humans can act to support safe passage of wildlife to new habitats. Wildlife corridors across the Trans-Canada highway have reduced wildlife motor-vehicle collisions by 80% thereby helping to maintain biodiversity in many species (70,71). Several Ontario Parks have also addressed climate change by installing ecopassages that are used by snakes, raccoons, muskrats, beavers, and chipmunks to safely cross roadways. In combination with road fencing, these ecopassages reduce mortalities associated with wildlife movement and migration (72).

Key message: Migration is a major coping tool for wildlife. As climates change, we need to provide safe passage to new ecosystems.

CAPACITY TO COPE WITH STRESSORS AND CHANGE

As life situations change, animals, and those responsible for their care, need to have capacity and resources to adapt to or cope with their new reality. "The components of effective adaptation to the future will be unpredictable and emergent rather than predictable and planned because of the unprecedented rate of social and environmental change and the complexity of interactions between co-occurring global threats and anthropogenic climate change" (4). This means animal health managers will need to protect and build capacity to cope with whatever may come with climate change.

Coping capacity is derived from the combined effects of individual, ecological, social, economic, and environmental factors and relationships that affect options for animals to positively respond to stressors and changes. Capacity to cope allows animals, populations, or systems to adjust to a disturbance, lessen damage, take advantage of opportunities, and adapt to the consequences (73). Vulnerability to the direct and indirect health impacts of climate change will vary widely by species, location, and management system, but it will ultimately be determined by exposure to the effects of climate change and capacity to adapt to or cope with those effects (74). Having capacity to cope with stressors and change is a major component of adapting to climate change, especially when not all adverse effects can be foretold.

Coping capacity can be an inherent part of an animal's physiology, life history, and ecology or can be augmented by human interventions in the animal's life circumstances. Health is built by empowering individuals and communities to increase access and control over their determinants of health, thereby promoting their capacity to live active and productive lives (75). Animal health and welfare provide the raw materials for resilience to a wide variety of threats (76). Protecting and sustaining the determinants of health is the bedrock of protecting coping capacity.

Building resiliency is part of coping capacity and can be seen as the process of adapting to adversity, trauma, and threats to grow and change and be prepared for the next trauma or threat (77). Resilience, despite having several debated definitions, is ultimately about healthy individuals and thriving ecological and social communities. A resilience-centered approach focuses on ensuring animals have the capacities and resources to promote sustainable well-being in the face of adversity and the challenges of daily living.

INNATE CAPACITY TO COPE

Animals' ability to cope with or adjust to changing climatic conditions can be classified as either "persisting in place" or "shifting in space" (78). Persist-in-place attributes enable species to survive under new conditions, whereas the shift-in-space responses require an ability to move to more favorable conditions. The inherent genetic diversity within populations has allowed for evolution in changing conditions. Unfortunately, for many species, the rate of climate change is happening faster than their genetic diversity allows them to evolve. For this reason, many investigators are turning to human-assisted genetic selection or modification to improve capacity to cope. Several livestock species are more heat tolerant than others (6,15,30,79) and crossbreeding programs can build resilience (31). For example, zebu and taurine cattle are more heat resistant than most breeds raised in North America, and Creole sows are more heat resistant than large white sows (80). Genetic diversity is also a critical component needed in aquaculture to breed for heat tolerance and create resilient ocean ecosystems (81). Maintaining biodiversity in ecosystems is essential to building resilience in animal populations, as is preserving genetic diversity within each species (82). For example, the limited genetic diversity in wild camels in China was identified as a point of vulnerability of this species to adapt to climate change (83).

Several candidate genes have been identified for modification to enhance capacity to cope with climate change in livestock species (84). Heat Shock Protein – 70 has been investigated for its ability to improve the heat tolerance of animals (14,15), along with the SLICK hair type gene in Holstein cattle to improve thermotolerance via production of silky hair of a shorter length (15,35,80).

Not all animals will have the same ability or opportunity to evolve to new climate realities, which creates species inequities. Some animals respond to changing environmental conditions through inherent physiological tolerance, developing new behavioral or physiological ways to adapt for survival, or migrating to areas that offer more favorable conditions (85). Specialist species, as opposed to generalist species, will have fewer options to adapt to conditions outside of their evolved norm without human intervention. Some species will be capable of adapting to the effects of climate change in small doses and over relatively narrow limits (see Response Example 5.4) if unimpeded by maladaptive human behaviors or actions. For example, habitat fragmentation could prevent some species from moving to more suitable areas, whereas engineering and breeding innovations may allow livestock production in otherwise unsuitable climates. Coping-focused climate adaptation actions need to understand both the inherent adaptive capacity of

---

**RESPONSE EXAMPLE 5.4   ANIMALS CAN ADAPT IF
WE LET THEM**

Animals adapt to climate change without human intervention (16). Heat loss can be increased through: behavior (e.g., moose seek out shaded, moist soil during the day to conduct body heat, and seals stop nursing pups to return to sea during high heat), and nest architecture (e.g., verdin orient their nests to face prevailing winds during the hottest months to increase nest survival, and the Eastern Three-Lined Skink adjusts its nest depth and timing of oviposition to buffer effects of increased temperature on offspring) (16). Other species have adapted to extreme weather events (e.g., feral cats were observed to travel up to 12 km to take advantage of easy hunting following wildfires, and sea turtles and saltmarsh sparrows rebuild their nests following weather-related damage) (16).

Key message: Animals have tremendous ability to adapt to different climates. Unfortunately, there is a limit at which animal adaptive capacities are no longer enough. Humans need to provide opportunities for animals to use their adaptation tools/behaviors.

the species and how human systems can compensate or compromise animals' inability to be resilient and bounce back from climate stressors and shocks. Health managers may need to triage populations, problems, and/or places with distinct climate sensitivity through vulnerability and impact assessments. An intersectoral approach is, therefore, fundamental to action on climate change and health.

BOLSTERING CAPACITY TO COPE

For livestock, several coping capacity solutions involve farmer interventions. Farmers can change or diversify the species they rear, adapt housing conditions to protect animals from extreme weather, select species and breeds that are better able to adapt to changing conditions, adjust their food and modify water-management practices, develop disaster-response plans, and adopt preventive veterinary practices suited to changing disease conditions (17).

Animal healthcare and veterinary services will need to change to ensure they can provide accessible and appropriate services. Although the goals of preventing disease, maintaining productivity, and sustaining health will remain, the form and scope of services will need to shift. Animal healthcare and veterinary services will need to: (i) provide services that mitigate climate impacts; (ii) reduce population vulnerability to lessen those impacts; (iii) enhance population resilience to avoid impacts; and (iv) address climate change risks at their sources (4). Programs will need to expand from only risk management to capacity-building for healthy, resilient animal populations and animal health systems.

MULTI-SOLVING

Animal health programs are climate change adaptation programs. Empowering people to protect and promote animal health and welfare improves livelihoods while building social and ecological resilience to endemic and emerging threats (19). This is an example of multi-solving. Protecting and promoting animal health in ways that are suitable and adaptable to local circumstances can concurrently help to build resilience against a changing future while dealing with today's challenges. Animal health managers need to create multi-solving strategies that protect multiple determinants of health at the same time. For example, habitat protection could provide food and physical security for wildlife while also preventing soil erosion for crop growing and providing cooler places for livestock. In a world of concurrent health, societal, and environmental problems, unique solutions for each problem are neither feasible nor effective (86).

MEETING EXPECTATIONS

PRODUCTIVITY

Societies' definitions of animal health are often based on the absence of certain diseases or the ability of an animal, herd, or population to produce desired products. For example, salmon fisheries regulators often focus on abundance for harvest as a surrogate for health (87). Production targets have often been incorporated into assessments of herd and individual livestock health. Being productive, therefore, is intertwined with health definitions for animals from which society derives economic or subsistence products. Animal health management of such species, therefore, must consider production expectations when managing climate change health impacts.

Publications on livestock and aquaculture species found in our review revealed that the effects of climate on animal health also impact productivity. For example, in dairy and beef cattle, heat stress decreases weight gain, production parameters, and feed conversion (88). Dairy cattle have very high water requirements for daily living, and even higher water intake needs to meet production goals; thus, they are more affected by heat stress compared to beef cattle (35). Heat stress leads to degraded mammary epithelial cell production and rumen fermentation in dairy cows (35). Reduced dry matter intake in dairy cows can put the animals at risk for negative-energy balance affecting hormone production and embryo survivability (35).

High temperatures suppress appetite, which, when combined with increased energy requirements, can lead to weight loss (6), which reduces milk production in dairy cows and meat production in other ruminants (20,21). Higher atmospheric CO_2 increases variability in pasture composition, water availability, and earlier or later spring and fall seasons (20), thereby compounding the impacts of heat stress. Precipitation also affects forage quality and quantity through long dry seasons or flooding that decrease rangeland for grazing or crop production (20,21). The introduction of more climate resilient crops and grasses, improvement of soil and water management, and changing land use and irrigation patterns are potential mitigation strategies for effects on animal production (6).

Social Pressures

Laws and regulations set the basis for societal response to many issues, including climate change. Agricultural subsidies can encourage or discourage actions that support animal health in the face of climate change and/or mitigate the climate change effects (23). Agricultural trade policies, along with production and income-assurance schemes, can produce incentives or disincentives for farmers to take adaptive actions. Policies to protect watersheds or wildlife habitat may reduce farmer flexibility to respond to climate change by reducing their ability to adapt land use and respond to extreme events. Many authors contemplated ways to mitigate the effects of agriculture on climate, and several studies proposed intensification of agricultural (6,30), diversification of agriculture production (6,31), and integration of crop-livestock systems (6,30,31). Ultimately, reducing climate change will enhance resilience and protect the determinants of health, but some shorter-term changes will be needed to mitigate immediate threats to production impacts.

The way in which people perceive climate change risk is informed by their social interactions and cultural worldviews (89), which in turn affects their willingness to respond to climate impacts on animals. Changed economic, recreational, and cultural opportunities can be expected as climate change alters fish and wildlife abundance, distribution, and safety. Impacts on rural, remote, and Indigenous community food security and cultural integrity are already being realized. Citizen expectations for animals to meet their aesthetic, abundance, and safety expectations can be expected to grow in the face of climate change impacts on animal health.

Science often sets our expectations for the abundance, distribution, and appearance of fish and wildlife. There are growing expectations to link wildlife health with conservation, but to do so through a health promotion rather than a disease-control perspective. This approach will necessitate the use of a determinants of health perspective. Animal welfare scientists are seeing a growing need to contribute their knowledge to refining expectations and adaptations to define and protect animal health (49).

CONCLUSION

We can reasonably expect that more harm will come to animals from loss or degradation of their determinants of health than from the diseases that may arise with climate change. Efforts to protect, maintain, and improve animal health and animal welfare need to provide a continuum of climate change actions that include: (i) providing services to mitigate climate change impacts; (ii) reducing population vulnerability to lessen those impacts; (iii) enhancing population resilience to avoid or cope with the impacts; and (iv) attacking climate change risks at their sources (4). This continuum of care cannot be created and maintained without strengthening links between animal health practice, policy, and research, with social, ecological, and economic sectors that influence the determinants of animal health.

Animals are being impacted by multiple anthropogenically-driven global crises simultaneously. Many of these concurrent threats are linked not only in their causes, but also in their solutions. A focus on all-hazards approach by attending to the determinants of health (access and quality) before harms emerge is a critical strategy to respond to climate change while also building readiness and resilience against concurrent global and local threats.

REFERENCES

1. Singer, M., 1996. A dose of drugs, a touch of violence, a case of AIDS: Conceptualizing the SAVA syndemic, Free Inquiry in Creative Sociology. 1; 28(1), pp. 13–24.
2. Stephen, C. and Wade, J., 2020. Missing in action: Sustainable climate change adaptation evidence for animal health. *Canadian Veterinary Journal-Revue Veterinaire Canadienne*, 61(9), pp. 966–970. PubMed PMID: WOS:000609170600006. English.

3. Wittrock, J., Duncan, C. and Stephen, C., 2019. A determinants of health conceptual model for fish and wildlife health. *Journal of Wildlife Diseases*, *55*(2), pp. 285–297. PubMed PMID: WOS:000464040500001. English.
4. Stephen, C. and Soos, C., 2021. The implications of climate change for veterinary services. *Rev Sci Tech*, *40*(2), pp. 421–430. PubMed PMID: 34542106. The implications of climate change for Veterinary Services. English.
5. Miraglia, M., Marvin, H.J.P., Kleter, G.A., Battilani, P., Brera, C., Coni, E., et al., 2009. Climate change and food safety: An emerging issue with special focus on Europe. *Food and Chemical Toxicology*, *47*(5), pp. 1009–1021. PubMed PMID: WOS:000265970600011. English.
6. Zhang, Y.Q.W., McCarl, B.A. and Jones, J.P.H., 2017. An overview of mitigation and adaptation needs and strategies for the livestock sector. *Climate*, *5*(4), pp. 15. PubMed PMID: WOS:000419196500020. English.
7. Christianson, K.R. and Johnson, B.M., 2020. Combined effects of early snowmelt and climate warming on mountain lake temperatures and fish energetics. *Arctic Antarctic and Alpine Research*, *52*(1), pp. 130–145. PubMed PMID: WOS:000529452300001. English.
8. Tirado, M.C., Clarke, R., Jaykus, L.A., McQuatters-Gollop, A. and Franke, J.M., 2010. Climate change and food safety: A review. *Food Research International*, *43*(7), pp. 1745–1765. PubMed PMID: WOS:000282860700003. English.
9. Bodmer, R., Mayor, P., Antunez, M., Chota, K., Fang, T., Puertas, P., et al., 2018. Major shifts in Amazon wildlife populations from recent intensification of floods and drought. *Conservation Biology*, *32*(2), pp. 333–344. PubMed PMID: WOS:000428319600009. English.
10. Reid, G.K., Gurney-Smith, H.J., Flaherty, M., Garber, A.F., Forster, I., Brewer-Dalton, K., et al., 2019. Climate change and aquaculture: Considering adaptation potential. *Aquaculture Environment Interactions*, *11*, pp. 603–624. PubMed PMID: WOS:000507378500001. English.
11. Meyers, E.M., 2021. Protecting a displaced species in an altered river: A case study of the endangered Sacramento River winter-run Chinook Salmon. *California Fish and Game*, *107*, pp. 172–188. PubMed PMID: WOS:000672513100009. English.
12. Abrahms, B., 2021. Human-wildlife conflict under climate change. *Science*, *373*(6554), pp. 484–485. PubMed PMID: WOS:000681716000017. English.
13. Davis, M.J., Woo, I., Ellings, C.S., Hodgson, S., Beauchamp, D.A., Nakai, G., et al. A climate-mediated shift in the estuarine habitat mosaic limits prey availability and reduces nursery quality for juvenile salmon. *Estuaries and Coasts*, 20. PubMed PMID: WOS:000706557200001. English.
14. Hassan, F.U., Nawaz, A., Rehman, M.S., Ali, M.A., Dilshad, S.M.R. and Yang, C.J., 2019. Prospects of HSP70 as a genetic marker for thermo-tolerance and immuno-modulation in animals under climate change scenario. *Animal Nutrition*, *5*(4), pp. 340–350. PubMed PMID: WOS:000502026500003. English.
15. Osei-Amponsah, R., Chauhan, S.S., Leury, B.J., Cheng, L., Cullen, B. and Clarke, I.J., et al., 2019. Genetic selection for thermotolerance in ruminants. *Animals*, *9*(11), pp. 18. PubMed PMID: WOS:000502299900094. English.
16. Buchholz, R., Banusiewicz, J.D., Burgess, S., Crocker-Buta, S., Eveland, L. and Fuller, L., 2019. Behavioural research priorities for the study of animal response to climate change. *Animal Behaviour*, *150*, pp. 127–137. PubMed PMID: WOS:000463590000011. English.
17. Calvosa, C., Chuluunbaatar, D. and Fara, K., 2009. Livestock and climate change. *International Fund for Agricultural Development, 2009.*

18. Schauberger, G., Hennig-Pauka, I., Zollitsch, W., Hortenhuber, S.J., Baumgartner, J., Niebuhr, K., et al., 2020. Efficacy of adaptation measures to alleviate heat stress in confined livestock buildings in temperate climate zones. *Biosystems Engineering*, *200*, pp. 157–175. PubMed PMID: WOS:000598490400012. English.

19. Njisane, Y.Z., Mukumbo, F.E. and Muchenje, V., 2020. An outlook on livestock welfare conditions in African communities - A review. *Asian-Australasian Journal of Animal Sciences*, *33*(6), pp. 867–878. PubMed PMID: WOS:000528253100001. English.

20. Rojas-Downing, M.M., Nejadhashemi, A.P., Harrigan, T. and Woznicki, S.A., 2017. Climate change and livestock: Impacts, adaptation, and mitigation. *Climate Risk Management*, *16*, pp. 145–163. PubMed PMID: WOS:000405852000012. English.

21. Hussain, M., Butt, A.R., Uzma, F., Ahmed, R., Irshad, S., Rehman, A., et al., 2020. A comprehensive review of climate change impacts, adaptation, and mitigation on environmental and natural calamities in Pakistan. *Environmental monitoring and assessment*, *192*(1), pp. 1–20.

22. Henry, B., Charmley, E., Eckard, R., Gaughan, J.B. and Hegarty, R., 2012. Livestock production in a changing climate: Adaptation and mitigation research in Australia. *Crop and Pasture Science*, *63*(3), pp. 191–202.

23. Antle, J.M., 2010. *Adaptation of Agriculture and the Food System to Climate Change: Policy Issues*. Washington, DC: Issue Brief to Resoucres for the Future.

24. Stevens, A.P., 2010. Introduction to the basic drivers of climate. *Nature Education Knowledge*. p. 10.

25. Sinclair, A.R. and Krebs, C.J., 2002. Complex numerical responses to top-down and bottom-up processes in vertebrate populations. *Philos Trans R Soc Lond B Biol Sci*, *357*(1425), pp. 1221–1231. PubMed PMID: 12396514. PMCID: PMC1693037. English.

26. White, T.C.R., 2008. The role of food, weather and climate in limiting the abundance of animals. *Biological Reviews*, *83*(3), pp. 227–248. PubMed PMID: WOS:000257796900001. English.

27. Rice, J., 1995. Food web theory, marine food webs, and what climate change may do to northern marine fish populations. *Canadian Special Publication of Fisheries and Aquatic Sciences*, pp. 561–568. PubMed PMID: 3793682. English.

28. Ullah, H., Nagelkerken, I., Goldenberg, S.U. and Fordham, D.A., 2018. Climate change could drive marine food web collapse through altered trophic flows and cyanobacterial proliferation. *Plos Biology*, *16*(1), p. 21. PubMed PMID: WOS:000423830300005. English.

29. Ghahramani, A., Howden, S.M., del Prado, A., Thomas, D.T., Moore, A.D., Ji, B., et al., 2019. Climate change impact, adaptation, and mitigation in temperate grazing systems: A review. *Sustainability*, *11*(24), p. 30. PubMed PMID: WOS:000506901400020. English.

30. Rivera-Ferre, M.G., Lopez-i-Gelats, F., Howden, M., Smith, P., Morton, J.F. and Herrero, M., 2016. Re-framing the climate change debate in the livestock sector: Mitigation and adaptation options. *Wiley Interdisciplinary Reviews-Climate Change*, *7*(6), pp. 869–892. PubMed PMID: WOS:000385827200006. English.

31. Henderson, B., Gerber, P. and Opio, C., 2011. Livestock and climate change, challenges and options. *CAB Reviews: Perspectives in Agriculture, Veterinary Science, Nutrition and Natural Resources*, *6*(016), pp. 1–11. English.

32. Holechek, J.L., Geli, H.M.E., Cibils, A.F. and Sawalhah, M.N., 2020. Climate change, rangelands, and sustainability of ranching in the western United States. *Sustainability*, *12*(12), p. 24. PubMed PMID: WOS:000553587300001. English.

33. Zinsstag, J., Crump, L., Schelling, E., Hattendorf, J., Maidane, Y.O., Ali, K.O., et al., 2018. Climate change and One Health. *Fems Microbiology Letters*, *365*(11), p. 9. PubMed PMID: WOS:000441113100003. English.

34. Noyes, P.D., McElwee, M.K., Miller, H.D., Clark, B.W., Van Tiem, L.A., Walcott, K.C., et al., 2009. The toxicology of climate change: Environmental contaminants in a warming world. *Environment International*, *35*(6), pp. 971–986. PubMed PMID: WOS:000267415500020. English.

35. Madhusoodan, A.P., Sejian, V., Rashamol, V.P., Savitha, S.T., Bagath, M. and Krishnan, G., et al., 2019. Resilient capacity of cattle to environmental challenges – An updated review. *Journal of Animal Behaviour and Biometeorology*, *7*(3), pp. 104–118. PubMed PMID: WOS:000467384700001. English.

36. Llonch, P., Haskell, M.J., Dewhurst, R.J. and Turner, S.P., 2017. Current available strategies to mitigate greenhouse gas emissions in livestock systems: an animal welfare perspective. *Animal*, *11*(2), pp. 274–284. English.

37. Møller, A.P., Rubolini, D. and Lehikoinen, E., 2008. Populations of migratory bird species that did not show a phenological response to climate change are declining. *Proceedings of the National Academy of Sciences*, *105*(42), pp. 16195–16200.

38. Turral, H., Burke, J. and Faurès, J-M, 2011. Climate change, water and food security: Food and Agriculture Organization of the United Nations (FAO). Available from https://www.fao.org/sustainable-food-value-chains/library/details/en/c/266046/. Accessed July 13, 2022.

39. Ebi, K.L. and Bowen, K., 2016. Extreme events as sources of health vulnerability: Drought as an example. *Weather and Climate Extremes*, *11*, pp. 95–102.

40. University, W.S., 2009. Animal health information for dairy producers affected by floods. ag animal health spotlight. Available from https://s3.wp.wsu.edu/uploads/sites/2147/2015/03/DAIRY-CATTLE-AND-FLOODS-JAN-2009.pdf. Accesed July 13, 2022.

41. Mukherjee, S., Mishra, A. and Trenberth, K.E., 2018. Climate change and drought: A perspective on drought indices. *Current Climate Change Reports*, *4*(2), pp. 145–163. PubMed PMID: WOS:000461110800006. English.

42. Surface runoff and the water cycle. United States Geological Survey; n.d.

43. Rahman, M.L., Shahjahan, M. and Ahmed, N., 2021. Tilapia farming in Bangladesh: Adaptation to climate change. *Sustainability*, *13*(14), p. 20. PubMed PMID: WOS:000676912600001. English.

44. Asian Disaster Reduction Center (ADRC). Information on disaster risk reduction of the member countries: Bangladesh. n.d. [Available from: https://www.adrc.asia/nationinformation.php?NationCode=50&Lang=en].

45. Ahmed, N. and Glaser, M., 2016. Can "Integrated Multi-Trophic Aquaculture (IMTA)" adapt to climate change in coastal Bangladesh? *Ocean & Coastal Management*, *132*, pp. 120–131. PubMed PMID: WOS:000385321500013. English.

46. Gamble, J., Balbus, J., Berger, M., Bouye, K., Campbell, V., Chief, K., et al., 2016. *Ch. 9: Populations of concern*. Washington, DC: US Global Change Research Program.

47. Ratnayake, H.U., Kearney, M.R., Govekar, P., Karoly, D. and Welbergen, J.A., 2019. Forecasting wildlife die-offs from extreme heat events. *Animal Conservation*, *22*(4), pp. 386–395. PubMed PMID: WOS:000476919200011. English.

48. Stephen, C. and Pollock, C.D., 2020. Climate Change: The Ultimate One Health Challenge. In Zinstagg et al. ed. *One Health: The Theory and Practice of Integrated Health Approaches*. London: CABI Press, p. 205.

49. Protopopova, A., Ly, L.H., Eagan, B.H. and Brown, K.M., 2021. Climate change and companion animals: Identifying links and opportunities for mitigation and adaptation strategies. *Integrative and Comparative Biology*, *61*(1), pp. 166–181.

50. Soravia, C., Ashton, B.J., Thornton, A. and Ridley, A.R., 2021. The impacts of heat stress on animal cognition: Implications for adaptation to a changing climate. *Wiley Interdisciplinary Reviews-Climate Change*, *12*(4), p. 29. PubMed PMID: WOS:000644630200001. English.

51. Logan, C.A., 2010. A review of ocean acidification and America's response. *Bioscience*, *60*(10), pp. 819–828. PubMed PMID: WOS:000283813700009. English.

52. Allison, E.H., Beveridge, M.C. and Van Brakel, M., 2009. Climate change, small-scale fisheries and smallholder aquaculture. *Fish, Trade and Development*. pp. 73–87. Royal Swedish Academy of Agriculture and Forestry. Available at https://digitalarchive.worldfishcenter.org/handle/20.500.12348/1330. Accessed July 13, 2022

53. Carvalho, J.S., Graham, B., Bocksberger, G., Maisels, F., Williamson, E.A., Wich, S., et al., 2021. Predicting range shifts of African apes under global change scenarios. *Diversity and Distributions*, *27*(9), pp. 1663–1679.

54. Nagelkerken, I. and Munday, P.L., 2016. Animal behaviour shapes the ecological effects of ocean acidification and warming: Moving from individual to community-level responses. *Global Change Biology*, *22*(3), pp. 974–989. PubMed PMID: WOS:000370491400004. English.

55. Santangeli, A., Lehikoinen, A., Bock, A., Peltonen-Sainio, P., Jauhiainen, L., Girardello, M., et al., 2018. Stronger response of farmland birds than farmers to climate change leads to the emergence of an ecological trap. *Biological Conservation*, *217*, pp. 166–172. PubMed PMID: WOS:000423005200018. English.

56. Varrin, R., Bowman, J. and Gray, P.A., 2007. The known and potential effects of climate change on biodiversity in Ontario's terrestrial ecosystems: Case studies and recommendations for adaptation. *Climate Change Research Report-Ontario Forest Research Institute*. (CCRR-09).

57. Bowles, D.C., Butler, C.D. and Morisetti, N., 2015. Climate change, conflict and health. *Journal of the Royal Society of Medicine*, *108*(10), pp. 390–395. PubMed PMID: WOS:000362725900005. English.

58. Dawoud, A.A. and Hamid, O.I.A. The impact of Darfur conflict on Animal health services, water and animal production in Wadi Salih and Zalingei localities in West Darfur. International Journal of Agr. & Env., *2013*(2), pp. 14–22.

59. Dudley, J.P., Ginsberg, J.R., Plumptre, A.J., Hart, J.A. and Campos, L.C., 2002. Effects of war and civil strife on wildlife and wildlife habitats. *Conservation Biology*, *16*(2), pp. 319–329.

60. Banholzer, S., Kossin, J. and Donner, S., 2014. The impact of climate change on natural disasters. *Reducing disaster: Early warning systems for climate change*. Springer, pp. 21–49.

61. Gallagher, C.A., Jones, B. and Tickel, J., 2020. Towards resilience: The One Health approach in disasters. 2020. In: *One Health: The theory and practice of integrated health approaches [Internet]*. CAB International, pp. 310–326.

62. Parrott, M.L., Wicker, L.V., Lamont, A., Banks, C., Lang, M., Lynch, M., et al., 2021. Emergency Response to Australia's Black Summer 2019-2020: The Role of a Zoo-Based Conservation Organisation in Wildlife Triage, Rescue, and Resilience for the Future. *Animals*, *11*(6), p. 22. PubMed PMID: WOS:000665291000001. English.

63. Garde, E., Pérez, G.E., Acosta-Jamett, G. and Bronsvoort, B.M., 2013. Challenges encountered during the veterinary disaster response: An example from Chile. *Animals (Basel)*, *3*(4), pp. 1073–1085. PubMed PMID: 26479753. English.

64. Mantyka-Pringle, C.S., Martin, T.G. and Rhodes, J.R., 2012. Interactions between climate and habitat loss effects on biodiversity: A systematic review and meta-analysis. *Global Change Biology*, *18*(4), pp. 1239–1252. PubMed PMID: WOS:000301533100004. English.

65. Segan, D.B., Murray, K.A. and Watson, J.E.M., 2016. A global assessment of current and future biodiversity vulnerability to habitat loss-climate change interactions. *Global Ecology and Conservation*, *5*, pp. 12–21. PubMed PMID: WOS:000413275900002. English.

66. De, K., Kumar, D., Thirumurugan, P., Sahoo, A. and Naqvi, S., 2017. Ideal Housing Systems for Sheep to Cope with Climate Change. *Sheep Production Adapting to Climate Change*. Springer, pp. 331–347.

67. Gilhooly, P.S., Nielsen, S.E., Whittington, J. and St Clair, C.C., 2019. Wildlife mortality on roads and railways following highway mitigation. *Ecosphere*, *10*(2), p. 16. PubMed PMID: WOS:000461577000020. English.

68. Lister, N.M., Brocki, M., Ament, R., 2015. Integrated adaptive design for wildlife movement under climate change. *Frontiers in Ecology and the Environment*, *13*(9), pp. 493–502. PubMed PMID: WOS:000364503100006. English.

69. Li, R., Xu, M., Wong, M.H.G., Qiu, S., Sheng, Q., Li, X., et al., 2015. Climate change-induced decline in bamboo habitats and species diversity: Implications for giant panda conservation. *Diversity and Distributions*, *21*(4), pp. 379–391.

70. Dickie, G., 2017. As Banff's famed wildlife overpasses turn 20, the world looks to Canada for conservation inspiration. *Canadian Geographic*.

71. Banff National Park Wildlife corridors - a 'moving' story. In: Canada P, editor. 2017.

72. Ecopassages help wildlife cross roads safely. In: Parks O, editor. 2016.

73. Burkett, V.R., 2013. Coping Capacity. In: Bobrowsky, P.T., eds. *Encyclopedia of Natural Hazards*. Dordrecht: Springer Netherlands, pp. 119–121.

74. Yohe, G.W., 2001. Mitigative capacity – The mirror image of adaptive capacity on the emissions side. *Climatic Change*, *49*(3), pp. 247–262. PubMed PMID: WOS:000167813000001. English.

75. Randolph, T.F., Schelling, E., Grace, D., Nicholson, C.F., Leroy, J.L., Cole, D.C., et al., 2007. Invited Review: Role of livestock in human nutrition and health for poverty reduction in developing countries. *Journal of Animal Science*, *85*(11), pp. 2788–2800. PubMed PMID: WOS:000250648400002. English.

76. Friggens, N.C., Blanc, F., Berry, D.P. and Puillet, L., 2017. Review: Deciphering animal robustness. A synthesis to facilitate its use in livestock breeding and management. *Animal*, *11*(12), pp. 2237–2251. PubMed PMID: WOS:000416121800016. English.

77. Palmiter, D., Alvord, M. and Dorlen, R., 2012. *Building your resilience*. American Psychological Association.

78. Thurman, L.L., Stein, B.A., Beever, E.A., Foden, W., Geange, S.R., Green, N., et al., 2020. Persist in place or shift in space? Evaluating the adaptive capacity of species to climate change. *Frontiers in Ecology and the Environment*, *18*(9), pp. 520–528. PubMed PMID: WOS:000564401800001. English.

79. Onyeneke, R.U., Nwajiuba, C.A., Emenekwe, C.C., Nwajiuba, A., Onyeneke, C.J., Ohalete, P., et al., 2019. Climate change adaptation in Nigerian agricultural sector: A systematic review and resilience check of adaptation measures. *AIMS Agriculture and Food*, *4*(4), pp. 967–1006.

80. Hoffmann, I., 2010. Climate change and the characterization, breeding and conservation of animal genetic resources. *Animal Genetics*, *41*, pp. 32–46. PubMed PMID: WOS:000276775100003. English.
81. Bricknell, I.R., Birkel, S.D., Brawley, S.H., Van Kirk, T., Hamlin, H., Capistrant-Fossa, K., et al., 2021. Resilience of cold water aquaculture: A review of likely scenarios as climate changes in the Gulf of Maine. *Reviews in Aquaculture*, *13*(1), pp. 460–503. PubMed PMID: WOS:000558670700001. English.
82. LeDee, O.E., Handler, S.D., Hoving, C.L., Swanston, C.W., Zuckerberg, B., 2021. Preparing wildlife for climate change: How far have we come? *Journal of Wildlife Management*, *85*(1), pp. 7–16. PubMed PMID: WOS:000584572300001. English.
83. Xue, Y.D., Li, J., Zhang, Y., Li, D.Q., Yuan, L., Cheng, Y., et al., 2021. Assessing the vulnerability and adaptation strategies of wild camel to climate change in the Kumtag Desert of China. *Global Ecology and Conservation*, *29*, p. 10. PubMed PMID: WOS:000701886400001. English.
84. Rovelli, G., Ceccobelli, S., Perini, F., Demir, E., Mastrangelo, S., Conte, G., et al., 2020. The genetics of phenotypic plasticity in livestock in the era of climate change: A review. *Italian Journal of Animal Science*, *19*(1), pp. 997–1014. PubMed PMID: WOS:000563076300001. English.
85. Peck, L.S., 2005. Prospects for surviving climate change in Antarctic aquatic species. *Frontiers in Zoology*, *2*(1), pp. 1–8.
86. Fried, L.P., Piot, P., Frenk, J.J., Flahault, A. and Parker, R., 2012. Global public health leadership for the twenty-first century: Towards improved health of all populations. *Global Public Health*, *7*, pp. S5–S15. PubMed PMID: WOS:000308034900002. English.
87. Wittrock, J., Anholt, M., Lee, M. and Stephen, C., 2019. Is fisheries and oceans Canada policy receptive to a new Pacific salmon health perspective? *Facets*, *4*, pp. 615–625. PubMed PMID: WOS:000502263100001. English.
88. Rust, J.M. and Rust, T., 2013. Climate change and livestock production: A review with emphasis on Africa. *South African Journal of Animal Science*, *43*(3), pp. 255–267. PubMed PMID: WOS:000327323900004. English.
89. McNeeley, S.M. and Lazrus, H., 2014. The cultural theory of risk for climate change adaptation. *Weather Climate and Society*, *6*(4), pp. 506–519. PubMed PMID: WOS:000343068900008. English.

6 Finding a Path through Complexity; Embedding the Science of Climate Change in the Study of Infectious Animal Diseases

Simone Vitali and Bethany Jackson

CONTENTS

DOI: 10.1201/9781003149774-6

KEY LEARNING OBJECTIVES

At the end of this chapter, you should be able to:

- Articulate, using appropriate examples, the complex and scale-dependent influences climate change may have on infectious-disease ecology, to facilitate integration of this knowledge into existing and future work with animals.
- Recognize the inevitability of ongoing change, the impossibility of knowing everything, and the centrality of fostering resilience within infectious disease systems.
- Explain the need for forward-facing animal health research that is iterative, adaptive, and integrates climate change to create meaningful management of infectious disease risks.

IMPLICATIONS FOR ACTION

- Identify organizations and resources in your region that support climate action within the animal health profession, empowering personal and professional change that will mitigate our contribution to the climate issue.
- Identify climate-sensitive infectious-disease systems by region, and focus efforts on sentinel programs that support systematic landscape restoration as a more efficient way to address risk than focusing on a single pathogen or disease association.
- Proactively engage policy makers and science communicators throughout the research process to ensure that actionable findings enter the political sphere where real change can and should occur.
- Institutions supporting research, particularly tertiary education, must develop metrics that recognize the value of clear and succinct communication that bridges the gap between science, politicians, and the broader animal health community.

HOW DOES CLIMATE CHANGE AFFECT INFECTIOUS DISEASE DYNAMICS? WELL...IT'S COMPLICATED

"It is often necessary to make a decision on the basis of knowledge sufficient for action but insufficient to satisfy the intellect." – Immanuel Kant

Biological systems are innately complex. All pathogens exist in a homeostatic balance with their host/s, vector/s, and the environment within which each component organism operates (1). These relationships are dynamic; a change in some aspect of any one component will evoke feedback of some kind from the others. This interconnectedness is central to some key complexities that must be understood if we are to grasp the impact of climate change on infectious disease systems.

CLIMATE CHANGE DOES NOT OCCUR IN A VACUUM

Human activity has become such a significant force for environmental change that the term "Anthropocene" has been coined for the current geological epoch (2). Anthropogenic climate change is just one of several processes through which humans are wielding this influence; others include habitat alteration (fragmentation, conversion, destruction, urbanization), overexploitation, pollution, tourism, and trade. Like climate change, these other anthropogenic drivers are a product of human social and economic development; like climate change, they exert their influence on the resilience of the environment and on the homeostatic balance of the biological and environmental components of infectious disease.

Climate change has been called a "threat multiplier" that interacts both directly and indirectly with these other eco-social variables. This interconnectedness makes it difficult, if not impossible, to isolate the influence of climate from that of other anthropogenic drivers (3,4). For example, "heat islands" generated in urbanized landscapes are exacerbated by climate change, increasing their potential as high-risk habitats for mosquito vectors and adding to the risk of vector-borne disease outbreaks (5).

The impact of other anthropogenic drivers on infectious disease is not necessarily summative or synergistic to the effects of climate change. For example, it is generally held that vector-borne diseases will begin to emerge at higher latitudes as global warming increasingly limits vector persistence in the tropics. However, this generalization does not consider the greater economic resilience of many high-latitude countries, which makes them better able to mitigate disease risks, thereby neutralizing the potential impact of climate change on disease expression (6).

In the same way as infectious-disease dynamics demonstrate complex interdependencies, the entire concept of animal health cannot be considered in isolation from the health of human populations and the ecosystems they all occupy. While this chapter will articulate how climate change can affect infectious disease, it is important to view climate-driven disease systems within a complex eco-social context if we are to develop effective response strategies (7). We must recognize the interdependence of human, animal, and ecosystem health as embodied in the "One Health" concept and accept the "critical foundational infrastructure" of intact and functional ecosystems to the well-being of all life on Earth if we are to adequately address 21st century global-health challenges (8).

IT'S A QUESTION OF SCALE ... AND EVERY SCALE MATTERS

Climate change exerts influence in a variety of ways at every order of magnitude, from the "broad grain" perspective of the biome and ecosystem through to the "fine grain" processes at the cellular and microbial level (3). These scales are hierarchically nested within one another: climatic shifts at the "continental" scale (10^6 km^2) bring changes in rainfall and temperature, which alter the vegetation and hydrological landscape at a "regional" scale (10^2–10^6 km^2), which in turn drives population changes and selective pressure on individuals at the "local" (10–10^3 km^2) and the "microscale" (<10 km^2). Measuring and integrating these variable, nested spatial impacts further complicate the modeling of disease-transmission dynamics but is fundamental to appropriately incorporating climate change into our understanding of where, when, and how infectious disease will occur (9).

Both climate data and animal-distribution models tend to supply broad grain information rather than fine detail (9,10). However, organisms do not experience climate on coarse scales, and the microclimates which they live in may diverge dramatically from the macroclimate of their region (10). This has important implications for infectious disease modeling. For example, Denmark experienced a significant outbreak of bluetongue disease virus in 2007–2008, despite predictions that the climate was too cold for the spread of the disease. Subsequent analysis of microclimate demonstrated that the habitats of midge and mosquito vectors (and the virus they were carrying) were warmer during the day than meteorological temperatures, leading to more rapid blood meal digestion and faster virus development than predicted using broad grain meteorological modeling (11). Given most pathogens and vectors are experiencing the selective pressures of climate change at this microscale, an understanding of microclimate is essential to epidemiological modeling of climate change.

In addition to affecting the way we model infectious disease, an appreciation of scale is integral to developing response strategies for disease management. The degree to which hazards manifest as disease will depend on a multitude of location-specific factors such as land use, sanitation, nutritional stressors, and the economic allocation to the response (12). Applying a blanket approach may ignore the local context of disease expression, which is critical to successful action. The involvement of multidisciplinary teams with local understanding and knowledge is the key to a successful transition of broad-scale trends to the local and microscale.

A CONSPICUOUS LACK OF LINEARITY

Given the web of interdependencies between the various components of disease ecology, it should not be surprising that the impact of climate change on disease expression is rarely, if ever, linear in nature. Many disease models are characterized by so-called "threshold behavior" in which large shifts in population abundance or species diversity can occur in response to relatively small changes in key parameters. For example, leptospirosis occurs in only a few sites in Salvador, Brazil, unless flooding occurs. In the presence of flooding, a dramatic and rapid increase occurs in the distribution and prevalence of leptospirosis (13).

The fitness of any given organism (host, pathogen, or vector) will demonstrate a threshold-like decline either side of its optimum thermal range (14). Within the life cycle of a pathogen or a vector, there may be many such nonlinear impacts, which may be conflicting in their effect on overall organism fitness. For example, although larval development time for the fluke parasite *Schistosoma mansoni* can be greatly accelerated by increased temperature, larval mortality also increases, so the overall impact on disease may depend on which of these effects dominates (15). This lack of linearity is also seen when examining the impacts of climate-driven environmental factors other than temperature, such as humidity, salinity, and acidification (1,16).

We must set aside the optimistic expectation that there will be consistent spatial or temporal changes in host or parasite fitness with climate change (14). Predicting the direction and magnitude of climate change impact on disease dynamics will require models that incorporate this nonlinearity (16).

MECHANISMS FOR CLIMATE CHANGE IMPACTS

Climate change acts on the environment, hosts, pathogens, and vectors, shifting the homeostatic balance between the various interactions according to the tolerance of each component. Figure 6.1 summarizes the climate-driven mechanisms that drive infectious-disease dynamics, while Table 6.1 provides some illustrations of these mechanisms as they are occurring today. Note that climate change does not inevitably promote an increase in disease occurrence; for any given disease, many of the mechanisms illustrated in Figure 6.1 will be occurring simultaneously, and each mechanism may exert a range of positive, negative, or null effects on the overall dynamics of disease expression.

The variations in atmospheric temperature that occur on a daily and seasonal basis shape the thermal tolerance or thermal "niche" of a given species, as well as its means of remaining within that thermal tolerance, which may be physiological (e.g. endothermy, hibernation), behavioral (e.g., migration), or both (6). If climate change creates a thermal environment that threatens survivorship or fitness, this environment will drive the species to move toward less thermally stressful locations. There is evidence for such climate-driven shifts in latitudinal and altitudinal distribution of species across a range of taxa (Figure 6.2), with both terrestrial and marine species tending to move poleward (17–19), fish species seeking deeper water (19), and insects moving to higher altitudes (20) with warming temperatures.

Challenge to an organism's thermal tolerance may also affect its fitness, survivorship, and recruitment within its current distribution, thereby causing shifts in abundance, resilience, and diversity of species, with flow-on effects for disease dynamics.

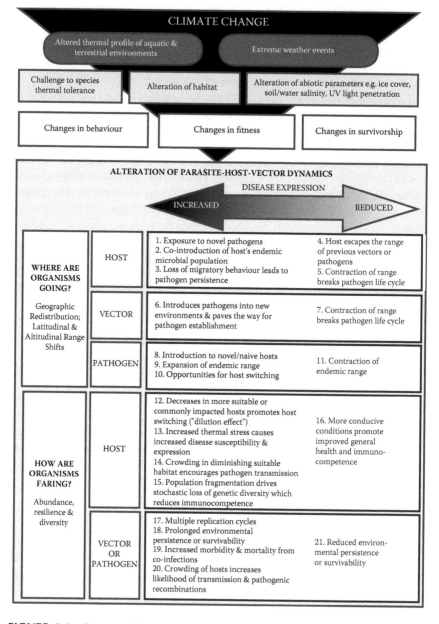

FIGURE 6.1 Conceptual framework identifying the mechanisms by which climate change influences infectious-disease expression.

The changes wrought by climate on a given species will have a variable impact on disease expression, depending on its relationships with the other biological and environmental "players" in the disease. For example, the natural host and reservoir for plague *(Yersinia pestis)* in Inner Mongolia is dominated

TABLE 6.1
Examples Illustrating Mechanisms by Which Climate Change Can Affect Infectious-Disease Expression. Mechanism Numbers in the Second Column Correspond to the Numbering in Figure 6.1

Example	Mechanism
Finland: Northward expansion of roe deer *(Capreolus capreolus)* populations exposes reindeer *(Rangifer tarandus)* as novel host to roe deer filarioid nematode *Setaria tundra*, causing outbreaks of peritonitis (30).	1,2, 8, 9
West Africa: Increasing rainfall increases suitable habitat for rodents, which are reservoir species for Lassa arenavirus, increasing the incidence of Lassa fever in humans (12).	2
North America: A milder climate reduces the need for monarch butterflies (*Danaus plexippus*) to migrate, leading to persistence of their protozoan parasites due to loss of migratory culling of infected hosts (31).	3
Costa Rica: *Norops polylepis* lizards inhabiting the edges of remnant forest plots have fewer mites than populations in the remnant interiors because the mites favor cooler and moister habitats (32).	4
Mongolia: Climate-driven habitat changes decrease reservoir population of Mongolian gerbils *Meriones unguiculatus*, reducing risk of plague *Yersinia pestis* outbreaks (21).	5, 11
Nepal: Dengue fever emerges at higher elevations following expanding range of *Aedes* vectors (20).	6
Ghana: A decline in the persistence of malaria occurs as temperatures exceed the optimum for year-round vector transmission (6,33).	7, 11
North America: Human cases of Lyme disease increase as deer numbers are reduced through habitat loss (24).	8,12
USA: Oyster parasite *Perkinsus marinus* extends into the northeast due to the increase in average winter low temperatures (15,34).	9
Canada: Larval and intermediate host survivability of *Varestrongylus* lungworm nematodes increases with warming temperatures, enabling host switching from migratory caribou *Rangifer tarandus* to endemic muskoxen *Ovibos moschatus* (35).	8, 10, 18
Mediterranean Sea: Mass mortalities of gorgonian corals correspond with warmer water temperatures, which stress the coral and improve conditions for pathogenic bacteria (36).	13, 18
Alaska: Larval development cycle of *Umingmakstrongylus pallikuukensis* lungworm has halved from 2 years to 1 year, leading to an increased frequency of disease outbreaks in its musk ox *(Ovibos moschatus)* host (37).	17
Tanzania: Extreme drought followed by the resumption of rain causes unusually high *Babesia* burdens in lions (*Panthera leo*) due to heavy infestations of vector ticks. A coinciding outbreak of canine distemper virus causes immunosuppression, which exacerbates the babesiosis, leading to mass mortality (38).	18, 19
Central Africa: Concentration of bat populations into smaller areas due to lengthy drought leads increases Ebola virus transmission and the risk of pathogenic recombinations (4).	14, 20
Australia: *Fasciola hepatica* fluke infestations of livestock may reduce in prevalence and distribution as summer rainfall and soil moisture decrease (3).	21
USA: Mortality in farmed salmonids due to *Flavobacterium psychrophilum* ("cold water disease") is greater at low water temperatures and may therefore be constrained by warming conditions (39).	16, 21

FIGURE 6.2 Changing climate drives latitudinal shifts in both hosts and pathogens (First Dog on the Moon, The Guardian www.theguardian.com/firstdog).

by the Mongolian gerbil (*Meriones unguiculatus*) in some foci, but in others, by the Daurian ground squirrel (*Spermophilus dauricus*). Climate-driven increases in temperature and precipitation are leading to increasing vegetation cover across Inner Mongolia, which promotes the persistence of ground squirrels but does not favor the gerbil, which avoids habitats with high vegetation cover. Consequently, climate change may decrease the risk of plague in foci where Mongolian gerbils are the primary reservoir, but may increase the risk where Daurian ground squirrels predominate (21). The degree and direction of the impact will also depend on the species' susceptibility to the changes being experienced; if the change in climate overcomes a key threshold of physiological tolerance, massive outbreaks may ensue in a short time. For example, unusually high relative humidity and temperature are postulated to have been a triggering factor in a mass mortality event of saiga (*Saiga tatarica*) from *Pasteurella* septicemia (22).

The permutations of climate-driven change in disease expression are at their most elaborate when the pathogen's life cycle involves intermediate hosts or vectors, each of which will have its own thermal niche and threshold tolerances that can be challenged by climate change, resulting in implications for disease dynamics. For example, the cycle of transmission of tickborne encephalitis virus (TBEV) requires synchronous feeding of the larval and nymph stages of the tick vector because viremia only persists in rodent hosts for two days (23). Because the co-feeding of larvae and nymphs is associated with rapid autumnal cooling, temperature rises due to climate change might decouple this relationship, thereby reducing the geographical range of TBEV (24).

It is often assumed that the effects of climate change (and other drivers) in decreasing biodiversity may also increase the vulnerability to disease of remaining species, a concept termed the "dilution effect." The dilution effect postulates that wildlife biodiversity acts as a buffer to the detrimental impacts of climate change on infectious disease because the number of hosts able to transmit a pathogen is "diluted" by abundant noncompetent hosts, thereby reducing chances of a successful transmission. However, as we have seen, transmission occurs via a complex interplay of environmental, host, and pathogen factors, creating challenges to ensuring a thorough and objective assessment of net costs and benefits (24). Systematic reviews of the literature suggest that while biodiversity loss at a local scale may exert a negative impact on human and wild animal health, the effect weakens as the spatial scale increases (25,26). As with any area of uncertainty, explicitly stating any assumptions about climatic impacts is an important fundamental step to maintaining flexibility and adaptability in disease management as more information is gathered.

In addition to the direct impacts of thermal change on hosts, parasites, and vectors (depending on their thermal tolerance and physiology), climate change drives other environmental shifts that exert influence on infectious disease expression. The salinity of both water and soil habitats are impacted by climate (1,27) as are a range of abiotic parameters of aquatic environments, including water acidity, oxygenation, and UV light penetration (1). Any such changes to an organism's environment may challenge its physiological tolerance, forcing geographic redistribution or changes in abundance and diversity (28). For example, climate-driven increase in salinity increases the spread, prevalence, and intensity of oyster infection with the protozoan *Perkinsus marinus* along the Gulf of Mexico (29).

INTERACTIONS OF CLIMATE CHANGE WITH OTHER ANTHROPOGENIC FORCES

Climate change is a "threat multiplier" for other anthropogenic alterations to the environment. Like other organisms, the thermal tolerance of humans is challenged by climate change. However, humans possess an unparalleled ability to "push back" when our tolerance thresholds are threatened. We are capable of large-scale manipulation of the environment, enabling us (and the domestic species in our care) to persist in more hostile conditions and expand into new territory. For example, as climate change diminishes the suitability of existing farmland, humans may respond by clearing and exploiting relatively undisturbed areas, thereby promoting habitat fragmentation and increasing the frequency and intensity of interactions between humans, domestic animals, and wildlife. Thus, anthropogenic climate change exacerbates other anthropogenic encroachments.

The concept of One Health has recently been reimagined as the "Berlin Principles." The principles recognize the profound influence of human activities on the health of humans, animals, and ecosystems, and in turn, on the emergence, exacerbation, and spread of disease (8). Figure 6.4 illustrates this connectivity through the example of Hendra Virus.

TEXT BOX 6.1   CLIMATE CHANGE AS A THREAT MULTIPLIER: THE STORY OF HENDRA VIRUS

Hendra virus (HeV) is a zoonotic paramyxovirus that emerged in Australia in 1994. The endemic reservoir hosts of HeV are Australian fruit bats of the *Pteropus* genus (40). The fruit bats' diet of nectar and fruit is ephemeral, and its distribution is patchy in native forests. In an undisturbed environment, this requires bats to forage widely and migrate frequently over large areas (41). However, with the loss of contiguous native forest due to human activities, fruit bats have become increasingly urbanized, relying om increasingly consistently available (but poorer quality) urban flowering sources (42). An increasing number of *Pteropus* bats are foregoing migration for the convenience and consistency of urban garden food sources, such that there are now continuously occupied colonies in most major city centers on the Australian east coast (42).

HeV circulates asymptomatically among fruit bats, with virus shedding followed by development of immunity, which wanes over time. There have been over 60 spillover events to horses, the majority of which have originated from continuously occupied urban camps (41). All cases of human infection have originated through contact with infected horses, which are the only mammalian species known to have been infected by bats. The virus is often fatal to both humans and horses, with approximate case fatality rates of 60% and 80%, respectively (43).

HeV spillover is related to climate by several different mechanisms acting at a variety of temporal and spatial scales (Figure 6.3), necessitating a considered and integrated approach to disease management, which avoids oversimplification (Figure 6.4).

TURNING KNOWLEDGE INTO ACTION

FIRST, A PROFESSIONAL STOCKTAKE

"Everyone thinks of changing the world, but no one thinks of changing himself."

– Leo Tolstoy

Before we look at ways to incorporate climate change into our work with infectious diseases and animal systems, it would be prudent to evaluate our personal and professional contributions to the problem. It is not within the scope of this chapter to review the many exceptional organizations working in this area. Instead, we encourage the reader to join or support local groups in line with those highlighted in the text box below (45–48). Collectively, we can take action now to minimize our professional impact on the very thing we are trying to mitigate.

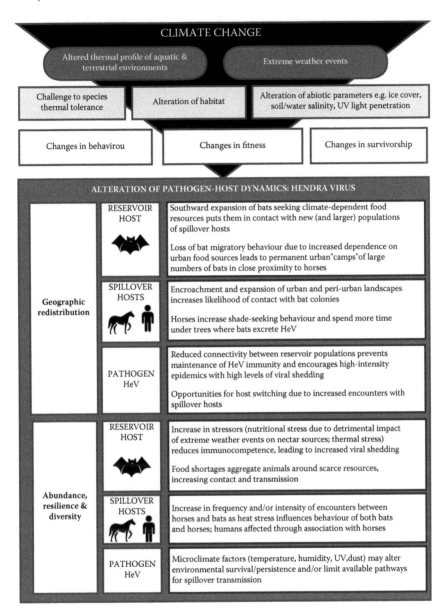

FIGURE 6.3 Hendra virus (HeV) – an illustration of the impact of climate change on disease dynamics (40–44).

FIGURE 6.4 It is necessary to resist simplistic solutions if we are to address infectious disease risk meaningfully in the context of climate change (First Dog on the Moon, The Guardian www.theguardian.com/firstdog).

TEXT BOX 6.2   WALKING THE TALK: REDUCING THE CARBON FOOTPRINT OF THE ANIMAL HEALTH SECTOR IN THE "SUNBURNT COUNTRY" (45)

Anthropogenic climate change is driving an increase in the frequency and severity of megafires in Australia. In the summer of 2019–2020, following two years of extreme drought in NSW and Queensland, Australia experienced one of the worst bushfire seasons on record (46). More than 24 million hectares of forest were lost (including rainforests that do not typically burn), an estimated 3 billion mammals, birds, and reptiles died (not accounting for invertebrates and soil microbiota), and 33 human lives were lost directly to fire (445 more if you consider impacts of smoke inhalation on pre-existing diseases) (47). These catastrophic events are no surprise to two organizations, "Farmers for climate action" and "Veterinarians for climate action," founded in 2015 and (perhaps presciently) 2019, respectively. These charitable organizations aim to galvanize professions affected by, and contributing to, climate change, to reduce their carbon footprint and the consequences for the people and animals in their communities and care. Both groups recognize that farmers and animal health practitioners are an integral part of the solution by virtue of the contribution livestock and pet industries make to climate change, and the opportunities therein to improve how we practice and how we educate our clients on their role in mitigation. The impact of climate change in rural Australia is particularly tangible, with disadvantage amplified due to the physical, cultural, and financial relationships people have with the land, and the relative lack of supportive services including mental health and veterinary clinics (48). The welfare consequences extend into the livestock and wildlife of these regions, particularly in extreme events where services are prioritized to the protection of infrastructure and human lives, and there are simply not enough resources to mitigate the impacts on animals and the environment as well.

Exercises for action

1. Read the opinion piece below from a country veterinarian in fire- and drought-affected Inverell, demonstrating the very personal impact of climate change on individuals dealing with the welfare fallout for humans and animals alike (49) https://www.smh.com.au/environment/climate-change/cattle-have-stopped-breeding-koalas-die-of-thirst-a-vet-s-hellish-diary-of-climate-change-20191220-p53m03.html
2. Explore the resources at Veterinarians for Climate Action (vfca.org.au), whose vision is to reduce carbon emissions in the veterinary and animal care community and promote climate action through targeted media and government submissions. Review their "Climate Smart" program, which provides resources and guidance for veterinary clinics to work toward carbon-neutral practice.
3. Explore the resources at Farmers for Climate Action (farmersforclimateaction.org.au), with their strategic focus on farmer education and training, advocacy to governments and industry, and partnerships between farmers, research institutions, and industry to drive action on climate change.
4. Identify similar organizations or groups in your region, and if these are lacking, why not start one?

PRACTICAL INTEGRATION STRATEGIES

Hopefully by this stage, the reader feels comfortable with the "how" and "why" of climate change interactions with infectious-disease ecology and examples that illustrate these complexities. We recognize, however, that bridging the gap between "knowing" and "doing" can be challenging in this space, particularly as climate change is just one of many interacting factors that act across multiple scales to drive infectious disease in socio-ecological systems.

Here we propose a series of extremely simple steps to develop a transparent appreciation of the climate change influences for a given infectious-disease system (or systems), and actions to mitigate any negative impacts (see Figure 6.5). These reflect the steps articulated within the process of disease risk analysis (50) with guidance centered on identifying climate change specific risks. We of course encourage readers to complement this approach with a broader reflection on the eco-social and economic drivers acting on the systems they are considering.

Step 1. Understand the Approach: What Should We Know before We Start?

At a global level, interdisciplinary collaboration has resulted in the development of reference documents that clearly articulate the link between the various anthropogenic drivers (including climate change) and the

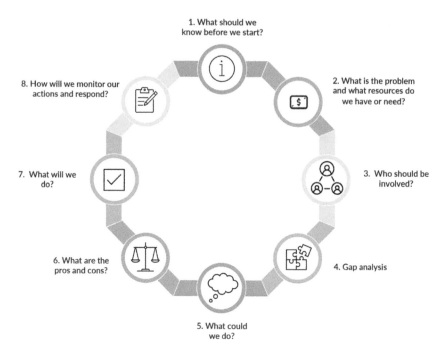

FIGURE 6.5 A step-wise approach to transparently evaluate climate change influences on a given infectious disease system (or systems).

development of disease, while also recognizing the inter-relatedness of the drivers themselves. The importance of thinking in terms of One Health is now well-recognized, with the "Tripartite Plus" agencies (World Health Organization; Food and Agriculture Organization of the United Nations; World Organization for Animal Health; United Nations Environment Program) being among the leaders in coordinating their programs within the theme of One Health (5).

TEXT BOX 6.3   EXERCISE FOR STEP 1

Readers are strongly encouraged to train their thinking to extend beyond the "one germ, one disease" paradigm. To this end, you should familiarize yourself with key recent texts on the Berlin Principles (8), infectious-disease risk in the context of One Health (5,51–53), and wild animal disease-risk analysis (50). Following this study, research your local, state, and federal systems for animal health and their inclusion or otherwise of climate change in procedures and policy.

Step 2. Understand the Question: What Is the Problem and What Resources Do We Have or Need?

Clearly articulating the question or problem you seek to address is always helpful. In the absence of something more specific, the default would be, "Is this infectious-disease system susceptible to climate influences, and, if so, how?" Consider the existing resources you have access to, inclusive of human capital, timeframes, and available budget. Approaches will be necessarily constrained by such factors, and early identification can help frame a more achievable approach (or identify where more resources are needed).

TEXT BOX 6.4   EXERCISE FOR STEP 2

What is the problem or system in question?
What resources do we have to hand?
What resources do we need?

Step 3. Find the Team: Who Should Be Involved?

The day of the polymath has not quite ended; however, with the advancement of specialized knowledge comes the loss of global abilities. We need specialists to help us navigate the exponential advancements in infectious-disease detection and modeling. However, we also need generalists who can integrate this specialized knowledge into broader objectives and understanding, as well as participants to provide critical perspective on the social, cultural, political, and jurisdictional implications at all stages of the process. Therefore, we must actively seek intersectoral collaborations that will fill our infectious disease and climate change toolkit, from the specialist to the generalist. The early identification and regular involvement of key stakeholders is fundamental to high quality decision making.

TEXT BOX 6.5   EXERCISE FOR STEP 3

Read the following articles and websites discussing One Health and collaborative, multidisciplinary successes:

- Errecaborde KM, Macy KW, Pekol A, Perez S, O'Brien MK, Allen I, et al. (2019) Factors that enable effective One Health collaborations – A scoping review of the literature. PLoS ONE 14(12): e0224660. https://doi.org/10.1371/journal.pone.0224660
- Mackenzie, J. S., & Jeggo, M. (2019). The One Health Approach-Why Is It So Important?. Tropical medicine and infectious disease, 4(2), 88. https://doi.org/10.3390/tropicalmed4020088
- One Health Basics (CDC), available at: https://www.cdc.gov/onehealth/basics/index.html (accessed 14/1/22).

Who is on your team, and what disciplines do they represent (refer to CDC website and recommended professionals and expertise in One Health approaches)?

Which disciplines are missing in your team?

How will you access this expertise?

Who are your polymaths?

Who are your science communicators, policy makers, policy advisors, and media representatives?

Who are the cultural and community liaisons who will provide the local context of those directly affected or directly responsible for implementation?

Step 4. Gap Analysis: What Do We Know? What Don't We Know? What Should We Know?

Using Figure 6.1, clarify aspects of the disease system you are studying that may be influenced by climate change. Be transparent; knowledge extrapolated from similar species, or the same species in a different region, may be the best information you currently have, but that does not mean it is sufficient to apply to the given system.

Critically, this step is an opportunity to acknowledge what is known and unknown in relation to climate influences on the host(s) and agent(s). This will broadly identify how climate might interact with the system in question, or if there are knowledge gaps that need to be addressed through research or expert opinion (or both).

TEXT BOX 6.6   EXERCISE FOR STEP 4

Is my system likely to be climate sensitive? Consider if the host(s) or agent(s) are dependent on factors likely to adjust with climate change, e.g. rainfall, humidity, temperature, extreme weather events, host densities, and redistribution.

Use Figure 6.1 to explore the climate links for your system.

Step 5. Determine the Options for Action: What *Could* We Do?

When considering what can be done, "nothing" should always be an option, as should possibilities that seem courageous, ambitious, politically sensitive, or radical. In this step, the argument is to aim high, consider all options, however unpalatable or improbable. Whether they are possible or not is a question for step 6.

TEXT BOX 6.7   EXERCISE FOR STEP 5

Reflect on the following questions as you derive a list of what could be done for the system or problem you are working with:

- If we had unlimited resources, what would we do to address the knowledge gaps and act on the problem?
- What are the priorities, why are they priorities, and who decides on the priorities?
- Is this more important than other issues currently facing the system in question?
- What is the simplest and most palatable thing to do? Why?
- What is the most difficult or unpalatable thing to do? Why?
- What are the trade-offs, barriers, and opportunities with the approaches suggested?

Step 6. Evaluate the Options for Action: What Are the Pros/Cons of Different Options?

This step is an opportunity to be explicit in terms of the feasibility of different options and the opportunities or barriers they afford. Clearly articulate the dependencies and subtleties of any risks, limitations, or opportunities, including political risks, financial risks, public risks, and so on.

TEXT BOX 6.8   EXERCISE FOR STEP 6

For the options identified above, consider the following:

- Are they feasible?
- What benefits do they bring?
- What barriers, limitations and risks do they present?
- Can we overcome these limitations, and if so, how?
- Do they foster resilience and reduce vulnerability long term?
- Do they close knowledge gaps?

Step 7. Decision Making: What *Will* We Do? And, Importantly, What Is Our Motivation to Do It?

Considering step 6, we prioritize our actions in a consultative manner, identifying the key drivers for our decisions (e.g. financial, preservation of genetic diversity, political will, public opinion).

TEXT BOX 6.9   EXERCISE FOR STEP 7

Identify your prioritized list of actions:

- Does this approach foster collaboration and longevity of impact?
- What or who is driving this decision?
- What or who stands to benefit?
- What or who stands to lose?

---

8. Review and Adapt: How Will We Monitor Our Actions and Respond?

As scientists and policy makers, we are reasonably driven to find the facts to support action. While inaction may be as dangerous as the wrong action, acting without measuring our success or otherwise is to deny a critical opportunity to learn and improve. In this step, explicitly derive objective measures of success, inclusive of how you will respond to these measures and who is responsible for managing and communicating the adaptive response process.

TEXT BOX 6.10   EXERCISE FOR STEP 8

What does success look like?
What will trigger revision of our actions?
Who will be responsible for decisions to adapt?
How will we communicate these decisions?

CASE STUDY: A MITE-Y COMMON FOE, SARCOPTIC MANGE EPIZOOTIC IN QUENDA (*ISOODON FUSCIVENTER*) OF WESTERN AUSTRALIA

We live in the age of panzootics, with fungal (51–53), parasitic (51), and viral threats (41) all considered major contributors to biodiversity loss and public health emergencies. These panzootics reflect the decline in resilience of complex biological systems, where anthropogenic changes are outpacing the capacity for adaptation. In the case of sarcoptic mange, a multihost and global parasitic disease whose severity is augmented by naïve exposure (54) and loss of host resilience (55), it is no surprise that we are seeing emergence across novel species and regions.

The southwest of Western Australia is a biodiversity hotspot, something to celebrate and lament given the history of land (mis)use in the region (56), as well as the impending impact of climate change (57–59). The region, which includes the state capital city of Perth, is predicted with high certainty to become hotter and drier, with increased frequency and intensity of bushfires, and decreasing autumn and winter rainfall (59). Native species which have a long evolutionary history tied to a Mediterranean climate with winter rainfall, are being driven southward to the tip of the continent by an expanding arid zone. For species reliant on ephemeral wetlands and rainfall, the existing impact of climate change has already precipitated the need to assist colonization to more suitable habitat and climate, if such species are to persist into the future (60,61).

In the Perth metropolitan region, the quenda (*Isoodon fusciventer:* a small marsupial from the bandicoot family) has to some degree adapted to urbanization of its habitat, incorporating human and pet food into its omnivorous diet (62). As a result, they come into conflict with the variety of domestic threats that also thrive in the urban matrix; namely cars, introduced predators, fire, and constant habitat modification to make way for roads and new housing developments (63). Quenda ecology and social networks in the urban environment have modified, with home ranges and densities fluctuating according to these urban resources, providing a fertile interface between humans, domestic pets, native and feral wildlife for zoonotic and epizootic emergence.

As testimony to the importance of wildlife care and rehabilitation facilities as providers of early-detection mechanisms for emerging diseases, a Perth facility (Darling Range Wildlife Shelter, DRWS) was the first to report an apparent epizootic of sarcoptic mange in quenda, which represented the first evidence of sustained transmission and disease in this species in the region, although sporadic cases had been reported previously (64). While not proven, it is likely that introduced foxes in the region would be the primary reservoir, with spillover a result of predator–prey cycles and/or shared urban resources. With 53 clinically affected quenda admitted to care between 2019 and 2021, representing a discrete geographic cluster (65), as well as anecdotal evidence of infestations in human and domestic species in the same area, the outbreak is necessarily drawing attention from local organizations and researchers recognizing the One Health implications. So how does climate change fit into this? A very good question, and one that has been highlighted as a research gap in our understanding of sarcoptic mange in wildlife (51,54).

Let us assume we have covered steps 1–3 in Figure 6.5, noting that we will be seeking local expertise, particularly in the areas of climate science and science policy, alongside the community groups, local government, state government, NGOs, human health, and other parties that already have a vested interest in this zoonotic disease outbreak. Step 4, the gap analysis, is where we will specifically address the ways in which climate change can interact with this infectious parasitic disease, directly or indirectly. Working from Figure 6.1 and using our knowledge of the epidemiology of sarcoptic mange, we can infer where we have sufficient knowledge to incorporate climate change into our understanding of the system and where we need more information. From the outset, we must be clear, climate change could have a negative, neutral, or positive impact on sarcoptic mange outbreaks, and these impacts will likely vary seasonally. Recall linearity is the last thing we should expect!

Changing climate is likely to drive dispersal and density of reservoir (foxes) or spillover (quenda) host populations, as well as their recruitment into urban environments or regions with resources that are dwindling elsewhere (specifically water). The *Sarcoptes* mite, like all organisms, has a preferred climate in which to persist on and off-host and complete its life cycle (62). Considering host-agent interactions, sarcoptic mange is known to have a seasonal pattern in some marsupial species, although this seasonality differs depending on species and location (66,67), reflecting climatic conditions that either optimize mite survival on and off host, and/ or minimize host immune responses due to poor resource availability. The mites prefer a low temperature and high humidity environment (62), so broadly the future of Perth's climate is looking somewhat grim if you are a *Sarcoptes* mite off host in the middle of summer. But does the quenda burrow microclimate offset this disadvantage? And what about spring, winter, and autumn? Where and when will we meet or not meet the needs of the mite? Where and when will climatic variables meet or not meet the needs of quenda or other spillover or reservoir hosts? We need data, yet we also need to start making reasonable hypotheses based on existing knowledge.

Mange cycles are typically maintained by direct contact and therefore density-dependent transmission, or an indirect/frequency dependent transmission through off-host survival in burrow systems and dens (51). Should direct contact be the primary route of transmission in the quenda epizootic, then the influence of climate change on host behavior, densities, dispersal, and resilience will be of greater importance perhaps than the influence of climate change on off-host survival through humidity and temperatures in burrow or shared den systems.

Using Figure 6.1, we consider the specific ways in which climate change may increase or decrease the occurrence of sarcoptic mange in urban and peri-urban areas of Perth, WA. Fundamentally, the questions reflect how climate change will interact with host health, distribution, behavior, and ecology (Figure 6.6), and how climate change will interact with mite survival and life cycles, on and off host.

FIGURE 6.6 Understanding the ecology of hosts and parasites, and how they experience climate, will assist in determining how climate change will affect disease expression (First Dog on the Moon, The Guardian www.theguardian.com/firstdog).

Key questions that arise are:
  i. How sensitive is the mite's biology, survival, and reproduction, on or off host, to the influence of climate change on temperature and humidity?
  ii. Is off-host survival an important factor in this system, and if so, where are foxes and quenda denning in the urban and peri-urban matrix, and how sensitive are these microclimates to coarse-scale climate change?
  iii. How sensitive are the susceptible hosts (especially reservoirs such as foxes) to impacts of climate change on biology/ecology, resources, bush-fires or extreme weather events, or thermoregulatory impacts of mange?
  iv. How likely is climate change to result in redistribution and/or changes in abundance and densities of reservoir and spillover hosts due to changes in preferred habitat and climatic zones?
  v. Do urban resources (food and water provisioning) counter potential redistribution or host health impacts of climate change on quenda or reservoir species such as foxes?

We might predict that decreasing rainfall (and humidity), coupled with in-creasing temperatures, would not favor off-host survival unless den or burrow systems protect against these external changes. The impacts are likely to differ across the urban, peri-urban, and bush environments, a question that warrants exploration as coarse-scale climate change will differentially affect these areas at the finer scale. For example, does the urban environment and its resources (food and water) offset the potential host health impacts of climate change? Do these same resources lead to higher host densities and behavioral changes that increase transmission risks (e.g. quenda have been observed using resources alongside domestic pets)? Will bush quenda be more impacted by climate change than urban counterparts owing to extreme weather events, bushfires, declining re-sources, and synergistic threats of predators and land-use modification, resulting in fade out of mange in these populations due to low densities and limited op-portunities for transmission? Or will these broader environmental changes due to climate impacts result in congregating species at key resources such as water points, thereby driving transmission up?

There will not be one answer, of course. However, each of these pathways by which climate change may engage with the broader threats to quenda health and the spread of sarcoptic mange warrants integration into any research or man-agement program so that our actions help derive answers to these questions. And, conversely, our answers can then drive future action.

CONCLUSION

"We shall never achieve harmony with the land, any more than we shall achieve absolute justice or liberty for people. In these higher aspirations the important thing is not to achieve but to strive."

– Aldo Leopold

The management approach to animal infectious disease will be more robust if it considers and incorporates the complexities and scale dependencies of climate change. At the broadest level, actions should be directed toward supporting ecosystem resilience, while reinforcing the capacity to detect and rapidly adapt to the inevitable changes to come. Given the innate interconnectedness of biological systems, researchers and managers should constantly work to structure investigations and translate findings in the context of the ecosystem, rather than just a single pathogen or disease. As well as being more palatable for decision-makers seeking holistic solutions, such an approach is likely to be more cost-effective and sustainable. The systematic approach should also acknowledge that human demands for land and natural resources are inescapable and therefore must be incorporated into the balance of actions for stabilizing biological systems (68).

Effectively embedding climate change in the management of infectious animal disease requires a continuum of engagement and effort, linking those directly affected by the disease (human, domestic animal, and/or wildlife communities), to those charged with investigation (researchers and organizations seeking answers to drive action), and those who influence and regulate global change (media, policy makers, state and federal animal health agencies, governments, and multinational organizations such as the UN, FAO, WOAH and WHO). If we are to achieve productive engagement across this continuum, it is essential to promote respectful, robust, and transparent collaboration and communication. Building strong personal cross-cultural relationships that acknowledge the demands, constraints, and reward systems of other parties is fundamental to bridging the gap between impact, investigation, and implementation (68,69). This type of pro-active "networking" is not a skill that comes naturally to many scientists, but it is vital to ensuring that decisions and actions are founded on evidence-based investigation.

For research to be effective as well as informative, there should be a balance between theoretical discovery and action-oriented research in funding agendas. The "publish or perish" paradigm rewards research that has high impact in the academic world, but if this same research remains inaccessible to those implementing management actions, it will remain lost in siloed science. Prioritizing science communication and policy outputs within major funding schemes is one widely employed method of encouraging scientists and scientific institutions to maintain an applied and collaborative approach to research. However, there are unexplored opportunities within the traditional journal-publication framework that would further incentivize the collective approach. Research metrics could reward the meaningful inclusion, from project inception to delivery, of collaborators who will make use of the research findings. Further, a simple policy and media brief could be a requirement of publication, to be reviewed alongside the manuscript and published online in an open-source fashion, thereby increasing accessibility to those who may use the findings of research.

It is clear that animal health practitioners will have to integrate climate change science into their management of infectious diseases, embedding this within the host-agent-environment triad and the myriad anthropogenic drivers (e.g. habitat modification, globalization, urbanization, and introduced species) that influence

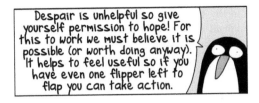

FIGURE 6.7   Neither despair nor scientific uncertainty can be allowed to paralyze us into inaction (First Dog on the Moon, The Guardian www.theguardian.com/firstdog).

animal disease systems. However, the immediacy of climate change challenges our desire as scientists to know everything before we do anything (Figure 6.7). We cannot afford to be paralyzed in the face of the inevitable uncertainty; we must take action that recognizes our knowledge gaps and identifies our assumptions. We will make mistakes and will need to be open and humble about these for the collective good. Ultimately, uncertainty cultivates opportunity and discovery; the relationships between infectious disease and climate change should be seen as providing us with abundant opportunities to innovate and adapt.

REFERENCES

1. Burge, C.A. and Hershberger, P.K., 2020. Climate change can drive marine diseases. *Marine Disease Ecology*. Oxford University Press, pp. 83–94.
2. Crutzen, P.J., 2002. Geology of mankind. *Nature (London)*, *415*(6867), pp. 23.
3. Black, P.F. and Butler, C.D., 2014. One Health in a world with climate change. *Revue Scientifique et Technique (International Office of Epizootics)*, *33*(2), pp. 465–473.
4. World Health Organization, 2015. Connecting global priorities: Biodiversity and human health: a state of knowledge review.
5. Intergovernmental Platform On Biodiversity and Ecosystem Services, 2020. Workshop Report on Biodiversity and Pandemics of the Intergovernmental Platform on Biodiversity and Ecosystem Services (IPBES). Zenodo; 2020/10/29.
6. Lafferty, K.D. and Mordecai, E.A., 2016. The rise and fall of infectious disease in a warmer world. *F1000 Research*, 5, pp. 2040.
7. Parham, P.E., Waldock, J., Christophides, G.K., Hemming, D., Agusto, F., Evans, K.J., et al., 2015. Climate, environmental and socio–economic change: Weighing up the balance in vector–borne disease transmission. *Philosophical Transactions of the Royal Society B*, *370*(1665), p. 20130551.
8. Gruetzmacher, K., Karesh, W.B., Amuasi, J.H., Arshad, A., Farlow, A., Gabrysch, S., et al., 2021. The Berlin principles on one health – bridging global health and conservation. *Science of The Total Environment*, *764*, p. 142919.
9. Goodin, D.G., Jonsson, C.B., Allen, L.J.S. and Owen, R.D., 2018. Integrating landscape hierarchies in the discovery and modeling of ecological drivers of zoonotically transmitted disease from wildlife. In: Hurst C.J., eds. *The Connections Between Ecology and Infectious Disease*. 5. Cham: Springer International Publishing, pp. 299–317.
10. Potter, K.A., Arthur Woods, H. and Pincebourde, S., 2013. Microclimatic challenges in global change biology. *Global Change Biology*, *19*(10), pp. 2932–2939.

11. Haider, N., Kirkeby, C., Kristensen, B., Kjær, L.J., Sørensen, J.H. and Bødker, R., 2017. Microclimatic temperatures increase the potential for vector–borne disease transmission in the Scandinavian climate. *Scientific Reports, 7*(1), p. 8175.

12. Gibb, R., Franklinos, L.H.V., Redding, D.W. and Jones, K.E., 2020. Ecosystem perspectives are needed to manage zoonotic risks in a changing climate. *BMJ (Online)*, p. 371.

13. Codeço, C.T., Lele, S., Pascual, M., Bouma, M. and Ko, A.I., 2008. A stochastic model for ecological systems with strong nonlinear response to environmental drivers: Application to two water–borne diseases. *Journal of the Royal Society Interface, 5*(19), pp. 247–252.

14. Rohr, J.R., Dobson, A.P., Johnson, P.T.J., Kilpatrick, A.M., Paull, S.H., Raffel, T.R., et al., 2011. Frontiers in climate change–disease research. *Trends in Ecology & Evolution, 26*(6), pp. 270–277.

15. Harvell, C.D., Mitchell, C.E., Ward, J.R., Altizer, S., Dobson, A.P., Ostfeld, R.S., et al., 2002. Climate warming and disease risks for terrestrial and marine biota. *Science, 296*(5576), pp. 2158–2162.

16. Dobson, A., Molnár, P.K. and Kutz, S., 2015. Climate change and Arctic parasites. *Trends in Parasitology, 31*(5), pp. 181–188.

17. La Sorte, F.A. and Jetz, W., 2010. Avian distributions under climate change: Towards improved projections. *Journal of Experimental Biology, 213*(6), pp. 862–869.

18. Chen, I.C., Hill, J.K., Ohlemuller, R., Roy, D.B. and Thomas, C.D., 2011. Rapid range shifts of species associated with high levels of climate warming. *Science, 333*(6045), pp. 1024–1026.

19. Dulvy, N.K., Rogers, S.I., Jennings, S., Stelzenmller, V., Dye, S.R. and Skjoldal, H.R., 2008. Climate change and deepening of the North Sea fish assemblage: A biotic indicator of warming seas. *Journal of Applied Ecology, 45*(4), pp. 1029–1039.

20. Acharya, K.P., Chaulagain, B., Acharya, N., Shrestha, K. and Subramanya, S.H., 2020. Establishment and recent surge in spatio–temporal spread of dengue in Nepal. *Emerging Microbes and Infections, 9*(1), pp. 676–679.

21. Xu, L., Schmid, B.V., Liu, J., Si, X., Stenseth, N.C. and Zhang, Z., 2015. The trophic responses of two different rodent–vector–plague systems to climate change. *Proceedings of the Royal Society B, 282*(1800), p. 20141846.

22. Kock, R.A., Orynbayev, M., Robinson, S., Zuther, S., Singh, N.J., Beauvais, W., et al., 2018. Saigas on the brink: multidisciplinary analysis of the factors influencing mass mortality events. *Science Advances, 4*(1), p. eaao2314.

23. Randolph, S., 2002. Predicting the risk of tick–borne diseases. *International Journal of Medical Microbiology, 291*, pp. 6–10.

24. Cable, J., Barber, I., Boag, B., Ellison, A.R., Morgan, E.R., Murray, K., et al., 2017. Global change, parasite transmission and disease control: Lessons from ecology. *Philosophical Transactions of the Royal Society of London. Series B, Biological Sciences, 372*, p. 1719.

25. Halliday, F.W. and Rohr, J.R., 2019. Measuring the shape of the biodiversity-disease relationship across systems reveals new findings and key gaps. *Nature Communications, 10*(1), p. 5032.

26. Buck, J.C. and Perkins, S.E., 2018. Study scale determines whether wildlife loss protects against or promotes tick–borne disease. *Proceedings of the Royal Society B, 285*(1878), p. 20180218.

27. Del Buono, D., 2021. Can biostimulants be used to mitigate the effect of anthropogenic climate change on agriculture? It is time to respond. *Science of the Total Environment, 751*, p. 141763.

28. Burge, C.A., Mark Eakin, C., Friedman, C.S., Froelich, B., Hershberger, P.K., Hofmann, E.E., et al., 2014. Climate change influences on marine infectious diseases: Implications for management and society. *Annual Review of Marine Science*, 6(1), pp. 249–277.
29. Soniat, T.M., Hofmann, E.E., Klinck, J.M. and Powell, E.N., 2008. Differential modulation of eastern oyster (*Crassostrea virginica*) disease parasites by the El–Niño–Southern Oscillation and the North Atlantic Oscillation. *International Journal of Earth Sciences*, 98(1), p. 99.
30. Laaksonen, S., Oksanen, A., Kutz, S., Jokelainen, P., Holma–Suutari, A. and Hoberg, E., 2017. 5. Filarioid nematodes, threat to arctic food safety and security. In: Paulsen, P., Bauer, A. and Smulders, F.J.M., eds. *Game Meat Hygiene*. The Netherlands: Wageningen Academic Publishers, pp. 101–120.
31. Bartel, R.A., Oberhauser, K.S., De Roode, J.C. and Altizer, S.M., 2011. Monarch butterfly migration and parasite transmission in eastern North America. *Ecology*, 92(2), pp. 342–351.
32. Schlaepfer, M.A. and Gavin, T.A., 2001. Edge effects on lizards and frogs in tropical forest fragments. *Conservation Biology*, 15(4), pp. 1079–1090.
33. Ryan, S.J., Carlson, C.J., Mordecai, E.A. and Johnson, L.R., 2019. Global expansion and redistribution of Aedes–borne virus transmission risk with climate change. *PLoS Neglected Tropical Diseases*, 13(3), p. e0007213.
34. Ward, J.R. and Lafferty, K.D., 2004. The elusive baseline of marine disease: Are diseases in ocean ecosystems increasing? *PLoS Biology*, 2(4), p. e120.
35. Hoberg, E.P. and Brooks, D.R., 2015. Evolution in action: Climate change, biodiversity dynamics and emerging infectious disease. *Philosophical Transactions of the Royal Society B*, 370(1665), p. 20130553.
36. Bally, M. and Garrabou, J., 2007. Thermodependent bacterial pathogens and mass mortalities in temperate benthic communities: A new case of emerging disease linked to climate change. *Global Change Biology*, 13(10), pp. 2078–2088.
37. Kutz, S.J., Hoberg, E.P., Polley, L. and Jenkins, E.J., 2005. Global warming is changing the dynamics of Arctic host–parasite systems. *Proceedings of the Royal Society B*, 272(1581), pp. 2571–2576.
38. Munson, L., Terio, K.A., Kock, R., Mlengeya, T., Roelke, M.E., Dubovi, E., et al., 2008. Climate extremes promote fatal co–infections during canine distemper epidemics in African lions. *PLoS ONE*, 3(6), p. e2545.
39. Holt, R.A., Amandi, A., Rohovec, J.S. and Fryer, J.L., 1989. Relation of water temperature to bacterial cold-water disease in coho salmon, chinook salmon, and rainbow trout. *Journal of Aquatic Animal Health*, 1(2), pp. 94–101.
40. Martin, G., Yanez–Arenas, C., Chen, C., Plowright, R.K., Webb, R.J. and Skerratt, L.F., 2018. Climate change could increase the geographic extent of Hendra virus spillover risk. *EcoHealth*, 15(3), pp. 509–525.
41. Plowright, R.K., Foley, P., Field, H.E., Dobson, A.P., Foley, J.E., Eby, P., et al., 2011. Urban habituation, ecological connectivity and epidemic dampening: the emergence of Hendra virus from flying foxes (*Pteropus* spp.). *Proceedings of the Royal Society B*, 278(1725), pp. 3703–3712.
42. Plowright, R.K., Eby, P., Hudson, P.J., Smith, I.L., Westcott, D., Bryden, W.L., et al., 2015. Ecological dynamics of emerging bat virus spillover. *Proceedings of the Royal Society B*, 282(1798), p. 20142124.
43. Yuen, K.Y., Fraser, N.S., Henning, J., Halpin, K., Gibson, J.S., Betzien, L., et al., 2021. Hendra virus: epidemiology dynamics in relation to climate change, diagnostic tests and control measures. *One Health (Amsterdam, Netherlands)*, 12, p. 100207.

44. Martin, G., Webb, R.J., Chen, C., Plowright, R.K. and Skerratt, L.F., 2017. Microclimates might limit indirect spillover of the bat borne zoonotic Hendra virus. *Microbial Ecology*, *74*(1), pp. 106–115.

45. Mackellar, D. and Brunsdon, J., 1990. *I Love a Sunburnt Country: The Diaries of Dorothea Mackellar*. North Ryde, NSW, Australia: Angus & Robertson, *1990*. p. 206.

46. Canadell, J.G., Meyer, C.P., Cook, G.D., Dowdy, A., Briggs, P.R., Knauer, J., et al., 2021. Multi–decadal increase of forest burned area in Australia is linked to climate change. *Nature Communications*, *12*(1), p. 6921.

47. Celermajer, D., Lyster, R., Wardle, G.M., Walmsley, R. and Couzens, E., 2021. The Australian bushfire disaster: How to avoid repeating this catastrophe for biodiversity. *Wiley Interdisciplinary Reviews Climate Change*, *12*(3).

48. Shorthouse, M. and Stone, L., 2018. Inequity amplified: Climate change, the Australian farmer, and mental health. *Medical Journal of Australia*, *209*(4), pp. 156–157.

49. Rhoades, G., 2019. Cattle have stopped breeding, koalas die of thirst: A vet's hellish diary of climate change2019. Available from: https://www.smh.com.au/ environment/climate-change/cattle-have-stopped-breeding-koalas-die-of-thirst-a-vet-s-hellish-diary-of-climate-change-20191220-p53m03.html.

50. Jakob–Hoff, R.M., MacDiarmid, S.C., Lees, C., Miller, P.S., Travis, D. and Kock, R.A., 2014. *Manual of Procedures for Wildlife Disease Risk Analysis*. Paris: World Organisation for Animal Health, *2014*. p. 149.

51. Escobar, L.E., Carver, S., Cross, P.C., Rossi, L., Almberg, E.S., Yabsley, M.J., et al., 2021. Sarcoptic mange: An emerging panzootic in wildlife. *Transboundary and Emerging Diseases*, p. tbed.14082.

52. Scheele, B.C., Pasmans, F., Skerratt, L.F., Berger, L., Martel, A., Beukema, W., et al., 2019. Amphibian fungal panzootic causes catastrophic and ongoing loss of biodiversity. *Science*, *363*(6434), pp. 1459–1463.

53. Fisher, M.C., Henk, D.A., Briggs, C.J., Brownstein, J.S., Madoff, L.C., McCraw, S.L., et al., 2012. Emerging fungal threats to animal, plant and ecosystem health. *Nature*, *484*(7393), pp. 186–194.

54. Skerratt, L.F., Death, C., Hufschmid, J., Carver, S. and Meredith, A., 2021. Guidelines for the treatment of Australian wildlife with sarcoptic mange, Part 2 - Literature review. *Brisbane*.

55. Martin, A.M., Fraser, T.A., Lesku, J.A., Simpson, K., Roberts, G.L., Garvey, J., et al., 2018. The cascading pathogenic consequences of Sarcoptes scabiei infection that manifest in host disease. *Royal Society Open Science*, *5*(4), p. 180018.

56. Bradshaw, C.J.A., 2012. Little left to lose: Deforestation and forest degradation in Australia since European colonization. *Journal of Plant Ecology*, *5*(1), pp. 109–120.

57. Schut, A.G.T., Wardell–Johnson, G.W., Yates, C.J., Keppel, G., Baran, I., Franklin, S.E., et al., 2014. Rapid characterisation of vegetation structure to predict refugia and climate change impacts across a global biodiversity hotspot. *PLoS ONE*, *9*(1), p. e82778.

58. Pettit, N.E., Naiman, R.J., Fry, J.M., Roberts, J.D., Close, P.G., Pusey, B.J., et al., 2015. Environmental change: Prospects for conservation and agriculture in a southwest Australia biodiversity hotspot. *E&S*, *20*(3), p. art10.

59. Steffen, W. and Hughes, L., 2011. The critical decade: Western Australian climate change impacts. *2011*.

60. Dade, M.C., Pauli, N. and Mitchell, N.J., 2014. Mapping a new future: Using spatial multiple criteria analysis to identify novel habitats for assisted colonization of endangered species: Multiple criteria analysis to guide assisted colonization. *Animal Conservation*, *17*, pp. 4–17.

61. Schmölz, K., Pinder, A., Kuchling, G. and Gollmann, G., 2021. Evaluating candidate wetlands for the assisted colonization of the western swamp turtle *Pseudemydura umbrina* in a changing climate: Macro-invertebrate food resources and turtle diet. *Aquatic Conservation: Marine and Freshwater Ecosystems*, *31*(7), pp. 1847–1858.

62. Arlian, L.G. and Morgan, M.S., 2017. A review of *Sarcoptes scabiei*: Past, present and future. *Parasites Vectors*, *10*(1), p. 297.

63. Ramalho, C.E., Ottewell, K.M., Chambers, B.K., Yates, C.J., Wilson, B.A., Bencini, R., et al., 2018. Demographic and genetic viability of a medium–sized ground–dwelling mammal in a fire prone, rapidly urbanizing landscape. *PLoS ONE*, *13*(2), p. e0191190.

64. Wicks, R.M., Clark, P. and Hobbs, R.P., 2007. Clinical dermatitis in a southern brown bandicoot (Isoodon obesulus) associated with the mite *Sarcoptes scabiei*. *Comparative Clinical Pathology*, *16*(4), pp. 271–274.

65. Botten, L., Ash, A. and Jackson, B., 2022. Characterising a sarcoptic mange epizootic in quenda (*Isoodon fusciventer*). *International Journal for Parasitology: Parasites and Wildlife*. In review. *18*, August 2022, pp. 172–179.

66. Speight, K., Whiteley, P., Woolford, L., Duignan, P., Bacci, B., Lathe, S., et al., 2017. Outbreaks of sarcoptic mange in free–ranging koala populations in Victoria and South Australia: A Case Series. *Australian Veterinary Journal*, *95*(7), pp. 244–249.

67. Wildlife Health Australia, 2021. Sarcoptic mange in Australian wildlife. *Wildlife Health Australia Fact Sheet*.

68. McAlpine, C., Lunney, D., Melzer, A., Menkhorst, P., Phillips, S., Phalen, D., et al., 2015. Conserving koalas: A review of the contrasting regional trends, outlooks and policy challenges. *Biological Conservation*, *192*, pp. 226–236.

69. Gibbons, P., Zammit, C., Youngentob, K., Possingham, H.P., Lindenmayer, D.B., Bekessy, S., et al., 2008. Some practical suggestions for improving engagement between researchers and policy–makers in natural resource management. *Ecological Management & Restoration*, *9*(3), pp. 182–186.

7 Zoonoses

*Nick H. Ogden, L. Robbin Lindsay,
and Michael A. Drebot*

CONTENTS

KEY LEARNING OBJECTIVES

- Zoonoses are infectious diseases in which pathogens are transmitted from animals to humans.
- Where and when risks from zoonoses occur often depends on weather and climate, which may affect how the pathogens are transmitted (particularly those transmitted by ticks and mosquitoes), the animals that are the natural hosts of the disease-causing organisms, or because of effects on the capacity of human society to control them.
- Climate change is likely to impact where and when zoonosis risks occur, causing diseases to emerge in regions at times and in places they have not previously been seen.

DOI: 10.1201/9781003149774-7

- A One Health approach that integrates responses by public health, animal health, and environmental health practitioners will be required to minimize risks from zoonoses that emerge or re-emerge with climate change.

IMPLICATIONS FOR CLIMATE ACTION

- To enhance our capacity to predict patterns of zoonosis occurrence, research to increase our understanding of the ecology and epidemiology of zoonoses will be required.
- With this enhanced knowledge, a wider scope of model-based risk assessments will allow us to project where and when zoonoses may emerge with climate change.
- Enhanced international event-based surveillance will allow early identification of emerging zoonoses around the world, while One Health surveillance in humans, animals, and vectors is required to detect emerging zoonoses and allow development of climate/weather-informed early-warning systems.
- The impact of emerging zoonoses will be minimized by research to develop novel methods to detect, prevent, and control them, as well as by early implementation of validated methods.

INTRODUCTION

This chapter focuses on the possible impact of climate change on zoonoses, i.e. infectious diseases for which pathogens are transmitted from animals to humans. Zoonoses, caused by bacteria, protozoa, macroparasites, viruses, and prions, are maintained by wildlife populations, domesticated livestock, and companion animals. Zoonoses comprise approximately 60% of known infectious diseases, but >70% of emerging infectious diseases of humans are zoonoses (if emergence of antimicrobial resistance is not considered), and a high proportion of these emerging zoonoses are maintained, or originated, in wildlife (1,2). Infectious diseases "emerge" by two main mechanisms (3):

1. Adaptive emergence by *de novo* emergence of a pathogen (or pathogen variant) that becomes infectious for and/or pathogenic in humans due to genetic change in the pathogen. Examples include influenza viruses and coronaviruses such as SARS-CoV and MERS-CoV (3).

2. Geographic range spread (of pathogens and, in some cases, their vectors) from one part of the world where they are endemic to another part of the world where they were previously absent. This may be long distance, such as the intercontinental spread of West Nile virus from (most likely) the Middle East to North America, or more short distance due to local geographic range expansion, such as the northward spread of Lyme disease risk into Canada from the United States.

Re-emergence is the re-appearance or epidemic expansion of an endemic disease. Both emergence and re-emergence of zoonoses involves spillover, which is the process whereby barriers are breached that have hitherto prevented an animal-reservoired pathogen from coming into contact with humans and infecting them. In general, only a small proportion of zoonoses are transmissible from human to human at all, and even less are transmitted with an efficiency such that they can be maintained by transmission among humans alone without animal reservoirs. Inefficient human-to-human transmission of zoonoses has been described as "stuttering chains" of transmission as these chains of transmission inevitably die out (4), although some are inherently transmissible from human to human (e.g. Ebola and SARS-CoV-2) and capable of causing human-to-human transmitted epidemics and pandemics. More frequent spillover events may facilitate selection and evolution of variants of zoonotic pathogens with more efficient human-to-human transmission (4). The likelihood that spillover occurs depends on characteristics of the pathogen and host, but also the likelihood of contact between infected hosts and humans, which may be facilitated by human activities (5). Examples include the changes in farming practices that allowed bat-reservoired Nipah virus to spill over into humans via livestock (6) and likely contributed to the emergence of Lyme disease in northeastern North America (7). Once capable of being transmitted human to human, more unrestrained transmission may allow the pathogen to evolve variants that are transmitted with increasing efficiency, as we have seen with SARS-CoV-2 (8).

Our concern is that climate change may drive the emergence and re-emergence of infectious diseases, including zoonoses (9). However, as well as direct effects of changes to climate and weather on zoonoses, several other drivers of infectious-disease emergence have been identified. These include socio-economic status of the human population where they occur, with lower status being associated with increased risk, while biodiversity of wildlife populations has complex associations with emergence – hot spots for emergence are those with high biodiversity but low biodiversity may increase risk from some zoonoses such as Lyme disease and West Nile virus (1,2,10). Characteristics of the pathogens themselves are also important drivers or facilitators of emergence in human populations. Pathogens that have wide, rather than narrow, animal-host ranges are more likely to be infectious for humans, while some RNA viruses are more likely to emerge and be transmissible human to human, likely due to their relatively high mutation rates and capacity for genetic recombination (1).

In this chapter, the possible impacts of climate change on emergence and re-emergence of zoonoses are reviewed by exploring mechanisms and evidence of climate and weather effects on zoonoses, using illustrative studies on predictive modeling and detection of zoonoses emergence, and attributing emergence to climate change. Finally, possible adaptive responses to reducing the risk and impact of zoonoses emergence and re-emergence are described.

WEATHER AND CLIMATE SENSITIVITY OF ZOONOSES

DIRECT EFFECTS ON PATHOGEN SURVIVAL

The survival of microbial and parasitic causal agents of zoonoses is intrinsically sensitive to the direct effects of all or one of temperature, desiccation, and ultraviolet light *in vitro*. Most life stages of macroparasites and bacteria will die if frozen (except in special storage conditions), over heated, or desiccated, while most viruses are killed by heat and UV light. In some circumstances, these effects may impact transmission of directly-transmitted pathogens; for example, transmission of SARS-CoV-2 is thought to be somewhat inhibited during summer months due to impacts of heat and UV light on virus survival in the environment (11). However, microbes and parasites are rarely directly exposed to the elements, mostly existing within the bodies of their reservoir hosts, intermediate hosts, or arthropod vectors. For many microbes and parasites, stages that are free-living have often evolved strategies that confer resistance against environmental damage, such as the highly resistant endospores of *Bacillus anthracis* (the agent of anthrax) and eggs of some roundworms such as *Baylisascaris procyonis* (12). However, this is not universal as *Toxacara canis* eggs are readily killed by freezing (13). Warming may inhibit transmission of some zoonoses; for example, transmission of avian influenza viruses among waterfowl reservoirs may be less efficient the warmer the water in which the birds swim (14), while *Echinococcus granulosus* eggs are killed by heating (15).

INDIRECT EFFECTS VIA IMPACTS ON MODES OF TRANSMISSION (INCLUDING BY ARTHROPOD VECTORS)

Apart from direct transmission due to close contact with infected animal reservoirs, zoonoses are also transmitted by arthropod vectors and by ingestion of/infection of contaminated food and water. Impacts of climate change on these modes of transmission may be key to impacts of climate change on zoonoses. The possible effects of climate, weather, and climate change on arthropod vectors have been explored extensively *in-vitro*, *in-vivo*, in modeling studies, and in reviews (16–19). Per-capita mortality rates, activity, length of seasons of activity, and rates of development from one life stage (and thus life-cycle length) to another are all affected by temperature to some extent, with warmer temperatures generally enhancing survival and abundance of vector populations. Increased humidity can also facilitate enhanced activity of vectors, while increased rainfall may increase habitat for immature stages of mosquitoes and other dipteran vectors. Extremes of temperature negatively impact the activity and survival of many vector species (18,19), while heavy rainfall can inhibit activity and may impact reproduction by displacing or flushing some aquatic life stages out of a favorable habitats such as artificial containers and sump pits in catch basins (20). These conflicting influences mean that the geographic footprint of vector populations may change in terms of latitude or altitude but not necessarily increase in size (18). Many zoonotic pathogens that are excreted in feces and/or urine of reservoir hosts are transmitted in water, with infection occurring by consumption of contaminated water for most. Some, such as *Leptospira* spp., may be acquired from contaminated water by contact with mucosae and broken skin. Warmer temperatures may enhance survival and multiplication of some zoonotic pathogens in water in some circumstances, but a significant anticipated impact of climate change is increased frequency of heavy rainfall events that increase human contact with zoonotic pathogens by contaminating drinking water systems or causing flooding (inter alia 21,22,23,24,25,26). Warmer temperatures are expected to increase the risk of food-borne zoonoses, particularly enteropathogens such as *Escherichia coli*, *Salmonella* spp. and *Campylobacter* spp. that are maintained within livestock production systems (21). Effects of climate warming on enteropathogen survival and multiplication are expected to occur along the farm-to-fork continuum and may particularly impact risk from *Campylobacter* spp. that are maintained in poultry production systems (27). It is feasible that the risk from some food-borne parasites (e.g. *Toxoplasma gondii*, cestodes), which may occur in meat from livestock and in food harvested from wildlife populations, may increase. However, such effects would likely be due to effects on transmission cycles among reservoir hosts (see below), as well as survival of eggs or other life stages in the environment.

INDIRECT EFFECTS VIA IMPACTS ON RESERVOIR HOSTS

The geographic, seasonal, and future occurrence and abundance of zoonotic pathogens may change due to effects of climate change on reservoir hosts and the communities in which they exist. A changing climate may drive range changes of individual wildlife reservoir host species, as has been suggested for the white-footed mouse, *Peromyscus leucopus,* in North America (28). Such range changes may be due to a warming climate increasing resource availability poleward, or possibly by negative effects on competitors, parasites, or predators. Overall impacts on zoonosis risk may be difficult to predict as range changes to one reservoir may impact another, as is being seen in southern Quebec, Canada. Here the white-footed mouse is expanding its range northward (possibly due to a warming climate) and displacing the deer mouse, *P. maniculatus* (for reasons currently unknown), but both are reservoirs of the agent of Lyme disease (28). Interrelated impacts of climate change, biodiversity change, and land-use change (including livestock production) may produce changes to reservoir host communities that impact the risk from zoonoses, including the rate of contact with humans (inter alia, 29,30,31,32). The complexity of the possible effects of climate change on wildlife communities may provide a significant challenge to prediction (30).

INDIRECT EFFECTS VIA IMPACTS ON HUMAN SOCIETY

Climate change will likely have impacts on zoonoses risks via impacts on human society. Climate change is likely to have impacts on agricultural productivity and food security, and thus economies, particularly in low and middle-income countries, impacting domestic capacity to control zoonoses and perhaps increasing economic migration and dispersal of zoonoses (33–36). Attempts to increase agricultural productivity may increase interactions between wildlife and livestock, increasing rates at which wildlife-reservoired zoonoses spill over into human populations via livestock (6,37). Negative socio-economic effects of climate change may increase human exposure and susceptibility to zoonotic infections (25,38), as well as the seasonality of that risk.

EFFECTS ON DISPERSION

It is generally considered that the impacts of climate change on the natural history of pathogens, parasites, and animal hosts will result in their geographic ranges moving poleward and to higher altitudes (reviewed in (18)). The capacity for the ranges of species to change depends on the extent to which mechanisms for dispersal exist, be that natural, such as dispersal of tick-borne zoonoses and influenza viruses by migratory birds (39,40), or by human agency, as has been suspected for West Nile virus (41). Furthermore, the capacity for dispersed zoonotic pathogens to become established in new locations depends on the degree of specialism of that species to particular communities – the greater the specialization, the narrower the niche and the fewer the suitable environments for its establishment; so, in general, habitat generalists are more capable of expanding or changing their range in response to climate change, while specialists with narrow niches may be more likely to become extinct associated with climate change (42). Not all zoonoses may expand their ranges poleward with climate change. Were Arctic fox rabies in northern Canada to expand, there may be more opportunity of spillover from Arctic foxes into Red foxes that can carry rabies south into Canada (43). However, the dynamics of Arctic rabies is complex, involving multiple species (including both host and prey populations) and multiple climatic determinants of species abundance and spatiotemporal dispersion; these underline that effects of climate change may be difficult to predict without a deeper understanding of the ecology of zoonoses (43) Figure 7.1.

MODEL-BASED PROJECTIONS FOR ZOONOSES, EVIDENCE FOR CHANGES IN PATTERNS, AND ATTRIBUTION TO CLIMATE CHANGE

Predicting if, where, and when risk from a zoonotic pathogen may change with climate change requires the use of models. Models aim to identify quantitative

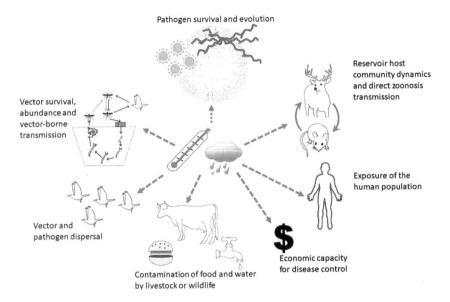

Pathogen survival and evolution

Vector survival, abundance and vector-borne transmission

Reservoir host community dynamics and direct zoonosis transmission

Exposure of the human population

Vector and pathogen dispersal

Contamination of food and water by livestock or wildlife

Economic capacity for disease control

FIGURE 7.1 Possible effects of weather, climate, and climate change on risk from zoonoses.

relationships between climate and occurrence of species. Having obtained these relationships, spatiotemporal occurrence under the current climate can be assessed, often for model validation purposes. Then output of global and/or regional climate models, developed by climatology research institutions, can be used to identify where and when zoonoses may be in the future with a changing climate. Two broad types of model are used to obtain zoonosis–climate associations. Process-based mathematical models use empirical data on effects of temperature and precipitation/humidity of lifecycle/transmission cycle processes to obtain relationships between weather/climate and survival and abundance of zoonotic pathogens (often using zoonosis vectors as a proxy for zoonosis occurrence). Nonclimatic determinants of transmission or population survival can be incorporated in these models also. These models require a considerable amount of detailed, quantitative knowledge of zoonotic pathogen ecology and transmission if they can usefully predict where and when the zoonoses occur, and that depth of knowledge is relatively uncommon. For this reason, a very common approach is to use data on presence of zoonoses in, for example, surveillance data, to estimate associations between environmental conditions (including climate) and the likelihood of occurrence of a species. These "correlative" approaches link occurrence data to environmental variables using geographic information systems, and then either statistical or machine-learning methods are used to determine the most likely "ecological niche" and, within that, the most likely "climatic niche" for the species. Again, once elucidated, the relationship between species distributions and climatic variables can be used to project future species distributions according to the climate projections obtained from climate models. These different model types, and their strengths and weaknesses, are reviewed by Sillero (44). The process-based mechanistic models have been mostly used to project future distributions of tick-borne zoonoses, including Lyme disease, and *Amblyomma americanum*-transmitted diseases such as Human Monocytic Ehrlichiosis (32,45,46). In contrast, correlative approaches have more commonly been used for mosquito-borne zoonoses, such as West Nile virus (47), plague (48), and directly-transmitted pathogens such as anthrax (49). The validity of models is established by comparing predicted current distributions to actual distributions using surveillance data (50,51) or data obtained by other methods such as "citizen science"-type observations of arthropod vectors (52,53).

Recent changes in spatiotemporal patterns of risk from zoonoses have only been detected for those zoonoses for which some form of systematic surveillance exists, such as tick-borne zoonoses, including Lyme disease in North America and northern Europe (reviewed in (18)) and Arctic Fox rabies (43). However, to both detect changes in risk and attribute such changes to a changing climate requires long-term datasets that are uncommon, such as the data on tick vectors in Canada that span the end of the 20th and beginning of the 21st centuries (52), and the historic data on plague in Asia and Europe (inter alia 54,55,56). It is even less common to have long-term surveillance datasets that detect changes in patterns of zoonoses over a long enough period and are sufficiently recent to be able to associate detected changes to recent anthropogenic climate change. To our knowledge, the only example to date is the attribution of emergence of Lyme disease in Canada (Figure 7.2). This emergence was detected by the long-term (circa 20-year) passive tick surveillance data, validated by field surveillance, with evidence that climate warming over the time period and in the locations where surveillance was conducted was most likely a climate anomaly driven by greenhouse gas emissions (57).

ONE HEALTH RESPONSES TO ZOONOSES RISKS

Responses to the threat of emerging and re-emerging zoonoses risks in general, and of specific zoonoses, encompass three main areas: assessment of where and when known zoonoses may emerge or re-emerge to maximise preparedness; surveillance for emerging and re-emerging zoonoses globally, nationally, and locally; and prevention and control responses to mitigate outbreaks and pandemics. These responses inherently implicate One Health approaches (58), whether or not they are driven by climate change. A One Health approach entails systematic engagement of those responsible for, and having expertise in, the relevant human, animal, and environmental health domains. It is necessary in order to understand the risk of zoonoses for public health because assessing risk involves understanding the ecology of wild and domesticated reservoir hosts (requiring skills of livestock health experts and wildlife disease ecologists), entomology (if transmitted by arthropod vectors), and the impact of the abiotic environment (climatologists, land managers) to understand what impacts there

Modelling to predict range spread   Surveillance for human cases and tick populations up to 2015

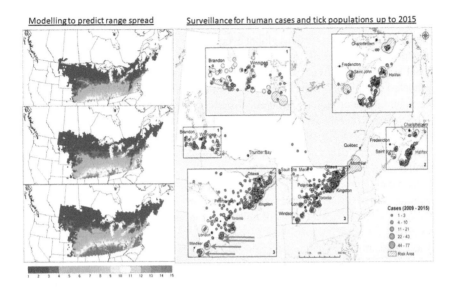

FIGURE 7.2   Left panel: Maps of values of the basic reproduction number (R_0) of the Lyme disease vector *I. scapularis* estimated from ANUSPLIN observed temperature (1971–2000: upper panel), and projected climate obtained from the climate model CRCM4.2.3 following the SRES A2 greenhouse gas emission scenario for 2011 to 2040 (middle panel) and 2041 to 2070 (bottom panel). The color scale indicates R_0 values. Temperature conditions that result in an R_0 of > 1 permit survival of *I. scapularis* populations. Note that within the zones where R_0 of *I. scapularis* is > 1, geographic occurrence of Lyme disease risk is also limited by other environmental factors. This figure is reproduced from Ogden et al. Environ Health Perspect 2014. Right panel: Surveillance for *I. scapularis* populations in central and eastern Canada conducted up to 2014, and surveillance for human Lyme disease cases from 2009 to 2015 (reproduced with permission from Gasmi et al. 2017). Regions where *I. scapularis* populations have been identified by field surveillance are shown as red hatched areas. In 2004, there were only four known *I. scapularis* populations in locations shown by the red arrows. Blue circles show municipalities where human Lyme disease cases had been identified, and the numbers of cases are indicated by the size of the circle.

may be on human/public health (which requires skills in infectious diseases in humans, public health managers/epidemiologists, human behavior, social sciences). One Health approaches are needed for surveillance for zoonoses, which frequently requires integrated programs in animal reservoirs, vectors, and humans (see below), while control programs may require interventions in domesticated animals, wildlife, vectors, and humans. One Health approaches are needed to ensure that these are effective and avoid unconsidered negative impacts (58).

Model-based risk assessments, as described above, provide our best estimates of current and future zoonoses risks at international, national, and local levels. These support jurisdiction-specific assessments of vulnerability and needs for adaptation to changing risks, thus maximizing our preparedness (59).

Surveillance needed to detect zoonoses risks includes global surveillance for known disease risks, as well as global event-based surveillance to detect emergence and spread of previously unknown zoonoses (such as COVID-19). Such international surveillance already exists and includes surveillance for human illness coordinated under the World Health Organisation's International Health Regulations, the voluntary international network of participating medical clinics known as the GeoSentinel Surveillance Network, and the joint FAO–OIE–WHO Global Early Warning System (GLEWS) (reviewed in (3)). Individual jurisdictions will need to undertake rapid risk assessments of detected zoonosis emergence events to determine their local importance and response needs.

Domestic surveillance for infections in humans, wild or domesticated animals, and in arthropod vectors, is of course needed for zoonoses known to be emerging or re-emerging. It would also be prudent to implement surveillance for zoonoses for which model-based risk assessments suggest the possibility of emergence, both for early detection and for model validation purposes. Novel molecular pathogen and vector detection methods that are capable of detecting wide ranges of pathogen and vector species (e.g. metabarcoding methods (60–62)) would be prudent. However, the combination of both conventional technique (e.g. electronic microscopy, cell culture, and serology) and molecular detection platforms has value when documenting and characterizing emerging pathogens and identifying incursions as part of a surveillance network (63,64). Serological methodologies provide a wider screening window since antibodies generally persist, while viremia /bacteremia may wane after several days, and detecting or culturing agents in biological or field-collected samples from animal hosts or humans can be challenging. Broad-based serological assays and antibody testing algorithms may provide initial clues to the nature of novel emerging pathogens because there is cross-reaction with known pathogens in many of these assays (65). Surveillance in environmental samples such as wastewater has been shown to be effective during the COVID-19 pandemic and is another technological addition to the surveillance arsenal (66).

Prevention and control of zoonoses is a huge field spanning the integrated systems needed to manage outbreaks by livestock, wildlife, and public/human health managers, nonpharmaceutical interventions (case detection and isolation, contact tracing and quarantine, restrictive closures, travel restrictions), medical countermeasures (vaccines, therapeutics), personal protective equipment, and the emergency stockpiling of equipment and therapeutics that have been the basis of previous pandemic planning (https://www.canada.ca/en/public-health/services/flu-influenza/pandemic-plans.html); these have had to be implemented to control the COVID-19 pandemic (https://www.canada.ca/en/public-health/services/diseases/coronavirus-disease-covid-19.html). This field is too wide to be adequately covered here, but in the context of weather and climate-sensitive diseases, weather/climate-informed early-warning systems are specific responses that are increasingly advocated (14,20). All of these responses are informed by model-based risk assessments of one form or another that allow us to be as prepared as we can possibly be.

CONCLUSION

It is likely that climate change will drive the emergence and re-emergence of zoonoses since most are, directly or indirectly, sensitive to changes in weather and climate. Mitigating risk from these infectious diseases will depend on a combination of model-based predictions of risk to enhance our preparedness, effective risk from zoonoses by implementing effective international, national, and regional surveillance, and by having effective response plans, capacity, and early-warning systems that adopt a One Health approach. Enhancing our knowledge of the ecology and epidemiology of zoonoses will be key to our efforts to effectively mitigate the risk of zoonoses and emerging infectious diseases. It is likely that there will be considerable review of our preparedness and responses to COVID-19, which are likely to impact and hopefully strengthen global capacity to detect and respond to emerging risks from zoonoses.

REFERENCES

1. Woolhouse, M.E. and Gowtage-Sequeria, S., 2005 Dec. Host range and emerging and reemerging pathogens. *Emerg Infect Dis*, *11*(12), pp. 1842–1847.
2. Jones, K.E., Patel, N.G., Levy, M.A., Storeygard, A., Balk, D., Gittleman, J.L. et al., 2008 Feb 21. Global trends in emerging infectious diseases. *Nature*, *451*(7181), pp. 990–993.
3. Ogden, N.H., AbdelMalik, P. and Pulliam, J., 2017 Oct 5. Emerging infectious diseases: Prediction and detection. *Can Commun Dis Rep*, *43*(10), pp. 206–211.
4. Lloyd-Smith, J.O., George, D., Pepin, K.M., Pitzer, V.E., Pulliam, J.R., et al., 2009 Dec 4. Epidemic dynamics at the human-animal interface. *Science*, *326*, pp. 1362–1367.
5. Plowright, R.K., Parrish, C.R., McCallum, H., Hudson, P.J., Ko, A.I., Graham, A.L., et al., 2017 Aug. Pathways to zoonotic spillover. *Nat Rev Microbiol*, *15*(8), pp. 502–510.
6. Pulliam, J.R., Epstein, J.H., Dushoff, J., Rahman, S.A., Bunning, M., Jamaluddin, A.A., et al., 2012 Jan 7. Agricultural intensification, priming for persistence and the emergence of Nipah virus: A lethal bat-borne zoonosis. *J R Soc Interface*, *9*(66), pp. 89–101.
7. Kilpatrick, A.M., Dobson, A.D.M., Levi, T., Salkeld, D.J., Swei, A., Ginsberg, S., et al., 2017 Jun 5. Lyme disease ecology in a changing world: Consensus, uncertainty, and critical gaps for improving control. *Philos Trans R Soc Lond B Biol Sci*, *372*, p. 20160117.
8. Otto, S.P., Day, T., Arino, J., Colijn, C., Dushoff, J., Li, M., et al., 2021 Jul 26. The origins and potential future of SARS-CoV-2 variants of concern in the evolving COVID-19 pandemic. *Curr Biol*, *31*(14), pp. R918–R929.
9. Ogden, N.H. and Gachon, P., 2019 Apr 4. Climate change and infectious diseases: What can we expect? *Can Commun Dis Rep*, *45*(4), pp. 76–80.
10. Keesing, F. and Ostfeld, R.S., 2021 Apr 27. Impacts of biodiversity and biodiversity loss on zoonotic diseases. *Proc Natl Acad Sci USA*, *118*(17), p. e2023540118.
11. World Meteorological Organization (WMO), 2021. First report of the WMO COVID-19 Task Team review on meteorological and air quality factors affecting the COVID-19 pandemic. WMO-No. 1262. Available from https://library.wmo.int/doc_num.php?explnum_id=10555.

12. Shafir, S.C., Sorvillo, F.J., Sorvillo, T. and Eberhard, M.L., 2011 Jul. Viability of *Baylisascaris procyonis* eggs. *Emerg Infect Dis*, *17*(7), pp. 1293–1295.

13. O'Lorcain, P., 1995 Jun. The effects of freezing on the viability of *Toxocara canis* and *T. cati* embryonated eggs. *J Helminthol*, *69*(2), pp. 169–171.

14. Morin, C.W., Semenza, J.C., Trtanj, J.M., Glass, G.E., Boyer, C. and Ebi, K.L., 2018 Dec. Unexplored opportunities: Use of climate- and weather-driven early warning systems to reduce the burden of infectious diseases. *Curr Environ Health Rep*, *5*(4), pp. 430–438.

15. Rausch, R., 1956 Nov. Studies on the helminth fauna of Alaska: XXX. The occurrence of *Echinococcus multilocularis* Leuckart, 1863, on the mainland of Alaska. *Am J Trop Med Hyg*, *5*(6), pp. 1086–1092.

16. Estrada-Pena, A., 2009 Jan 1. Tick-borne pathogens, transmission rates and climate change. *Front Biosci (Landmark Ed)*, *14*, pp. 2674–2687.

17. Ciota, A.T. and Keyel, A.C., 2019 Nov 1. The role of temperature in transmission of zoonotic arboviruses. *Viruses*, *11*(11), pp. 1013.

18. Ogden, N.H., Beard, C.B., Ginsberg, H. and Tsao, J., 2020 Jul 16. Possible effects of climate change on Ixodid ticks and the pathogens they transmit: What has been projected and what has been observed. *J. Med. Entomol*, *58*(4), pp. 1536–1545.

19. Agyekum, T.P., Botwe, P.K., Arko-Mensah, J., Issah, I., Acquah, A.A., Hogarh, J.N., et al., 2021 Jul 7. A systematic review of the effects of temperature on *Anopheles* mosquito development and survival: Implications for malaria control in a future warmer climate. *Int J Environ Res Public Health*, *18*(14), p. 7255.

20. Ogden, N.H., Lindsay, L.R., Ludwig, A., Morse, A.P., Zheng, H. and Zhu, H., 2019 May 2. Weather-based forecasting of mosquito-borne disease outbreaks in Canada. *Can Commun Dis Rep*, *45*(5), pp. 127–132.

21. Fleury, M., Charron, D.F., Holt, J.D., Allen, O.B. and Maarouf, A.R., 2006 Jul. A time series analysis of the relationship of ambient temperature and common bacterial enteric infections in two Canadian provinces. *Int J Biometeorol*, *50*(6), pp. 385–391.

22. Semenza, J.C., Herbst, S., Rechenburg, A., Suk, J.E., Hoser, C., Schreiber, C. et al., 2012 Apr. Climate change impact assessment of food- and waterborne diseases. *Crit Rev Environ Sci Technol*, *42*, pp. 857–890.

23. Chhetri, B.K., Takaro, T.K., Balshaw, R., Otterstatter, M., Mak, S., Lem, M., et al., 2017 Oct. Associations between extreme precipitation and acute gastrointestinal illness due to cryptosporidiosis and giardiasis in an urban Canadian drinking water system (1997-2009). *J Water Health*, *15*(6), pp. 898–907.

24. Chhetri, B.K., Galanis, E., Sobie, S., Brubacher, J., Balshaw, R., Otterstatter, M., et al., 2019 Dec 30. Projected local rain events due to climate change and the impacts on waterborne diseases in Vancouver, British Columbia, Canada. *Environ Health*, *18*(1), p. 116.

25. Hacker, K.P., Sacramento, G.A., Cruz, J.S., de Oliveira, D., Nery, N., Jr, Lindow, J.C., et al., 2020 Feb. Influence of rainfall on *Leptospira* infection and disease in a tropical urban setting, Brazil. *Emerg Infect Dis*, *26*(2), pp. 311–314.

26. Chadsuthi, S., Chalvet-Monfray, K., Wiratsudakul, A., Modchang, C., 2021 Jan 15. The effects of flooding and weather conditions on leptospirosis transmission in Thailand. *Sci Rep*, *11*(1), p. 1486.

27. Smith, B.A., Meadows, S., Meyers, R., Parmley, E.J. and Fazil, A., 2019 Jan. Seasonality and zoonotic foodborne pathogens in Canada: Relationships between climate and *Campylobacter*, *E. coli* and *Salmonella* in meat products. *Epidemiol Infect*, *147*, p. e190.

28. Simon, J.A., Marrotte, R.R., Desrosiers, N., Fiset, J., Gaitan, J., Gonzalez, A., et al., 2014 Aug. Climate change, habitat fragmentation, ticks and the white-footed mouse drive occurrence of *B. burgdorferi*, the agent of Lyme disease, at the northern limit of its distribution. *Evol Appl, 7*(7), pp. 750–764.

29. Brierley, L., Vonhof, M.J., Olival, K.J., Daszak, P., Jones, K.E., 2016 Feb. Quantifying global drivers of zoonotic bat viruses: A process-based perspective. *Am Nat, 187*(2), pp. E53–E64.

30. Altizer, S., Ostfeld, R.S., Johnson, P.T.J., Kutz, S. and Harvell, C.D., 2013 Aug 2. Climate change and infectious diseases: From evidence to a predictive framework. *Science, 341*(6145), pp. 514–519.

31. Brooks, D.R., Hoberg, E.P. and Boeger, W.A., 2019. *The Stockholm Paradigm: Climate Change and Emerging Disease.* Chicago: University of Chicago Press.

32. Li, S., Gilbert, L., Vanwambeke, S.O., Yu, J., Purse, B.V. and Harrison, P.A., 2019 Jun. Lyme disease risks in Europe under multiple uncertain drivers of change. *Environ Health Perspect, 127*(6), p. 67010.

33. Adger, W.N., Pulhin, J.M., Barnett, J., Dabelko, G.D., Hovelsrud, G.K., Levy, M., et al., 2014. Human security In: Field, C.B., Barros, V.R., Dokken, D.J., Mach, K.J., Mastrandrea, M.D., Bilir, T.E., Chatterjee, M., Ebi, K.L., Estrada, Y.O., Genova, R.C., Girma, B., Kissel, E.S., Levy, A.N., MacCracken, S., Mastrandrea, P.R., White, L.L., eds., *Climate Change 2014: Impacts, Adaptation, and Vulnerability. Part A: Global and Sectoral Aspects. Contribution of Working Group II to the Fifth Assessment Report of the Intergovernmental Panel on Climate Change.* Cambridge: Cambridge University Press, pp. 755–791.

34. Godde, C.M., Mason-D'Croz, D., Mayberry, D.E., Thornton, P.K. and Herrero, M., 2021 Mar. Impacts of climate change on the livestock food supply chain; a review of the evidence. *Glob Food Sec, 28*, p. 100488.

35. Leal Filho, W., Azeiteiro, U.M., Balogun, A.L., Setti, A.F.F., Mucova, S.A.R., Ayal, D., et al., 2021 Jul 20. The influence of ecosystems services depletion to climate change adaptation efforts in Africa. *Sci Total Environ, 779*, p. 146414.

36. Jones, C.M. and Welburn, S.C., 2021 Feb 26. Leishmaniasis beyond East Africa. *Front Vet Sci, 8*, p. 618766.

37. Meurens, F., Dunoyer, C., Fourichon, C., Gerdts, V., Haddad, N., Kortekaas, J., et al., 2021 Jun. Animal board invited review: Risks of zoonotic disease emergence at the interface of wildlife and livestock systems. *Animal, 15*(6), p. 100241.

38. Arthur, R.F., Gurley, E.S., Salje, H., Bloomfield, L.S., Jones, J.H., 2017 May 5. Contact structure, mobility, environmental impact and behaviour: The importance of social forces to infectious disease dynamics and disease ecology. *Philos Trans R Soc Lond B Biol Sci, 372*(1719), p. 20160454.

39. Ogden, N.H., Lindsay, L.R., Hanincová, K., Barker, I.K., Bigras-Poulin, M., Charron, D.F., et al., 2008 Mar. Role of migratory birds in introduction and range expansion of *Ixodes scapularis* ticks and of *Borrelia burgdorferi* and *Anaplasma phagocytophilum* in Canada. *Appl Environ Microbiol, 74*(6), pp. 1780–1790.

40. Blagodatski, A., Trutneva, K., Glazova, O., Mityaeva, O., Shevkova, L., Kegeles, E., et al., 2021 May 20. Avian influenza in wild birds and poultry: Dissemination pathways, monitoring methods, and virus ecology. *Pathogens, 10*(5), p. 630.

41. Nett, R.J., Campbell, G.L. and Reisen, W.K., 2009 Oct. Potential for the emergence of Japanese encephalitis virus in California. *Vector Borne Zoonotic Dis, 9*(5), pp. 511–517.

42. Nicotra, A.B., Beever, E.A., Robertson, A.L., Hofmann, G.E. and O'Leary, J., 2015 Oct. Assessing the components of adaptive capacity to improve conservation and management efforts under global change. *Conserv Biol, 29*(5), pp. 1268–1278.

43. Simon, A., Beauchamp, G., Bélanger, D., Bouchard, C., Fehlner-Gardiner, C., Lecomte, N., et al., 2021 Sep. Ecology of Arctic rabies: 60 years of disease surveillance in the warming climate of northern Canada. *Zoonoses Public Health*, *68*(6), pp. 601–608.

44. Sillero, N., 2011 Apr 24. What does ecological modelling model? A proposed classification of ecological niche models based on their underlying methods. *Ecol Modell*, *222*, pp. 1343–1346.

45. Ogden, N.H., Radojevic, M., Wu, X., Duvvuri, V.R., Leighton, P.A. and Wu, J., 2014 Jun. Estimated effects of projected climate change on the basic reproductive number of the Lyme disease vector *Ixodes scapularis*. *Environ Health Perspect*, *122*(6), pp. 631–638.

46. Sagurova, I., Ludwig, A., Ogden, N.H., Pelcat, Y., Dueymes, G. and Gachon, P., 2019 Oct. Predicted northward expansion of the geographic range of the tick vector *Amblyomma americanum* in North America under future climate conditions. *Environ. Health Perspect*, *127*(10), pp. 107014.

47. Chen, C.C., Jenkins, E., Epp, T., Waldner, C., Curry, P.S. and Soos, C., 2013 Jul 22. Climate change and West Nile virus in a highly endemic region of North America. *Int J Environ Res Public Health*, *10*(7), pp. 3052–3071.

48. Nakazawa, Y., Williams, R., Peterson, A.T., Mead, P., Staples, E. and Gage, K.L., 2007 May. Climate change effects on plague and tularemia in the United States. *Vector Borne Zoonotic Dis*, *7*(4), pp. 529–540.

49. Walsh, M.G., de Smalen, A.W. and Mor, S.M., 2018 Jun 18. Climatic influence on anthrax suitability in warming northern latitudes. *Sci Rep*, *8*(1), p. 9269.

50. Ogden, N.H., St-Onge, L., Barker, I.K., Brazeau, S., Bigras-Poulin, M. and Charron, D.F., et al., 2008 May 22. Risk maps for range expansion of the Lyme disease vector, *Ixodes scapularis*, in Canada now and with climate change. *Int J Health Geogr*, *7*, p. 24.

51. Clow, K.M., Leighton, P.A., Ogden, N.H., Lindsay, L.R., Michel, P. and Pearl, D.L., et al., 2017 Dec 27. Northward range expansion of *Ixodes scapularis* evident over a short timescale in Ontario, Canada. *PLoS One*, *12*(12), p. e0189393.

52. Leighton, P., Koffi, J., Pelcat, Y., Lindsay, L.R. and Ogden, N.H., 2012 Apr. Predicting the speed of tick invasion: An empirical model of range expansion for the Lyme disease vector *Ixodes scapularis* in Canada. *J Appl Ecol*, *49*, pp. 457–464.

53. Kampen, H., Medlock, J.M., Vaux, A.G., Koenraadt, C.J., van Vliet, A.J. and Bartumeus, F., et al., 2015 Jan 8. Approaches to passive mosquito surveillance in the EU. *Parasit Vectors*, *8*, p. 9.

54. Kausrud, K.L., Begon, M., Ari, T.B., Viljugrein, H., Esper, J. and Büntgen, U., et al., 2010 Aug 27. Modeling the epidemiological history of plague in Central Asia: Palaeoclimatic forcing on a disease system over the past millennium. *BMC Biol*, *8*, p. 112.

55. Samia, N.I., Kausrud, K.L., Heesterbeek, H., Ageyev, V., Begon, M. and Chan, K.S., et al., 2011 Aug 30. Dynamics of the plague-wildlife-human system in Central Asia are controlled by two epidemiological thresholds. *PNAS*, *108*(35), pp. 14527–14532.

56. Earn, D.J.D., Ma, J., Poinar, H., Dushoff, J. and Bolker, B.M., 2020 Nov 3. Acceleration of plague outbreaks in the second pandemic. *PNAS*, *117*(44), pp. 27703–27711.

57. Ebi, K.L., Ogden, N.H., Semenza, J.C. and Woodward, A., 2017 Aug 7. Detecting and attributing health burdens to climate change. *Environ Health Perspect*, *125*(8), p. 085004.

58. Ogden, N.H., Wilson, J.R.U., Richardson, D.M., Hui, C., Davies, S.J., Kumschick, S., et al., 2019 Mar 13. Emerging infectious diseases and biological invasions – a call for a One Health collaboration in science and management. *R Soc Open Sci*, 6, p. 181577.
59. Berry, P., Enright, P.M., Shumake-Guillemot, J., Villalobos Prats, E. and Campbell-Lendrum, D., 2018 Nov 23. Assessing health vulnerabilities and adaptation to climate change: A review of international progress. *Int J Environ Res Public Health*, 15(12), p. 2626.
60. Mechai, S., Bilodeau, G., Lung, O., Roy, M., Steeves, R., Gagné, N., et al., 2021 Jul 16. Mosquito identification from bulk samples using DNA metabarcoding: A protocol to support mosquito-borne disease surveillance in Canada. *J Med Entomol*, 58(4), pp. 1686–1700.
61. Schuele, L., Lizarazo-Forero, E., Strutzberg-Minder, K., Schütze, S., Löbert, S., Lambrecht, C., et al., 2021 Aug 4. Application of shotgun metagenomics sequencing and targeted sequence capture to detect circulating porcine viruses in the Dutch-German border region. *Transbound Emerg Dis*. Online ahead of print.
62. Izumi, T., Morioka, Y., Urayama, S.I., Motooka, D., Tamura, T., Kawagishi, T., et al., 2021 Jul. DsRNA sequencing for RNA virus surveillance using human clinical samples. *Viruses*, 13(7), p. 1310.
63. Kuleš, J., Potocnakova, L., Bhide, K., Tomassone, L., Fuehrer, H.P., Horvatić, A., et al., 2017 May. The challenges and advances in diagnosis of vector-borne diseases: Where do we stand? *Vector Borne Zoonotic Dis*, 17(5), pp. 285–296.
64. Bird, B.H. and Mazet, J.A.K., 2018 Feb 15. Detection of emerging zoonotic pathogens: An integrated One Health approach. *Annu Rev Anim Biosci*, 6, pp. 121–139.
65. Van Horn, C.J., 2018 Nov. Hantavirus pulmonary syndrome—the 25th anniversary of the Four Corners outbreak. *Emerg Infect Dis*, 24(11), pp. 2056–2060.
66. Rasero, F.J.R., Moya Ruano, L.A., Del Real, P.R., Gómez, L.C. and Lorusso, N., 2021 Sep 25. Associations between SARS-CoV-2 RNA concentrations in wastewater and COVID-19 rates in days after sampling in small urban areas of Seville: A time series study. *Sci Total Environ*, 806, p. 150573.

8 Interactions between Climate Change and Contaminants

Julia E. Baak, Rose M. Lacombe, Emily S. Choy, Kyle H. Elliott, and John E. Elliott

CONTENTS

KEY LEARNING OBJECTIVES

1. Climate change and contaminants act as cumulative stressors on wildlife health.
2. Climate change increases the release and cycling of contaminants in the environment.
3. Climate change will alter contaminant levels in wildlife, causing various impacts on individual health.

DOI: 10.1201/9781003149774-8

IMPLICATIONS FOR ACTION

1. Standardized, long-term monitoring programs are needed to identify how changes in climate are impacting contaminant concentrations in wildlife.
2. Supplementing existing long-term wildlife-monitoring programs with climate change, contaminants, and wildlife health measures will allow us to assess how multiple stressors may cumulatively impact wildlife health.
3. Future research should focus on the interactive effects of climate change and contaminant levels at an ecosystem-level scale to better understand how contaminants released from climate change can move through the environment and impact wildlife.

INTRODUCTION

The loss of biodiversity is an urgent issue in the Anthropocene as we approach a tipping point or planetary boundary beyond which recovery is difficult (1,2). Two of the greatest risks to wildlife, of the nine identified planetary boundaries, are climate change and toxic contamination (3,4). Understanding the interactive effects of climate change and chemical pollution on wildlife is a great challenge that requires multidisciplinary approaches that integrate ecotoxicology, environmental chemistry, and remote sensing.

The effects of climate change on contaminant cycling are complex and interactive; they include different processes, such as transformations (5–7), biological uptake (8,9), transport by air, snow or ice, and ocean currents (10–13), biotransport (14,15), and biomagnification (16). Moreover, these trends can differ geographically depending on differences in climate, contamination, and wildlife.

The influence of climate change on wildlife can be separated into direct mechanisms (heat stress) and indirect mechanisms, such as bottom-up processes (changing diet) and top-down processes (changing predators; 17–20). Each of these mechanisms can interact with toxic contamination. For example, since most contaminants are obtained via diet, climate-related bottom-up processes that influence diet may lead to higher levels of contaminants if the new prey, itself, feeds at a higher trophic position. At the same time, in the face of environmental perturbations, contaminant exposures may exacerbate effects of climate change through interfering with the ability of wildlife to respond to indirect and direct threats due to climate change.

In this chapter, we examine the interactive effects of climate change and contaminants on wildlife health. First, we examine the abiotic effects of climate change on contaminants, as well as biotic effects on wildlife via diet changes.

Next, we examine how contaminants may influence resilience to climate change, with an emphasis on endocrine disruption. Finally, we discuss two case studies in Arctic and alpine regions where climate change is occurring more rapidly than anywhere else on the globe. While this chapter focuses primarily via a wildlife lens, the general trends and circumstances are also relevant to domestic species.

CHANGE IN ABIOTIC FACTORS INFLUENCING CONTAMINANT LEVELS

GLACIERS, SEA-ICE AND PERMAFROST

The impacts of climate change are greatest in Arctic and alpine habitats, where warming is twice the global average and critical cold habitats are in rapid decline (8,21,22). We have lost over 50% of the world's glaciers, 30% of Arctic sea ice, and a large portion of permafrost, with ice-free summers predicted to occur before 2050 (23). This dramatic ice loss can lead to the mobilization of high levels of contaminants that have been deposited in cold regions in glaciers, sea ice, and permafrost. At the same time, many contaminants produced in warmer climates are brought by air and water currents to colder climates where they are then deposited, a process known as the "Grasshopper Effect" (24–27). Such contamination is of concern because of the high levels that biomagnify through the food web and bioaccumulate in top predators (28). Because many bio-magnifying contaminants are lipophilic (i.e., dissolve in fats), they pose a risk to Arctic species that have large fat reserves as an adaptation for living in a colder climate, but also to northern communities who harvest Arctic predators as important traditional foods (29).

Glacial runoff from melts can interact with cold condensation, contaminating surrounding watersheds due to the minimal volatilization loss from cold runoff waters, limited catchment retention owing to low soil organic matter within such catchments, and the rapid channeling of the runoff from the glacier surface (30–32). Alpine areas in western North America are one region where contaminants, such as persistent organic pollutants (POP) and mercury, from Asia or industrial regions of North America, can be transported atmospherically and deposited (30,31,33), potentially contaminating wildlife and people that eat fish from alpine lakes (34,35). For example, the 1960–1980 global cold cycle trapped contaminants in glaciers, but these contaminants would have likely melted out by 1980 (12,34,35). Nonetheless, local processes coupled with ongoing POPs deposition in cold environments mean that significant levels of POPs remain in alpine environments worldwide (12,32,34,36). Thus, understanding the role of melting glaciers in contaminant levels in biota will be of increasing importance in a changing climate (Case Study 8.1).

Sea ice plays a key role in Arctic marine ecosystems, and many of the most threatened Arctic species are pagophilic, which means they depend on ice for some or all of their annual cycle (53,54). Sea ice can provide a platform for hunting,

CASE STUDY 8.1   ALPINE GLACIERS IN WESTERN CANADA

Contaminants that are resistant to degradation, including many POPs, occur in remote alpine habitats like the Alps and Rockies, partially due to scavenging by snow crystals, possibly via adsorption or entrapment (24,37,38). High levels in glacial runoff have been particularly evident in western North America, partially due to atmospheric transport from Asian sources (24,34,39). Half of the world's glaciers outside of the Antarctic and Greenland ice sheets have melted since 1950 (40), meaning pollutants trapped in alpine glaciers can melt out and accumulate in nearby water bodies. Indeed, most of the "legacy" pollutants in alpine lakes originated from the organo-chlorine era (1950–1970) and have since melted out of glaciers (35,41,42).

Pollutants released from glacial melt can bioaccumulate and biomag-nify through lacustrine and riverine food webs and reach high levels in fish and their predators (43,44). Indeed, elevated pollutant levels in fish and mussel tissues, including insecticides such as toxaphene and industrial pollutants such as Polychlorinated biphenyls (PCBs) occur even in locations far removed from contaminant sources (45–47). For example, zebra mussels (*Dreissena polymorpha*) from Lake Iseo, Italy, had high levels of Dichlorodiphenyltrichloroethane (DDT) from a glacial-melting event, which led to mistimed sperm release relative to egg release (46). Likewise, trout from glacier-fed lakes in the Rockies had high levels of DDT (200 ng/g), which can then biomagnify in their predatorspredators, such as osprey (Pandion haliaetus). Indeed, DDT levels are expected to be above 20ug/g for osprey feeding exclusively on DDT levels are expected to be above 20 ug/g for osprey (*Pandion haliaetus*) feeding exclusively on trout in glacial-fed lakes, well above the levels associated with a 15% reduction in eggshell thickness and effects on embryonic viability (5 ug/g; 34,35,44,48,49). Thus, climate-induced glacial melting can lead to high contaminant levels in wildlife, including piscivorous birds that are feeding at a similar trophic level as humans.

Although melting glaciers were historically a source of contaminants, particularly during the melt following the 1950–1980 cool period after the organochlorine era (see above, and Figure 8.1 in 35), current levels in alpine glaciers are very low compared to global hotspots. Moreover, the portions of alpine glaciers that are now melting out predate the Industrial Revolution (50). For example, legacy pollutant levels in osprey were not associated with glacier coverage in an alpine region (34). Although high levels in some lakes may be due to historical melting, as well as the concentration of levels deposited over large glacier areas into small downstream lakes, ongoing melting is unlikely to lead to the high levels of pollutants seen in the late 20th century. Recent investigations of contaminant release have shifted, however, to other chemicals of concern, such as pharmaceuticals, plastics, and surfactants (51,52).

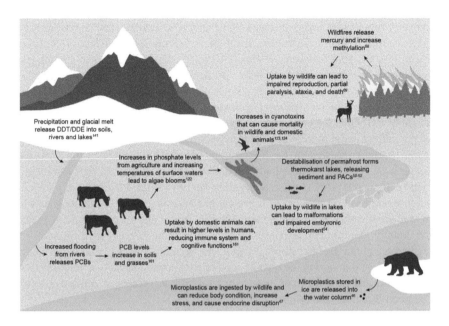

FIGURE 8.1 Example of how climate change can influence contaminant movements across different environmental compartments, and the interactive affects this movement can have on the health of wildlife and domestic animals. For each example, we highlight one contaminant that is well-studied in that context, but the same mechanism is often implicated for many different contaminants.

reproduction, travel, refuge from predators and during melt, acts to inject freshwater and nutrients that enable blooms in primary productivity (53–55). However, a dramatic decline in Arctic sea-ice cover is underway as a result of climate change, with ice-free summers predicted to occur before 2050 (21,56). Reduced ice cover can increase wildlife exposures to other anthropogenic stressors, such as increased industrial development, shipping activity, and contaminants (21,57). Indeed, sea ice can act as a sink, source, and transport medium for many contaminants (see also Case Study 8.2; 58–60). For example, microplastics can be deposited in sea ice during formation, and then transported long distances to remote regions such as the Arctic (60,61). As climate change exacerbates sea-ice melt, these contaminants can then be remobilized, thus increasing their availability in the environment (Figure 8.1; 62). Once ingested by biota, microplastics can cause a myriad of impacts, including reduced growth, reduced body condition, increased stress, endocrine disruption, reduced reproductive success, and even death (reviewed in 63). Thus, understanding how sea ice acts as a source of contaminants to wildlife is an important step to better understand how climate change may impact contaminant concentrations in wildlife.

Permafrost thaw and the melting of land ice is the largest change in the Arctic attributable to rising temperatures in soil and terrestrial environments (80), and this change is occurring across the circumpolar Arctic (81). For example, using satellite imagery, Lewkowicz and colleagues (82) found a 60-fold increase in retrogressive thaw slumps (large catastrophic thaw features) on Banks Island (Northwest Territories, Canada) between 1984 and 2015. The destabilization and slumping of permafrost, forming thermokarst lakes, is releasing sediment to downstream lakes and waterways as well as to the Arctic Ocean from coastal erosion (82–84). This release can increase both primary (released from permafrost) and secondary (released from sediment erosion, altered hydrological flow, or increased runoff caused by permafrost thaw) sources of contaminants to terrestrial and marine environments in the Arctic (reviewed in 6). For example, thermokarst lakes in the Canadian Arctic were found to contain higher concentrations of metals and polycyclic aromatic compounds (PACs) released by thaw slumps (85). PACs are highly toxic to wildlife and have been found to cause cardiac and skeletal malformations, as well as impaired embryonic development in fish, frogs, and birds (Figure 8.1; 86). Importantly, thermokarst responses to climate warming are not uniform and vary regionally within the circumpolar Arctic in relation to local landscape factors (87,88). Thus, monitoring the quantity and rate that contaminants are released from permafrost, and how this may pose a risk to wildlife species, is important to examine on a region- and species-specific basis.

LAKES

Freshwater ecosystems have been altered by climate change, with many lakes drying up via evaporation and thermokarst development, and others affected by altered precipitation patterns (32,89–91). Moreover, the ice-free period is increasing in many polar regions, influencing the stratification and biogeochemistry of lakes (92–95). Together, these processes increase the light available in the water column, influencing levels of dissolved organic carbon (DOC) and contaminant levels in lakes (96,97). Specifically, DOC can bind to methylmercury (MeHg), the most bioavailable and toxic form of mercury, and facilitate its dissolution into the water, increasing the amount of bioavailable MeHg to wildlife (98,99). MeHg is a powerful neurotoxin that can cause reproductive impairment (e.g., the suppression of sex hormones, the underdevelopment of gonads), neurological effects (e.g., partial paralysis, ataxia) and have fatal effects in vertebrates (Figure 8.1; 100–102). Importantly, MeHg can biomagnify the food web; thus, organisms at all levels may suffer severe health effects (28). The complex interactions between DOC and mercury cycling make it difficult to understand how physical changes to lakes might influence the breadth of contaminants' impact on the environment and wildlife.

WILDFIRES

In terrestrial ecosystems, wildfires play an important role in the life cycle of many plants in forest systems. However, the frequency of wildfires in some regions, exacerbated by particular forest-management regimes, has increased in recent decades. Moreover, regions such as the Arctic tundra, where forests historically have been rare (103–105), are also expected to experience an increase in wildfire frequency (105–107). Increased forest fires have complex effects on contaminant cycling. As climate change increases the frequency and intensity of wildfires, the burning of peatlands is expected to increase, thus releasing mercury into the environment, especially as methylation rates of mercury increase with temperature (Figure 8.1; 100). Wildfire suppression can also lead to outbreaks of pests, such as the mountain pine beetle (*Dendroctonus ponderosae*) in western North America, which are further exacerbated by climate change when warmer winters lead to reduced beetle deaths during overwintering (108). Increased use of pesticides, such as monosodium methanearsonate (MSMA), to control these beetles can alter the abundance and distribution of beetle species and their predators, such as woodpeckers (108). Understanding how wildfires impact the introduction of contaminants into the environment is crucial to managing wildlife health and populations.

PRECIPITATION

Climate change–induced shifts in precipitation patterns can impact the transport and remobilization of contaminants. For example, Lu et al. (109) conducted a leaching experiment on a tailings deposit in northern Norway to examine how temperature and precipitation changes due to climate change may impact heavy metal leaching from the tailings. The authors found that higher precipitation rates alone cause metal leaching from the tailing, but that increasing temperature and precipitation rates synergistically increase metal leaching, causing higher amounts of contaminants to be released from the tailings deposit. Further, Zhu et al. (110) show that increasing precipitation rates in the mid-Atlantic Ocean result in higher contaminant concentrations in surface waters, where contaminants likely come from both surface and sediment runoffs and sewage overflows. Additionally, increased precipitation rates can cause further sea ice, glacier, and permafrost melt, thus exacerbating the release of contaminants from those stores (see section *Glaciers, sea-ice and permafrost*). Leached contaminants, such as cadmium, cause a range of health issues in wildlife. For example, in Colorado, USA, cadmium from ore-mining has contaminated soils and is taken up and highly biomagnified by willows (*Salix* spp.), an important winter food source for the white-tailed ptarmigan (*Lagopus leucurus*; 111). As the ptarmigans age and continue to feed on willows, cadmium will bioaccumulate in their tissues, especially in kidneys, weakening skeletal integrity, causing renal failure, and even leading to death once the cadmium has reached its toxicity threshold (100 mg/kg; 111). These interactions emphasize the importance of assessing multiple factors when examining the interactions between climate change and contaminant concentrations, especially when trying to assess risks to wildlife.

DIET CHANGES

Warming ocean temperatures and sea-ice loss due to climate change have resulted in significant prey shifts in several Arctic marine predators (17,112,113),

with potential impacts on contaminants. For example, Øverjordet et al. (114) examined mercury concentrations in two seabird species, black-legged kittiwakes
(*Rissa tridactyla*) and dovekies (*Alle alle*), in Svalbard and found that mercury levels
in black-legged they were lower in years that they fed at lower trophic levels. The
authors suggested that this situation was a result of lower sea-ice concentrations,
where in years with less sea ice, black-legged kittiwakes would have less access to
relatively high trophic-level prey (e.g., ice-associated Arctic cod; *Boreogadus saida*)
that generally have higher mercury concentrations (16,115). Moreover, changes in
sea-ice or prey populations can cause various Arctic biota to have prolonged periods
of fasting (i.e., increased reliance on fat stores), which can lead to the release of
more fat-soluble contaminants into the body. Sea-ice concentrations can also influence the diet of terrestrial mammals. For example, Andersen et al. (116) examined
POP concentrations in Arctic foxes (*Vulpes lagopus*) from Svalbard, Norway, in
1997–2013, and found that POP concentrations increased with higher proportions of
marine diet (e.g., ringed seal; Pusa hispida) than terrestrial diet (e.g., Svalbard
reindeer; Rangifer tarandus platyrhynchus). Specifically, the authors found that
concentrations of beta- Hexachlorocyclohexane (β-HCH; an isomer of the insecticide lindane), increased with increasing sea-ice cover, suggesting higher β-HCH
levels in marine prey. This shows that climate-induced diet changes can influence
contaminant levels in complex ways (Case Study 8.2).

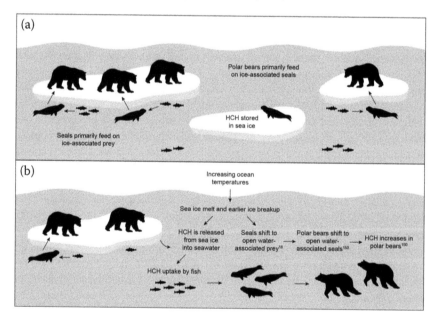

FIGURE 8.2 Example of how climate change can influence hexachlorocyclohexane
(HCH) in an Arctic food web, where panel A represents "normal" climate conditions 9/25
and panel B represents climate-induced increases in ocean temperatures. We highlight HCH
because it is well-studied in this context, but the same mechanism is often implicated for
many different contaminants.

CASE STUDY 8.2   DIET CHANGES IN RINGED SEALS AND POLAR BEARS

Ringed seals and polar bears (*Ursus maritimus*) are ecologically, economically, culturally, and nutritionally significant in the Arctic, and they are a primary food source in many Arctic communities (64,65). However, as an ice-dependent species, climate change-induced changes in ocean temperatures and sea-ice concentrations can impact ringed seal and polar bear diets, and ultimately, contaminant concentrations in their tissues. Indeed, there is evidence that ringed seals and polar bears across the Arctic are changing their diet, likely as a result of warming temperatures and changes in sea ice (18,66–69). For example, in western Hudson Bay (Nunavut, Canada), ringed-seal diet shifted from ice-associated Arctic cod to a sub-Arctic species, capelin (*Mallotus villosus*), through 1991 to 2006, likely as a result of changes in water temperatures and sea ice in this region (18). As most contaminants in these predators are likely obtained from diet, shifts in diet can play an important role in contaminant levels in this species.

Indeed, Braune et al. (16) examined changes in the feeding ecology of polar bears from Hudson Bay from 1991 to 2007 and found that several chlorinated and brominated contaminants (e.g., β-HCH, PCBs, Polybrominated diphenyl ethers PBDEs) increased over time due to shifts in diet to more open water-associated species because of earlier sea-ice breakup (Figure 8.2). Similarly, Gaden et al. (70) observed that ringed seals from the Canadian Arctic had higher concentrations of organochlorines (OCs) in years of earlier ice break-up, likely related to shifts in diet. Looking at the food web as a whole, McKinney et al. (71) examined temporal changes in OCs in a Canadian Arctic marine food web, including seals and polar bears, and found that contaminants were generally higher in ice-free seasons, but also that the presence of more contaminated, transient, or sub-Arctic species during ice-free seasons may also alter contaminant concentrations in the food web. Thus, climate change can cause shifts in diet that may alter contaminant levels, but may also cause more contaminated species to shift northward, which may increase environmental levels in the region (72).

Currently, knowledge on contaminants in ringed seal prey is limited, and results vary spatially and temporally. For example, Pedro et al. (73) and Braune et al. (16) sampled Arctic cod and capelin from different regions in Nunavut, Canada, and found that mercury concentrations were consistently higher in Arctic cod than capelin. However, when examining POPs, this relationship is less clear. While Braune et al. (16) found that PCB concentrations were also higher in Arctic cod than capelin, Pedro et al. (73) found the opposite: PCBs levels in capelin were higher than in Arctic cod from the same region. These differences are likely related to the migratory behavior of these fish species, and as climate change continues to shift species distributions northward (74), understanding the role of species shifts on contaminants throughout the Arctic food web will be of increasing importance.

Climate change-induced changes in contaminant concentrations in ringed seals and polar bears can also impact Indigenous communities who rely on these predators for sustenance. Indeed, in eastern Greenland, Sonne et al. (75) estimated that subsistence hunters that regularly consume ringed seals and polar bears may exceed the tolerable daily intake of PCBs and chlordane pesticides by approximately five-fold. Moreover, Quinn et al. (76) modeled exposure to PCBs in Indigenous women and estimated that women who consume an entirely traditional diet (i.e., seal blubber) have 15–150 times higher body burdens of PCBs than women who consume imported foods. PCBs can cause severe health effects in humans, including cancer, reduced immune-system function, and lower cognitive abilities (77). While it can be difficult to model how climate change and contaminant levels in food webs will change over time, models that consider emissions, transport, and fate of contaminants under changing climate conditions will be useful to examine how these cumulative effects may impact Arctic biota and the people that harvest them (78). Hickie et al. (79) modeled the accumulation of POPs in ringed seal blubber using data from the Canadian Arctic and found that their ability to eliminate many POPs, along with a fast juvenile growth rate and population turnover, allows this species to respond rapidly to contaminant changes. Thus, ringed seals are a useful indicator species to monitor spatial and temporal trends in POPs and other contaminants, while also providing data for risk assessments for Indigenous communities in the north (79).

IMMUNOCOMPETENCE AND ENDOCRINE DISRUPTION

Contaminants can disrupt endocrine systems and cause altered reproductive behaviors in wildlife (117). Two mechanisms by which climate change is impacting wildlife populations are first, by a mismatch in the timing of breeding and environmental cues (17,118), and second, increased parasitism and disease (19,119). For example, PCBs and PBDEs both mimic thyroid hormones and bind readily to the carrier proteins in gulls (120). This may compromise the ability of individuals to respond to environmental change and lead to reduced reproductive success (121). Infectious diseases and parasites have caused the population collapse of many wildlife species, including white-nose syndrome in North American Myotis bats and Batrachochytrium dendrobatidis on amphibians (119). At the same time, many contaminants can alter the immune status of wildlife, leaving them at greater risk for these diseases (122,123).

Endocrine Disruption

Primary productivity can respond rapidly to increases in energy associated with climate change, while animals at higher trophic levels may use other cues to determine optimal timing of breeding (17,118). For example, in Arctic seabirds, a phytoplankton bloom immediately following the departure of sea ice sets a timeline that dictates the food available for offspring (55). Seabirds must be able to time their egg-laying approximately one month ahead of time to match the period of highest energy demands (i.e., chick-rearing) with the peak in prey availability (when the most energy can be obtained for the least energy spent; 17). Climate change can cause ice to melt more rapidly than seabirds are able to respond, creating a mismatch between the time for chick-rearing and the peak of food availability, leading to unsustainable levels of energy expenditure (17). Contaminants such as per and poly-fluoroalkyl acids can disrupt thyroid hormone homeostasis in seabirds (16,124), which are associated with avian thermoregulation. It is predicted that warming temperatures due to climate change may affect avian thermoregulation and endocrine regulation, especially processes regulated by thyroid hormones (125). Therefore, the effects of climate change on endocrine and thyroid function may be exacerbated by contaminants. Moreover, the brain is particularly sensitive to contamination (126,127), potentially influencing behavioral responses to climate change. For example, neonicotinoid pesticides alter the behavior of migratory songbirds, which may exacerbate mismatches between historical migratory tendencies and current peaks in food availability (the mismatch hypothesis; 128,129). A major factor in determining the resilience of wildlife to climate change is their behavioral plasticity (or flexibility) to accommodate changes in food availability. However, contaminants may disrupt hormones (and related biomarkers) that allow such behavioral plasticity.

Immune Disruption

Many human and wildlife parasites and diseases will increase in response to climate change, although others are expected to go extinct (130). For example, avian cholera, human and avian malaria, and Rift Valley Fever have increased their distribution with climate change (130), while increased mosquitoes and blackflies on warm days has led to increased mortality in seabirds and birds of prey (19,131). Many contaminants cause immune disruption at environmentally relevant levels in the lab (132); however, this effect is less studied in natural environments. Moreover, wildlife living in highly toxic environments, such as the Baltic Sea, Saint Lawrence Estuary, and Great Lakes, had high rates of lesions and diseases during the organochlorine era and shortly thereafter (133–135). Thus, contaminant-induced immune deficiencies in wildlife may reduce their plasticity when responding to novel diseases facilitated by climate change.

CARCINOGEN EXPOSURE AND NEOPLASIA RISKS

Contaminant concentrations in the environment are continuously increasing, by exacerbated weather patterns and natural disasters, or by shifts in contaminant bioavailability and cycling (136,137). Animals are increasingly exposed to these environmental contaminants, many of which are carcinogenic, which lead to the deteriorating health of populations. Contaminants like polycyclic aromatic hydrocarbons (PAH) and HCH have adverse effects on immune system function and can cause the development of neoplasms (abnormal tissue growth) in animals, which may become malignant (137–140). Contaminants may further act as endocrine disruptors or cause oxidative damage to DNA in vertebrates leading to the development of various forms of cancer (138).

For example, PAHs were linked to liver lesions in winter flounder (*Pleuronectes americanus*) and cancerous tumors in Saint Lawrence Estuary belugas (*Delphinapterus leucas*) in habitats with contaminated sediments (141–143). Cancer was the main cause of death in the Saint Lawrence Estuary belugas, and researchers identified 21 types of cancers, the most prevalent consisting of intestinal tumors (143). The authors attribute the high rate of cancer in belugas (27%; comparable to cancer rates in humans) to their diet and feeding strategies: belugas disturb sediments to feed on invertebrates, which have been shown to be highly contaminated by PAHs produced by aluminum-smelting factories in proximity to the Saint Lawrence Estuary (143,144). Further, wildfires are another important source of PAHs, and their frequency is expected to increase as climate change progresses (106,145). Thus, climate change is an additional environmental stressor that can work to weaken immune systems in wildlife, making animals more vulnerable to disease. Climate change and pollutants, therefore, work synergistically as threats to animal health (139,140).

TOXIC ALGAE BLOOMS

Marine heatwaves doubled between 1982 and 2016, and they are expected to increase rapidly in the near future (146). Similarly, inland waters are warming at 0.13°C per decade with 99% showing warming trends (147). These warming trends in marine and terrestrial environments, are especially extreme events, are negatively affecting aquatic animals, causing increases in coral bleaching (148), die-offs (149), and the need to search over long distances for dwindling food (150). Moreover, these responses can persist long after the heat waves subside (151). One of the greatest impacts of warming waters is increased toxic-algae blooms. For example, 68% of lakes have experienced more algae blooms since satellite imagery started, while only 8% declined (152). Similarly, harmful algal blooms in marine environments are expected to increase with ocean stratification (153).

In terrestrial environments, increased nutrient loading has led to a global increase in freshwater algal blooms (147). Higher levels of nutrients (particularly phosphate) from intensive agriculture, coupled with warmer surface waters and increased stratification and eutrophication, can lead to increases in cyanobacteria, the main algae involved in harmful freshwater algae blooms. For example, overflowing hog farms following Hurricane Florence led to large blooms in North Carolina (154). Cyanobacteria produce a variety of cyanotoxins that are harmful to fish and humans alike, and can cause large-scale mortality of fish, domestic livestock, and birds (Figure 8.1; 155,156).

In the marine environment, roughly 200 species of phytoplankton cause harmful algal blooms, and future impacts are difficult to predict. Indeed, poisoning from "red tide" was recorded in 731 in Japan and the 1500s in the Gulf of Mexico (157). Yet, some forms of harmful algal blooms have only been described in the last 40 years (157). Toxins produced during these algal blooms, such as domoic acid and saxotoxins, can cause mass die-offs of fish, seabirds and marine mammals, including large-scale mortality of endangered species (158–160). No change in marine harmful algal blooms has been detected globally, although trends are difficult to detect because of the many taxa involved, regional differences in reporting, awareness of novel toxins, and because many reports are anecdotal or based on impacts to poisoned species (157). Changes in temperature, salinity, stratification, light, storm intensity, and ocean nitrification and acidification are all likely to influence future algae bloom trends and wildlife health in complex and unpredictable ways (153).

DOMESTIC ANIMALS

This chapter has illustrated the variety and magnitude of climate change and contaminants interactions by focusing on wildlife. Domestic animals, however, will also be affected, in turn affecting humans. Some of the impacts described for wildlife will be the same for domestic animals. For example, just as precipitation and warming will increase wildlife exposure to freshwater toxic algae, contaminants in drinking water or immunosuppressive chemicals, domestic animal exposures will increase as well. These will have implications for human health when domestic animals enter the human food chain or when agriculture waste products enter soil and waterways. There may also be effects that are unique to domestic animals. For example, as climate change impacts the prevalence of certain infectious diseases, so too might the use of veterinarian use of veterinary drugs and chemicals, with subsequent risks of increased meat contamination (161). Livestock foods may be increasingly plagued by mycotoxins, which are expected to become more prevalent in crops with climate change (162). Additionally, while increased flooding of agriculture areas can redistribute environmental pathogens and contaminants, but contaminants, it may also disperse fuels and chemicals that are usually safety stored on farms or nearby areas. For example, increased flooding in urban and industrial areas has been shown to increase contaminant concentrations in domestic cows' milk on nearby farms, likely due to higher levels in the grasses on which they feed (163). Domestic animal health-adaptation strategies must, therefore, include consideration of direct effects of contaminants on animal and public health, as well as how changes in exposure to chemicals contaminants may interact with and modify other climate change related health effects.

SUMMARY AND FUTURE DIRECTIONS

Climate change and toxic contaminants interact in complex and unpredictable ways. As climate change continues, contaminants will continue to be released by ice melt, wildfires, erosion, precipitation, and more. As wildlife respond to these changes by shifting their diet or range, contaminant concentrations may be altered further. These changes can result in a variety of impacts, including neurological effects, altered stress and hormone levels, endocrine disruption, and mortality. While it can be difficult to model how climate change and contaminant levels will change and interact over time, models that consider sources, transport, and fate of contaminants under changing climate conditions will be useful to examine how these cumulative effects may impact wildlife and the people that harvest them. Although wildlife species can be used as indicators of changes in climate or contaminants, studies rarely examine both. Much more work is needed to better understand how wildlife health will continue to be impacted by contaminants in the context of exacerbating climate change. In the short term, bioremediation solutions can reduce the overall quantity of pollutants that wildlife are exposed to during their lifecycle. Future research should focus on examining the interactive effects of climate change and contaminants at an ecosystem-level scale to more accurately assess how these two threats interact and impact wildlife and the environment.

REFERENCES

1. Rockström, J., Steffen, W., Noone, K., Persson, Å, Chapin, III, F.S., Lambin, E., et al., 2009. Planetary boundaries: Exploring the safe operating space for humanity. *Ecology and Society*, *14*(2).
2. Steffen, W., Richardson, K., Rockström, J., Cornell, S.E., Fetzer, I., Bennett, E.M., et al., 2015. Planetary boundaries: Guiding human development on a changing planet. *Science*, *347*(6223).
3. Armitage, J.M., Quinn, C.L. and Wania, F., 2011. Global climate change and contaminants—an overview of opportunities and priorities for modelling the potential implications for long-term human exposure to organic compounds in the Arctic. *Journal of Environmental Monitoring*, *13*(6), pp. 1532–1546.
4. Canham, R., González-Prieto, A.M. and Elliott, J.E., 2021. Mercury exposure and toxicological consequences in fish and fish-eating wildlife from anthropogenic activity in Latin America. *Integrated Environmental Assessment and Management*, *17*(1), pp. 13–26.
5. MacMillan, G.A., Girard, C., Chételat, J., Laurion, I. and Amyot, M., 2015. High methylmercury in Arctic and subarctic ponds is related to nutrient levels in the warming eastern Canadian Arctic. *Environmental Science & Technology*, *49*(13), pp. 7743–7753.
6. Vonk, J.E., Tank, S.E., Bowden, W.B., Laurion, I., Vincent, W.F., Alekseychik, P., et al., 2015. Reviews and syntheses: Effects of permafrost thaw on Arctic aquatic ecosystems. *Biogeosciences*, *12*(23), pp. 7129–7167.
7. Wang, X., Sun, D. and Yao, T., 2016. Climate change and global cycling of persistent organic pollutants: A critical review. *Science China Earth Sciences*, *59*(10), pp. 1899–1911.
8. Post, E., Forchhammer, M.C., Bret-Harte, M.S., Callaghan, T.V., Christensen, T.R., Elberling, B., et al., 2009. Ecological dynamics across the Arctic associated with recent climate change. *Science*, *325*(5946), pp. 1355–1358.

9. Hudelson, K.E., Muir, D.C., Drevnick, P.E., Köck, G., Iqaluk, D., Wang, X., et al., 2019. Temporal trends, lake-to-lake variation, and climate effects on Arctic char (Salvelinus alpinus) mercury concentrations from six High Arctic lakes in Nunavut, Canada. *Science of The Total Environment, 678*, pp. 801–812.
10. Fernández, P. and Grimalt, J.O., 2003. On the global distribution of persistent organic pollutants. *CHIMIA International Journal for Chemistry, 57*(9), pp. 514–521.
11. von Waldow, H., MacLeod, M., Jones, K., Scheringer, M. and Hungerbühler, K., 2010. Remoteness from emission sources explains the fractionation pattern of polychlorinated biphenyls in the northern hemisphere. *Environmental Science & Technology, 44*(16), pp. 6183–6188.
12. Steinlin, C., Bogdal, C., Luthi, M.P., Pavlova, P.A., Schwikowski, M., Zennegg, M., et al., 2016. A temperate alpine glacier as a reservoir of polychlorinated biphenyls: Model results of incorporation, transport, and release. *Environmental Science & Technology, 50*(11), pp. 5572–5579.
13. Zdanowicz, C.M., Proemse, B.C., Edwards, R., Feiteng, W., Hogan, C.M., Kinnard, C., et al., 2018. Historical black carbon deposition in the Canadian High Arctic: A >250-year long ice-core record from Devon Island. *Atmospheric Chemistry and Physics, 18*(16), pp. 12345–12361.
14. Blais, J.M., Kimpe, L.E., McMahon, D., Keatley, B.E., Mallory, M.L., Douglas, M.S., et al., 2005. Arctic seabirds transport marine-derived contaminants. *Science, 309*(5733), p. 445.
15. Choy, E.S., Gauthier, M., Mallory, M.L., Smol, J.P., Douglas, M.S., Lean, D., et al., 2010. An isotopic investigation of mercury accumulation in terrestrial food webs adjacent to an Arctic seabird colony. *Science of the Total Environment, 408*(8), pp. 1858–1867.
16. Braune, B.M., Gaston, A.J., Elliott, K.H., Provencher, J.F., Woo, K.J., Chambellant, M., et al., 2014. Organohalogen contaminants and total mercury in forage fish preyed upon by thick-billed murres in northern Hudson Bay. *Marine Pollution Bulletin, 78*(1–2), pp. 258–266.
17. Gaston, A.J., Gilchrist, H.G., Mallory, M.L. and Smith, P.A., 2009. Changes in seasonal events, peak food availability, and consequent breeding adjustment in a marine bird: A case of progressive mismatching. *The Condor, 111*(1), pp. 111–119.
18. Chambellant, M., Stirling, I. and Ferguson, S.H., 2013. Temporal variation in western Hudson Bay ringed seal Phoca hispida diet in relation to environment. *Marine Ecology Progress Series, 481*, pp. 269–287.
19. Gaston, A.J. and Elliott, K.H., 2013. Effects of climate-induced changes in parasitism, predation and predator-predator interactions on reproduction and survival of an Arctic marine bird. *Arctic*, pp. 43–51.
20. Blunden, J. and Arndt, D.S., 2016. State of the climate in 2015. *Bulletin of the American Meteorological Society, 97*(8), pp. Si–S275.
21. Overland, J.E. and Wang, M., 2013. When will the summer Arctic be nearly sea ice free? *Geophysical Research Letters, 40*(10), pp. 2097–2101.
22. McKinney, M.A., Pedro, S., Dietz, R., Sonne, C., Fisk, A.T., Roy, D., et al., 2015. A review of ecological impacts of global climate change on persistent organic pollutant and mercury pathways and exposures in arctic marine eco-systems. *Current Zoology, 61*(4), pp. 617–628.
23. Wang, M. and Overland, J.E., 2012. A sea ice free summer Arctic within 30 years: An update from CMIP5 models. *Geophysical Research Letters, 39*(18).

24. Wania, F. and Mackay, D., 1996. Tracking the distribution of persistent organic pollutants. *Environmental Science and Technology*, *30*(9).

25. Durnford, D., Dastoor, A., Figueras-Nieto, D. and Ryjkov, A., 2010. Long range transport of mercury to the Arctic and across Canada. *Atmospheric Chemistry and Physics*, *10*(13), pp. 6063–6086.

26. Vorkamp, K. and Rigét, F.F., 2014. A review of new and current-use contaminants in the Arctic environment: Evidence of long-range transport and indications of bioaccumulation. *Chemosphere*, *111*, pp. 379–395.

27. Foster, K.L., Braune, B.M., Gaston, A.J. and Mallory, M.L., 2019. Climate influence on legacy organochlorine pollutants in Arctic seabirds. *Environmental Science & Technology*, *53*(5), pp. 2518–2528.

28. Atwell, L., Hobson, K.A. and Welch, H.E., 1998. Biomagnification and bioaccumulation of mercury in an arctic marine food web: Insights from stable nitrogen isotope analysis. *Canadian Journal of Fisheries and Aquatic Sciences*, *55*(5), pp. 1114–1121.

29. Kuhnlein, H.V. and Chan, H.M., 2000. Environment and contaminants in traditional food systems of northern indigenous peoples. *Annual review of nutrition*, *20*(1), pp. 595–626.

30. Blais, J.M., Schindler, D.W., Muir, D.C., Sharp, M., Donald, D., Lafreniere, M., et al., 2001. Melting glaciers: A major source of persistent organochlorines to subalpine Bow Lake in Banff National Park, Canada. *Ambio*, pp. 410–415.

31. Blais, J.M., Schindler, D.W., Sharp, M., Braekevelt, E., Lafrenière, M., McDonald, K., et al., 2001. Fluxes of semivolatile organochlorine compounds in Bow Lake, a high-altitude, glacier-fed, subalpine lake in the Canadian Rocky Mountains. *Limnology and Oceanography*, *46*(8), pp. 2019–2031.

32. Milner, A.M., Khamis, K., Battin, T.J., Brittain, J.E., Barrand, N.E., Füreder, L., et al., 2017. Glacier shrinkage driving global changes in downstream systems. *Proceedings of the National Academy of Sciences*, *114*(37), pp. 9770–9778.

33. Jaffe, D., McKendry, I., Anderson, T. and Price, H., 2003. Six 'new' episodes of trans-Pacific transport of air pollutants. *Atmospheric Environment*, *37*(3), pp. 391–404.

34. Elliott, J.E., Levac, J., Guigueno, M.L.F., Shaw, D.P., Wayland, M., Morrissey, C.A., et al., 2012. Factors influencing legacy pollutant accumulation in alpine osprey: Biology, topography, or melting glaciers? *Environmental Science & Technology*, *46*(17), pp. 9681–9689.

35. Guigueno, M.F., Elliott, K.H., Levac, J., Wayland, M. and Elliott, J.E., 2012. Differential exposure of alpine ospreys to mercury: Melting glaciers, hydrology or deposition patterns? *Environment International*, *40*, pp. 24–32.

36. Luo, W., Gao, J., Bi, X., Xu, L., Guo, J., Zhang, Q., et al., 2016. Identification of sources of polycyclic aromatic hydrocarbons based on concentrations in soils from two sides of the Himalayas between China and Nepal. *Environmental Pollution*, *212*, pp. 424–432.

37. Blais, J.M., Schindler, D.W., Muir, D.C., Kimpe, L.E., Donald, D.B. and Rosenberg, B., 1998. Accumulation of persistent organochlorine compounds in mountains of western Canada. *Nature*, *395*(6702), pp. 585–588.

38. Franz, T.P. and Eisenreich, S.J., 1998. Snow scavenging of polychlorinated biphenyls and polycyclic aromatic hydrocarbons in Minnesota. *Environmental Science & Technology*, *32*(12), pp. 1771–1778.

39. Atlas, E. and Giam, C., 1981. Global transport of organic pollutants: Ambient concentrations in the remote marine atmosphere. *Science*, *211*(4478), pp. 163–165.

40. Slater, T., Lawrence, I.R., Otosaka, I.N., Shepherd, A., Gourmelen, N., Jakob, L., et al., 2020. Earth's ice imbalance.

41. Levy, W., Pandelova, M., Henkelmann, B., Bernhöft, S., Fischer, N., Antritter, F., et al., 2017. Persistent organic pollutants in shallow percolated water of the Alps Karst system (Zugspitze summit, Germany). *Science of The Total Environment*, *579*, pp. 1269–1281.

42. Li, J., Yuan, G.-L., Wu, M.-Z., Sun, Y., Han, P. and Wang, G.-H., 2017. Evidence for persistent organic pollutants released from melting glacier in the central Tibetan Plateau, China. *Environmental Pollution*, *220*, pp. 178–185.

43. Kidd, K.A., Schindler, D., Hesslein, R.H. and Muir, D., 1995. Correlation between stable nitrogen isotope ratios and concentrations of organochlorines in biota from a freshwater food web. *Science of the Total Environment*, *160*, pp. 381–390.

44. Henny, C.J., Kaiser, J.L., Grove, R.A., Bentley, V.R. and Elliott, J.E., 2003. Biomagnification factors (fish to osprey eggs from Willamette River, Oregon, USA) for PCDDs, PCDFs, PCBs and OC pesticides. *Environmental Monitoring and Assessment*, *84*(3), pp. 275–315.

45. Muir, D., Grift, N., Ford, C., Reiger, A., Hendzel, M. and Lockhart, W., 1990. Evidence for long-range transport of toxaphene of remote arctic and subarctic waters from monitoring of fish tissues.

46. Bacchetta, R. and Mantecca, P., 2009. DDT polluted meltwater affects reproduction in the mussel Dreissena polymorpha. *Chemosphere*, *76*(10), pp. 1380–1385.

47. Grenier, P., Elliott, J.E., Drouillard, K.G., Guigueno, M.F., Muir, D., Shaw, D.P., et al., 2020. Long-range transport of legacy organic pollutants affects alpine fish eaten by ospreys in western Canada. *Science of the Total Environment*, *712*, pp. 135889.

48. Elliott, J.E., Wilson, L.K., Henny, C.J., Trudeau, S.F., Leighton, F.A., Kennedy, S.W., et al., 2001. Assessment of biological effects of chlorinated hydrocarbons in osprey chicks. *Environmental Toxicology and Chemistry: An International Journal*, *20*(4), pp. 866–879.

49. Elliott, K.H., Braune, B.M. and Elliott, J.E., 2021. Beyond bulk δ15N: Combining a suite of stable isotopic measures improves the resolution of the food webs mediating contaminant signals across space, time and communities. *Environment International*, *148*, pp. 106370.

50. Cogley, J.G., 2009. Geodetic and direct mass-balance measurements: Comparison and joint analysis. *Annals of Glaciology*, *50*(50), pp. 96–100.

51. Battaglin, W.A., Bradley, P.M., Iwanowicz, L., Journey, C.A., Walsh, H.L. and Blazer, V.S., 2018. Pharmaceuticals, hormones, pesticides, and other bioactive contaminants in water, sediment, and tissue from Rocky Mountain National Park, 2012–2013. *Science of the Total Environment*, *643*, pp. 651–673.

52. Ambrosini, R., Azzoni, R.S., Pittino, F., Diolaiuti, G., Franzetti, A. and Parolini, M., 2019. First evidence of microplastic contamination in the supraglacial debris of an alpine glacier. *Environmental Pollution*, *253*, pp. 297–301.

53. Granskog, M.A., Kaartokallio, H. and Shirasawa, K., 2003. Nutrient status of Baltic Sea ice: Evidence for control by snow-ice formation, ice permeability, and ice algae. *Journal of Geophysical Research: Oceans*, *108*(C8).

54. Rausch, R.L., George, J.C. and Brower, H.K., 2007. Effect of climatic warming on the Pacific walrus, and potential modification of its helminth fauna. *Journal of Parasitology*, *93*(5), pp. 1247–1251.

55. Laidre, K.L., Heide-Jørgensen, M.P., Nyeland, J., Mosbech, A. and Boertmann, D., 2008. Latitudinal gradients in sea ice and primary production determine Arctic seabird colony size in Greenland. *Proceedings of the Royal Society B: Biological Sciences*, *275*(1652), pp. 2695–2702.

56. Comiso, J.C., Parkinson, C.L., Gersten, R. and Stock, L., 2008. Accelerated decline in the Arctic sea ice cover. *Geophysical Research Letters*, *35*(1). Online, no page numbers https://agupubs.onlinelibrary.wiley.com/doi/10.1029/2007GL031972

57. Pizzolato, L., Howell, S.E., Derksen, C., Dawson, J. and Copland, L., 2014. Changing sea ice conditions and marine transportation activity in Canadian Arctic waters between 1990 and 2012. *Climatic Change*, *123*(2), pp. 161–173.

58. Korsnes, R., Pavlova, O. and Godtliebsen, F., 2002. Assessment of potential transport of pollutants into the Barents Sea via sea ice—an observational approach. *Marine Pollution Bulletin*, *44*(9), pp. 861–869.

59. Ma, J., Hung, H., Tian, C. and Kallenborn, R., 2011. Revolatilization of persistent organic pollutants in the Arctic induced by climate change. *Nature Climate Change*, *1*(5), pp. 255–260.

60. Kanhai, L.D.K., Gardfeldt, K., Krumpen, T., Thompson, R.C. and O'Connor, I., 2020. Microplastics in sea ice and seawater beneath ice floes from the Arctic Ocean. *Scientific Reports*, *10*(1), pp. 5004.

61. Peeken, I., Primpke, S., Beyer, B., Gütermann, J., Katlein, C., Krumpen, T., et al., 2018. Arctic sea ice is an important temporal sink and means of transport for microplastic. *Nature Communications*, *9*(1), pp. 1505.

62. Obbard, R.W., Sadri, S., Wong, Y.Q., Khitun, A.A., Baker, I. and Thompson, R.C., 2014. Global warming releases microplastic legacy frozen in Arctic Sea ice. *Earth's Future*, *2*(6), pp. 315–320.

63. Zhang, S., Wang, J., Liu, X., Qu, F., Wang, X., Wang, X., et al., 2019. Microplastics in the environment: A review of analytical methods, distribution, and biological effects. *TrAC Trends in Analytical Chemistry*, *111*, pp. 62–72.

64. Pelly, D.F. and Franks, C.E.S., 2002. Sacred hunt: A portrait of the relationship between seals and Inuit. *The American Review of Canadian Studies*, *32*(4), pp. 723–725.

65. Dowsley, M., 2010. The value of a polar bear: Evaluating the role of a multiple-use resource in the Nunavut mixed economy. *Arctic Anthropology*, *47*(1), pp. 39–56.

66. McKinney, M.A., Peacock, E. and Letcher, R.J., 2009. Sea ice-associated diet change increases the levels of chlorinated and brominated contaminants in polar bears. *Environmental Science & Technology*, *43*(12), pp. 4334–4339.

67. Young, B. and Ferguson, S., 2013. Seasons of the ringed seal: Pelagic open-water hyperphagy, benthic feeding over winter and spring fasting during molt. *Wildlife Research*, *40*(1), pp. 52–60.

68. Lowther, A.D., Fisk, A., Kovacs, K.M. and Lydersen, C., 2017. Interdecadal changes in the marine food web along the west Spitsbergen coast detected in the stable isotope composition of ringed seal (Pusa hispida) whiskers. *Polar Biology*, *40*(10), pp. 2027–2033.

69. Boucher, N.P., Derocher, A.E. and Richardson, E.S., 2020. Spatial and temporal variability in ringed seal (Pusa hispida) stable isotopes in the Beaufort Sea. *Ecology and Evolution*, *10*(10), pp. 4178–4192.

70. Gaden, A., Ferguson, S.H., Harwood, L., Melling, H., Alikamik, J. and Stern, G., 2012. Western Canadian Arctic ringed seal organic contaminant trends in relation to sea ice break-up. *Environmental Science & Technology*, *46*(8), pp. 4427–4433.

71. McKinney, M.A., McMeans, B.C., Tomy, G.T., Rosenberg, B., Ferguson, S.H., Morris, A., et al., 2012. Trophic transfer of contaminants in a changing arctic marine food web: Cumberland Sound, Nunavut, Canada. *Environmental Science & Technology*, 46(18), pp. 9914–9922.
72. Hallanger, I.G., Warner, N.A., Ruus, A., Evenset, A., Christensen, G., Herzke, D., et al., 2011. Seasonality in contaminant accumulation in Arctic marine pelagic food webs using trophic magnification factor as a measure of bioaccumulation. *Environmental Toxicology and Chemistry*, 30(5), pp. 1026–1035.
73. Pedro, S., Fisk, A.T., Tomy, G.T., Ferguson, S.H., Hussey, N.E., Kessel, S.T., et al., 2017. Mercury and persistent organic pollutants in native and invading forage species of the canadian arctic: Consequences for food web dynamics. *Environmental Pollution*, 229, pp. 229–240.
74. Perry, A.L., Low, P.J., Ellis, J.R. and Reynolds, J.D., 2005. Climate change and distribution shifts in marine fishes. *Science*, 308(5730), pp. 1912–1915.
75. Sonne, C., Dietz, R. and Letcher, R.J., 2013. Chemical cocktail party in East Greenland: A first time evaluation of human organohalogen exposure from consumption of ringed seal and polar bear tissues and possible health implications. *Toxicological & Environmental Chemistry*, 95(5), pp. 853–859.
76. Quinn, C.L., Armitage, J.M., Breivik, K. and Wania, F., 2012. A methodology for evaluating the influence of diets and intergenerational dietary transitions on historic and future human exposure to persistent organic pollutants in the Arctic. *Environment International*, 49, pp. 83–91.
77. Carpenter, D.O., 2006. Polychlorinated biphenyls (PCBs): Routes of exposure and effects on human health. *Reviews on Environmental Health*, 21(1), pp. 1–24.
78. Kenny, T.-A., 2019. Climate change, contaminants, and country food: Collaborating with communities to promote food security in the Arctic. *Predicting Future Oceans*: Elsevier;. p. pp. 249–263.
79. Hickie, B.E., Muir, D.C.G., Addison, R.F. and Hoekstra, P.F., 2005. Development and application of bioaccumulation models to assess persistent organic pollutant temporal trends in arctic ringed seal (*Phoca hispida*) populations. *Science of The Total Environment*, 351-352, pp. 413–426.
80. Liljedahl, A.K., Boike, J., Daanen, R.P., Fedorov, A.N., Frost, G.V., Grosse, G., et al., 2016. Pan-Arctic ice-wedge degradation in warming permafrost and its influence on tundra hydrology. *Nature Geoscience*, 9(4), pp. 312–318.
81. Biskaborn, B.K., Smith, S.L., Noetzli, J., Matthes, H., Vieira, G., Streletskiy, D.A., et al., 2019. Permafrost is warming at a global scale. *Nature Communications*, 10(1), pp. 1–11.
82. Lewkowicz, A.G. and Way, R.G., 2019. Extremes of summer climate trigger thousands of thermokarst landslides in a High Arctic environment. *Nature Communications*, 10(1), pp. 1–11.
83. Lantuit, H. and Pollard, W., 2008. Fifty years of coastal erosion and retrogressive thaw slump activity on Herschel Island, southern Beaufort Sea, Yukon Territory, Canada. *Geomorphology*, 95(1-2), pp. 84–102.
84. Kokelj, S., Tunnicliffe, J., Lacelle, D., Lantz, T., Chin, K. and Fraser, R., 2015. Increased precipitation drives mega slump development and destabilization of ice-rich permafrost terrain, northwestern Canada. *Global and Planetary Change*, 129, pp. 56–68.
85. Thienpont, J.R., Eickmeyer, D.C., Kimpe, L.E. and Blais, J.M., 2020. Thermokarst disturbance drives concentration and composition of metals and polycyclic aromatic compounds in lakes of the western Canadian Arctic. *Journal of Geophysical Research: Biogeosciences*, 125(12), p. e2020JG005834.

86. Wallace, S.J., De Solla, S.R., Head, J.A., Hodson, P.V., Parrott, J.L., Thomas, P.J., et al., 2020. Polycyclic aromatic compounds (PACs) in the Canadian environment: Exposure and effects on wildlife. *Environmental Pollution*, *265*, p. 114863.

87. AMAP, 2017. Snow, Water, Ice and Permafrost in the Arctic (SWIPA) 2017.

88. Loranty, M.M., Berner, L.T., Taber, E.D., Kropp, H., Natali, S.M., Alexander, H.D., et al., 2018. Understory vegetation mediates permafrost active layer dynamics and carbon dioxide fluxes in open-canopy larch forests of northeastern Siberia. *Plos One*, *13*(3), pp. e0194014.

89. Stokes, C., Popovnin, V., Aleynikov, A., Gurney, S. and Shahgedanova, M., 2007. Recent glacier retreat in the Caucasus Mountains, Russia, and associated increase in supraglacial debris cover and supra-/proglacial lake development. *Annals of Glaciology*, *46*, pp. 195–203.

90. Carroll, M.L., Townshend, J., DiMiceli, C., Loboda, T. and Sohlberg, R., 2011. Shrinking lakes of the Arctic: Spatial relationships and trajectory of change. *Geophysical Research Letters*, *38*(20). Online, no page nmbers.

91. Finger Higgens, R., Chipman, J., Lutz, D., Culler, L., Virginia, R. and Ogden, L., 2019. Changing lake dynamics indicate a drier Arctic in Western Greenland. *Journal of Geophysical Research: Biogeosciences*, *124*(4), pp. 870–883.

92. Vincent, A.C., Mueller, D.R. and Vincent, W.F., 2008. Simulated heat storage in a perennially ice-covered high Arctic lake: Sensitivity to climate change. *Journal of Geophysical Research: Oceans*, *113*(C4).

93. Brown, L. and Duguay, C., 2011. The fate of lake ice in the North American Arctic. *The Cryosphere*, *5*(4), pp. 869–892.

94. Surdu, C.M., Duguay, C.R. and Fernández Prieto, D., 2016. Evidence of recent changes in the ice regime of lakes in the Canadian High Arctic from spaceborne satellite observations. *The Cryosphere*, *10*(3), pp. 941–960.

95. Priet-Mahéo, M., Ramón, C., Rueda, F. and Andradóttir, H., 2019. Mixing and internal dynamics of a medium-size and deep lake near the Arctic Circle. *Limnology and Oceanography*, *64*(1), pp. 61–80.

96. Outridge, P., Sanei, H., Stern, G., Hamilton, P. and Goodarzi, F., 2007. Evidence for control of mercury accumulation rates in Canadian High Arctic lake sediments by variations of aquatic primary productivity. *Environmental Science & Technology*, *41*(15), pp. 5259–5265.

97. Cory, R.M., Ward, C.P., Crump, B.C. and Kling, G.W., 2014. Sunlight controls water column processing of carbon in arctic fresh waters. *Science*, *345*(6199), pp. 925–928.

98. Driscoll, C.T., Blette, V., Yan, C., Schofield, C., Munson, R. and Holsapple, J., 1995. The role of dissolved organic carbon in the chemistry and bioavailability of mercury in remote Adirondack lakes. *Water, Air, and Soil Pollution*, *80*(1), pp. 499–508.

99. Ravichandran, M., 2004. Interactions between mercury and dissolved organic matter—a review. *Chemosphere*, *55*(3), pp. 319–331.

100. Turetsky, M.R., Harden, J.W., Friedli, H.R., Flannigan, M., Payne, N., Crock, J., et al., 2006. Wildfires threaten mercury stocks in northern soils. *Geophysical Research Letters*, *33*(16).

101. Scheuhammer, A.M., Meyer, M.W., Sandheinrich, M.B. and Murray, M.W., 2007. Effects of environmental methylmercury on the health of wild birds, mammals, and fish. *Ambio*, pp. 12–18.

102. Hong, Y.-S., Kim, Y.-M. and Lee, K.-E., 2012. Methylmercury exposure and health effects. *Journal of Preventive Medicine and Public Health*, *45*(6), pp. 353.

103. Wein, R.W., 1976. Frequency and characteristics of arctic tundra fires. *Arctic*, *29*(4), pp. 213–222.

104. Chipman, M., Hudspith, V., Higuera, P., Duffy, P., Kelly, R., Oswald, W., et al., 2015. Spatiotemporal patterns of tundra fires: Late-Quaternary charcoal records from Alaska. *Biogeosciences*, *12*(13), pp. 4017–4027.

105. Hu, F.S., Higuera, P.E., Duffy, P., Chipman, M.L., Rocha, A.V., Young, A.M., et al., 2015. Arctic tundra fires: Natural variability and responses to climate change. *Frontiers in Ecology and the Environment*, *13*(7), pp. 369–377.

106. French, N.H., Jenkins, L.K., Loboda, T.V., Flannigan, M., Jandt, R., Bourgeau-Chavez, L.L., et al., 2015. Fire in arctic tundra of Alaska: Past fire activity, future fire potential, and significance for land management and ecology. *International Journal of Wildland Fire*, *24*(8), pp. 1045–1061.

107. Young, A.M., Higuera, P.E., Duffy, P.A. and Hu, F.S., 2017. Climatic thresholds shape northern high-latitude fire regimes and imply vulnerability to future climate change. *Ecography*, *40*(5), pp. 606–617.

108. Morrissey, C.A., Dods, P.L. and Elliott, J.E., 2008. Pesticide treatments affect mountain pine beetle abundance and woodpecker foraging behavior. *Ecological Applications*, *18*(1), pp. 172–184.

109. Lu, J. and Yuan, F., 2019. The effect of temperature and precipitation on the leaching of contaminants from Ballangen tailings deposit, Norway. *WIT Transactions on Ecology and the Environment*, *231*, pp. 75–89.

110. Zhu, L., Jiang, C., Panthi, S., Allard, S.M., Sapkota, A.R. and Sapkota, A., 2021. Impact of high precipitation and temperature events on the distribution of emerging contaminants in surface water in the Mid-Atlantic, United States. *Science of the Total Environment*, *755*, pp. 142552.

111. Larison, J.R., Likens, G.E., Fitzpatrick, J.W. and Crock, J.G., 2000. Cadmium toxicity among wildlife in the Colorado Rocky Mountains. *Nature*, *406*(6792), pp. 181–183.

112. Gaston, A.J., Woo, K. and Hipfner, J.M., 2003. Trends in forage fish populations in northern Hudson Bay since 1981, as determined from the diet of nestling thick-billed murres Uria lomvia. *Arctic*, *56*(3), pp. 227–233.

113. Yurkowski, D.J., Hussey, N.E., Ferguson, S.H. and Fisk, A.T., 2018. A temporal shift in trophic diversity among a predator assemblage in a warming Arctic. *Royal Society Open Science*, *5*(10), p. 12.

114. Øverjordet, I.B., Gabrielsen, G.W., Berg, T., Ruus, A., Evenset, A., Borgå, K., et al., 2015. Effect of diet, location and sampling year on bioaccumulation of mercury, selenium and cadmium in pelagic feeding seabirds in Svalbard. *Chemosphere*, *122*, pp. 14–22.

115. Pedro, S., Boba, C., Dietz, R., Sonne, C., Rosing-Asvid, A., Hansen, M., et al., 2017. Blubber-depth distribution and bioaccumulation of PCBs and organochlorine pesticides in Arctic-invading killer whales. *Science of the Total Environment*, *601*, pp. 237–246.

116. Andersen, M.S., Fuglei, E., König, M., Lipasti, I., Pedersen, ÅØ, Polder, A., et al., 2015. Levels and temporal trends of persistent organic pollutants (POPs) in arctic foxes (*Vulpes lagopus*) from Svalbard in relation to dietary habits and food availability. *Science of the Total Environment*, *511*, pp. 112–122.

117. Colborn, T., Vom Saal, F.S. and Soto, A.M., 1993. Developmental effects of endocrine-disrupting chemicals in wildlife and humans. *Environmental health perspectives*, *101*(5), pp. 378–384.

118. Edwards, M. and Richardson, A.J., 2004. Impact of climate change on marine pelagic phenology and trophic mismatch. *Nature*, *430*(7002), pp. 881–884.

119. Harvell, C.D., Mitchell, C.E., Ward, J.R., Altizer, S., Dobson, A.P., Ostfeld, R.S., et al., 2002. Climate warming and disease risks for terrestrial and marine biota. *Science, 296*(5576), pp. 2158–2162.

120. Ucan-Marin, F., Arukwe, A., Mortensen, A.S., Gabrielsen, G.W. and Letcher, R.J., 2010. Recombinant albumin and transthyretin transport proteins from two gull species and human: Chlorinated and brominated contaminant binding and thyroid hormones. *Environmental Science & Technology, 44*(1), pp. 497–504.

121. Jenssen, B.M., 2006. Endocrine-disrupting chemicals and climate change: A worst-case combination for arctic marine mammals and seabirds? *Environmental Health Perspectives, 114*(Suppl 1), pp. 76–80.

122. Hooper, M.J., Ankley, G.T., Cristol, D.A., Maryoung, L.A., Noyes, P.D. and Pinkerton, K.E., 2013. Interactions between chemical and climate stressors: A role for mechanistic toxicology in assessing climate change risks. *Environmental Toxicology and Chemistry, 32*(1), pp. 32–48.

123. Noyes, P.D., McElwee, M.K., Miller, H.D., Clark, B.W., Van Tiem, L.A., Walcott, K.C., et al., 2009. The toxicology of climate change: environmental contaminants in a warming world. *Environment International, 35*(6), pp. 971–986.

124. Melnes, M., Gabrielsen, G.W., Herzke, D., Sagerup, K. and Jenssen, B.M., 2017. Dissimilar effects of organohalogenated compounds on thyroid hormones in glaucous gulls. *Environmental Research, 158*, pp. 350–357.

125. Ruuskanen, S., Hsu, B-Y and Nord, A., 2021. Endocrinology of thermoregulation in birds in a changing climate. *Molecular and Cellular Endocrinology, 519*, pp. 111088.

126. Iwaniuk, A.N., Koperski, D.T., Cheng, K.M., Elliott, J.E., Smith, L.K., Wilson, L.K., et al., 2006. The effects of environmental exposure to DDT on the brain of a songbird: Changes in structures associated with mating and song. *Behavioural Brain Research, 173*(1), pp. 1–10.

127. Guigueno, M.F. and Fernie, K.J., 2017. Birds and flame retardants: A review of the toxic effects on birds of historical and novel flame retardants. *Environmental Research, 154*, pp. 398–424.

128. Jones, T. and Cresswell, W., 2010. The phenology mismatch hypothesis: Are declines of migrant birds linked to uneven global climate change? *Journal of Animal Ecology, 79*(1), pp. 98–108.

129. Eng, M.L., Stutchbury, B.J. and Morrissey, C.A., 2019. A neonicotinoid insecticide reduces fueling and delays migration in songbirds. *Science, 365*(6458), pp. 1177–1180.

130. Rohr, J.R., Dobson, A.P., Johnson, P.T., Kilpatrick, A.M., Paull, S.H., Raffel, T.R., et al., 2011. Frontiers in climate change–disease research. *Trends in Ecology & Evolution, 26*(6), pp. 270–277.

131. Franke, A., Lamarre, V. and Hedlin, E., 2016. Rapid nestling mortality in Arctic peregrine falcons due to the biting effects of black flies. *Arctic*, pp. 281–285.

132. Luster, M.I., Portier, C., Pait, D.G., Rosenthal, G.J., Germolec, D.R., Corsini, E., et al., 1993. Risk assessment in immunotoxicology: II. Relationships between immune and host resistance tests. *Toxicological Sciences, 21*(1), pp. 71–82.

133. Helle, E., Olsson, M. and Jensen, S., 1976. PCB levels correlated with pathological changes in seal uteri. *Ambio*, pp. 261–262.

134. Gilbertson, M., Kubiak, T., Ludwig, J. and Fox, G., 1991. Great lakes embryo mortality, edema, and deformities syndrome (GLEMEDS) in colonial fish-eating birds: Similarity to chick-edema disease. *Journal of Toxicology and Environmental Health, Part A Current Issues, 33*(4), pp. 455–520.

135. Martineau, D., De Guise, S., Fournier, M., Shugart, L., Girard, C., Lagace, A., et al., 1994. Pathology and toxicology of beluga whales from the St. Lawrence Estuary, Quebec, Canada. Past, present and future. *Science of the Total Environment*, *154*(2–3), pp. 201–215.

136. Alava, J.J., Cheung, W.W., Ross, P.S. and Sumaila, U.R., 2017. Climate change–contaminant interactions in marine food webs: Toward a conceptual framework. *Global Change Biology*, *23*(10), pp. 3984–4001.

137. Nogueira, L.M., Yabroff, K.R. and Bernstein, A., 2020. Climate change and cancer. *CA: A Cancer Journal for Clinicians*, *70*(4), pp. 239–244.

138. McAloose, D. and Newton, A.L., 2009. Wildlife cancer: A conservation perspective. *Nature Reviews Cancer*, *9*(7), pp. 517–526.

139. Pesavento, P.A., Agnew, D., Keel, M.K. and Woolard, K.D., 2018. Cancer in wildlife: Patterns of emergence. *Nature Reviews Cancer*, *18*(10), pp. 646–661.

140. Baines, C., Lerebours, A., Thomas, F., Fort, J., Kreitsberg, R., Gentes, S., et al., 2021. Linking pollution and cancer in aquatic environments: A review. *Environment International*, *149*, p. 106391.

141. Malins, D.C., McCain, B.B., Brown, D.W., Chan, S.L., Myers, M.S., Landahl, J.T., et al., 1984. Chemical pollutants in sediments and diseases of bottom-dwelling fish in Puget Sound, Washington. *Environmental Science & Technology*, *18*(9), pp. 705–713.

142. Chang, S., Zdanowicz, V. and Murchelano, R., 1998. Associations between liver lesions in winter flounder (*Pleuronectes americanus*) and sediment chemical contaminants from north-east United States estuaries. *ICES Journal of Marine Science*, *55*(5), pp. 954–969.

143. Martineau, D., Lemberger, K., Dallaire, A., Labelle, P., Lipscomb, T.P., Michel, P., et al., 2002. Cancer in wildlife, a case study: Beluga from the St. Lawrence estuary, Québec, Canada. *Environmental Health Perspectives*, *110*(3), pp. 285–292.

144. Dalcourt, M., Béland, P., Pelletier, E. and Vigneault, Y., 1992. Caractérisation des communautés benthiques et étude des contaminants dans des aires fréquentées par le béluga du Saint-Laurent: Ministère des Pêches et des Océans.

145. Mansilha, C., Carvalho, A., Guimarães, P. and Espinha Marques, J., 2014. Water quality concerns due to forest fires: Polycyclic aromatic hydrocarbons (PAH) contamination of groundwater from mountain areas. *Journal of Toxicology and Environmental Health, Part A*, *77*(14-16), pp. 806–815.

146. Frölicher, T.L., Fischer, E.M. and Gruber, N., 2018. Marine heatwaves under global warming. *Nature*, *560*(7718), pp. 360–364.

147. Woolway, R.I. and Maberly, S.C., 2020. Climate velocity in inland standing waters. *Nature Climate Change*, *10*(12), pp. 1124–1129.

148. Leggat, W.P., Camp, E.F., Suggett, D.J., Heron, S.F., Fordyce, A.J., Gardner, S., et al., 2019. Rapid coral decay is associated with marine heatwave mortality events on reefs. *Current Biology*, *29*(16), pp. 2723–2730. e4.

149. Piatt, J.F., Sydeman, W.J. and Wiese, F., 2007. Introduction: A modern role for seabirds as indicators. *Marine Ecology Progress Series*, *352*, pp. 199–204.

150. Osborne, O.E., Hara, P.D., Whelan, S., Zandbergen, P., Hatch, S.A. and Elliott, K.H., 2020. Breeding seabirds increase foraging range in response to an extreme marine heatwave. *Marine Ecology Progress Series*, *646*, pp. 161–173.

151. Suryan, R.M., Arimitsu, M.L., Coletti, H.A., Hopcroft, R.R., Lindeberg, M.R., Barbeaux, S.J., et al., 2021. Ecosystem response persists after a prolonged marine heatwave. *Scientific Reports*, *11*(1), pp. 1–17.

152. Ho, J.C., Michalak, A.M. and Pahlevan, N., 2019. Widespread global increase in intense lake phytoplankton blooms since the 1980s. *Nature*, *574*(7780), pp. 667–670.

153. Wells, M.L., Karlson, B., Wulff, A., Kudela, R., Trick, C., Asnaghi, V., et al., 2020. Future HAB science: Directions and challenges in a changing climate. *Harmful Algae*, *91*, p. 101632.

154. Paul, S., Ghebreyesus, D. and Sharif, H.O., 2019. Brief communication: Analysis of the fatalities and socio-economic impacts caused by Hurricane Florence. *Geosciences*, *9*(2), p. 58.

155. Carmichael, W.W., 2001. Health effects of toxin-producing cyanobacteria: "The CyanoHABs". *Human and Ecological Risk Assessment: An International Journal*, *7*(5), pp. 1393–1407.

156. Breinlinger, S., Phillips, T.J., Haram, B.N., Mareš, J., Yerena, J.A.M., Hrouzek, P., et al., 2021. Hunting the eagle killer: A cyanobacterial neurotoxin causes vacuolar myelinopathy. *Science*, *371*(6536). p. eaax9050. https://doi.org/10.1126/science.aax9050

157. Hallegraeff, G.M., Anderson, D.M., Belin, C., Bottein, M-YD, Bresnan, E., Chinain, M., et al., 2021. Perceived global increase in algal blooms is attributable to intensified monitoring and emerging bloom impacts. *Communications Earth & Environment*, *2*(1), pp. 1–10.

158. Flewelling, L.J., Naar, J.P., Abbott, J.P., Baden, D.G., Barros, N.B., Bossart, G.D., et al., 2005. Red tides and marine mammal mortalities. *Nature*, *435*(7043), pp. 755–756.

159. Lefebvre, K.A., Robertson, A., Frame, E.R., Colegrove, K.M., Nance, S., Baugh, K.A., et al., 2010. Clinical signs and histopathology associated with domoic acid poisoning in northern fur seals (*Callorhinus ursinus*) and comparison of toxin detection methods. *Harmful Algae*, *9*(4), pp. 374–383.

160. Fire, S.E., Pruden, J., Couture, D., Wang, Z., Bottein, M-YD, Haynes, B.L., et al., 2012. Saxitoxin exposure in an endangered fish: Association of a shortnose sturgeon mortality event with a harmful algal bloom. *Marine Ecology Progress Series*, *460*, pp. 145–153.

161. Cooper, K.M., McMahon, C., Fairweather, I. and Elliott, C.T., 2015. Potential impacts of climate change on veterinary medicinal residues in livestock produce: An island of Ireland perspective. *Trends in Food Science & Technology*, *44*(1), pp. 21–35.

162. Tirado, M.C., Clarke, R., Jaykus, L., McQuatters-Gollop, A., Frank, J., 2010. Climate change and food safety: A review. *Food Research International*, *43*(7), pp. 1745–1765.

163. Lake, I.R., Foxall, C.D., Lovett, A.A., Fernandes, A., Dowding, A., White, S., et al., 2005. Effects of river flooding on PCDD/F and PCB levels in cows' milk, soil, and grass. *Environmental Science & Technology*, *39*(23), pp. 9033–9038.

9 Climate Change-related Hazards and Disasters: An Unrelenting Threat to Animal and Ecosystem Health

Christa A. Gallagher and Jimmy Tickel

CONTENTS

KEY LEARNING OBJECTIVES

- Recognize the inherent value of animals to society and the critical role animals play in the maintenance of the planet's healthy ecosystems.
- Understand that through the human-animal bond, individuals and communities are connected to, and responsible for, animal health, safety, and wellness.

DOI: 10.1201/9781003149774-9

- Realize that animals are as vulnerable to natural and anthropogenic disasters as humans and require human consideration and assistance to protect them in disasters.
- The One Health and "All of society" approaches are key to reciprocal planetary health and safety in disasters.

IMPLICATIONS FOR ACTION

- The impacts to animals in disasters should be considered and worked into each pre-and post-event phase of the disaster-management cycle.
- The One Health and related multidisciplinary and multisectoral systems approaches should be adopted and used for emergency and disaster management to provide symbiotic planning and response to humans, animals, and the environment.
- One Health professionals should collaboratively leverage and expand upon existing emergency-management system structure to prepare for and respond to animals in disasters.
- A highly organized incident-management system should be used for optimal efficiency and effectiveness of emergency response in disasters.

INTRODUCTION

Weather-related hazards and disasters have long plagued societies and ecosystems, but their escalation over past decades is becoming more evident and concerning to the scientific community and general public. Based on the Centre for Research on the Epidemiology of Disaster's EM-DAT records (1), the World Meteorological Organization (WMO) recently reported that the number of weather-, climate- and water-related disasters has increased by a factor of five from 1970 to 2019. These natural hazards accounted for 50% of all disasters (including directly associated technological hazards), 45% of all reported human deaths (91% of which occurred in developing countries), and 74% of all reported economic losses. This toll translated into 2.06 million human deaths and US$ 3.6 trillion in economic losses (2). Over these last 50 years, extreme events such as heat waves, heavy precipitation, droughts, and tropical cyclones have occurred in every region of the world and have necessitated societal development of decision-making, policy, and capacity to mitigate these events. What's more, anthropogenic drivers such as population growth, land-use change, urbanization, industrialization, and social inequity are interacting and amplifying in dynamic and unanticipated ways with devastating consequences to living species and environments. The Intergovernmental Panel on Climate Change's 2021 report unequivocally discloses that human activities have warmed the Earth's land, oceans, and atmosphere and contributed to rapid and widespread changes to the biosphere (3). Collectively, these disasters have significant adverse social, economic, and ecological impacts to humans and animals and the environment we share, directly and indirectly affecting species and ecosystem health and survival in the short and long term. The lamentable projection shared among many experts is that climate change will continue to spawn more frequent and higher intensity weather, climate, and water extremes in vast parts of the world, which may be unprecedented.

PROFILING HAZARDS AND DISASTERS

Natural and anthropogenic hazards are precursors that (may) morph into disasters if their impacts exceed certain thresholds and cause devastating consequences (see Box 9.1 from (4)). Hazards become disasters when profound and often crippling effects affect local populations that lack the capacity and/or the capability to effectively deal with them. Disasters are largely characterized and defined through the lens of ensuing human impact, most often depicted as loss of life and/or economic damage. The Centre for Research on the Epidemiology of Disasters (CRED) largely defines disasters by their humanitarian statistics, and events are so-named when at least one of the following criteria is met: 10 or more people are reported killed; 100 or more people are reported affected; a state of emergency is declared; or there is a call for international assistance (5). It is noteworthy that animals are not specifically mentioned in the definition of a disaster or included within its statistical criteria. This fact is especially concerning since animal health both depends on and directly affects human health.

BOX 9.1   KEY DEFINITIONS

Hazard: A process, phenomenon, or human activity that may cause loss of life, injury, or other health impacts, property damage, social and economic disruption, or environmental degradation.

Disaster: A serious disruption of the functioning of a community or a society at any scale due to hazardous events interacting with conditions of exposure, vulnerability, and capacity, leading to one or more of the following: human, material, economic, and environmental losses and impacts … The effect may test or exceed the capacity of a community or society to cope using its own resources.

Natural hazards, which make up the majority of disasters, can be geophysical, meteorological, hydrological, climatological (like hurricanes and earthquakes), extraterrestrial, or biological phenomena (like infectious disease) (6). Over the last five decades, 61% of disasters have been associated with floods and cyclones (44% riverine and general floods and 17% tropical cyclones) (2). Anthropogenic hazards (including human activities such as technological or industrial incidents) can lead to what is commonly termed "man-made disasters," which include examples such as chemical spills, explosions, environmental degradation, and nuclear accidents (6). These events can be unintentional and represent unfortunate accidents, or they can be the result of intentional human behavior demonstrated by calculated and willful acts of terrorism. Importantly, climate change can influence any of these types of hazards to become substantial disaster events. Also, significant anthropogenic links have been made to many types of natural hazards, including heat waves, extreme rainfall, and sea-level events, while less is scientifically known about the human signal on drought events or small-scale weather events such as thunderstorms and tornadoes (2). The WMO 2021 report found 60 disasters in the technological category (including aviation or boat-transportation accidents), where natural hazards were listed as a primary or main contributor (2). Additionally, the United Nations and other global experts report that climate change acts as a threat multiplier to pressurize social conflicts and exacerbate fragile societal and governmental structures, especially as related to food and water insecurity (7–9). Both examples highlight the inextricable link and even existent cycling between societal activity and climate change.

A disaster's onset is variable and can be slow and insidious, like a drought or sea-level rise, or a sudden event, like an earthquake or volcanic eruption. Disasters can be classified according to their frequency. Disasters can be frequent and expected to occur at certain intervals, like cyclones; can be infrequent, such as the 100-year storm predictions; or the frequency may be unknown, such as for wildfires or incursions of infectious disease in people and animals. The frequency of these variables is dependent upon an area's geographic location, existing climate and climate change influences, and human and animal interaction with the hazards. Over the last 50 years, floods were the most common of the weather-, climate-, and water-related disaster types recorded.

Lastly, and most notably, disasters can be described by their impacts. They can be typed as small-scale and large-scale disasters; the former indicating an event that impacts a local community(s) and requires assistance beyond the impacted community, whereas the latter happens on a grander scope and calls for national or international support (10). The post-disaster impacts on communities, animals, and the environment occur on a continuum from the immediate, short-term to longer-lasting or delayed effects, and they vary widely in their severity. In general, immediate effects could include illness and death in people and animals resulting from physical and emotional trauma, and sheer physical destruction of infrastructure and natural habitats (11,12). Longer-lasting and more indirect effects include, but are not limited to: continued or delayed onset illness and death from exposure to hazardous materials, dense living conditions, and medical and public health breakdowns, population displacement, contaminated ecosystems, destroyed or disrupted industries and economies, and food insecurity (13,14). More on the specific impacts of animals from disasters will follow later in the chapter.

CONSIDERING ANIMALS IN DISASTERS

Humans and animals are deeply and intimately connected. Animals are woven into the fabric of every society, as has been the case throughout history. The simple truth is that humans need animals to enhance their daily lives, and we rely upon them for survival. Likewise, many animals depend on us in the same manner or are at least directly influenced by our behaviors and activities. If we take the time to consider the many ways that animals are a part of contemporary human existence, we reach the following conclusion: there is very little separation of species. Though we can frame humankind's need to keep animals healthy and safe in many ways, we offer four substantial reasons why we need to protect animals from injury, illness, and death in times of disasters.

1. **Animals are a valued resource:** Animal-sourced proteins provide much-needed nutrition globally. Economic and social status gains are made every day through the sale and trade of animals and animal products. In many parts of the world, horses, llamas, elephants, and camels are a major form of transportation. Societies clothe themselves with animal fiber, use animal manure as a renewable energy source, and burn dung for heating and cooking. Scientifically, animals are a critical resource for translational studies that function to maintain health in humans and other animal species. Non-human primates were especially pivotal animal models in the development of the SARS-CoV-2 treatments and vaccines (15). Specifically related to disasters, dogs and horses are irreplaceable in search-and-rescue activities, serving to locate lost, injured, or trapped individuals, because they can be even more effective than modern technology.
2. **Humans are emotionally bonded to animals:** The human-animal bond has existed for over 10,000 years, and many experts believe, through advancements in societal attitudes, this bond has only strengthened over time. In many homes around the world, pets are considered part of the family. Through their immeasurable comfort and companionship, animals provide support to human emotional and mental health and even contribute to improvements in physical health status and social functioning (16). The recent COVID-19 pandemic has highlighted the significance of the human-animal connection as animals helped people to survive the disaster (17). Moreover, animals are tied to some human activities like hunting, fishing, and animal-related tourism.

We are connected to the natural world through observations and interactions with wildlife and exotic (zoo) species. Globally, animals have long been connected to societies through culture and religion. The human-animal bond is directly relative to disasters because people will fail to evacuate without their animals (18) and attempt re-entry into dangerous disaster zones to care for or save their pets or livestock (19).

3. **Animal health is essential for ecosystem health and survival:** An ecosystem is a complex community of living organisms (including humans and animals) interacting with their physical landscape and weather. Inherent to (healthy) ecosystems is interaction and interdependence among and between all members to maintain structure and function over time in the face of stressors and change. Animals are crucial components to ecosystems, and without them, ecosystems would cease to provide the services that support life to humans and other living species. Powerful agricultural examples are livestock breeds that provision humans with food and fiber, and domestic honeybees that pollinate the world's fruit and vegetable crops. Conversely, animals acquiring diseases in disasters have potential for transmitting disease directly and indirectly to humans (zoonotic transmission). Animals both represent, as well as support, essential biodiversity without which planetary life would quickly collapse and cease to exist.

4. **Humans have a responsibility to provide for the care and welfare of animals:** Animals are just as vulnerable to the effects of disasters as humans, and perhaps more so since many species depend on individuals and communities to care for them. Companion animals, livestock, and captive animal species (housed within zoological and laboratory facilities) are all reliant on human care for their health and well-being, and even wildlife species at times may require or receive provisioning or medical aid from humans. For many experts involved in the realm of animal disaster, the position of helping animals in dire emergency situations is a fundamental ethical and moral societal responsibility; without human intercession, animals will suffer. It is simply the "right thing to do" considering the immense value of animals and represents a positive means to bring planetary stewardship into practice and sustainability. The USA's National Animal Rescue and Sheltering Coalition (NARSC) has identified five primary support functions for the health, safety, and wellness of animals in disasters: (1) assessment; (2) evacuation and transportation; (3) search and rescue; (4) sheltering, and (5) veterinary care (20). Contributions made to these five functions will subsequently lessen the impact of disastrous event on humans. For example, maintaining health in food animal species will help to ensure communal food safety and security, and rescuing and securing displaced animals may lessen harmful animal interactions like animal bites or vehicular accidents.

EXAMINING THE ANIMAL IMPACTS

Animals can suffer from the primary impact directly caused by a disaster and become displaced, injured, sick, or killed. Unfortunately, the adverse effects do not stop there; they may progress to a delayed order of effects to include secondary and tertiary effects. Secondary effects occur because of the primary effect, and tertiary effects are the long-term or permanent effects set off as a result of the primary event. Using an animal example of deer confronting a tropical hurricane, the primary effects could include impact injury to the deer from high winds and drowning from flooding. Secondary effects could include the environmental contamination of drinking water with associated water-borne disease, leading to further illness and death. Habitat damage and biodiversity loss of deer and other reliant species would represent tertiary ecosystem effects. So the effects to animals and their habitats in disasters can be variable in time and space as effects are amplified and compounded within relevant ecosystems.

Since there are many different types of animals and respective societal purposes, there are many diverse ways disasters can impact animal species. It is therefore necessary to include many diverse animal professionals in the planning and response for animals in disasters. In general, impacts to animals from weather-related disasters include injury/loss of life and suffering through separation or lack of access to needed resources that humans and animals use to conduct their daily lives (examples include loss of access to shelter, food, water, services, mobility, power, and social interaction). Companion animals can be hurt or killed if their caretakers fail to evacuate them to safety. This was a poignant case in Hurricane Katrina in 2005 in the United States when an estimated tens of thousands of pets died. Livestock, wildlife, and exotics (in zoos or sanctuaries) are particularly vulnerable to physical harm because they are exposed to the elements and experience health and life-threatening changes during meteo-/geological disasters. They cannot typically be evacuated due to overall feasibility related to sheer numbers of animals, resources, and capability. Some of these populations of livestock, exotic, and laboratory species can be secured or evacuated pre-event; these should remain options whenever possible. Wildlife is often severely impacted by disasters because their health and well-being are intimately connected to the environment. In extreme circumstances for vulnerable populations, disaster harms to wildlife species could actually threaten populations to the point of extinction or extirpation. Longer-term effects may be alterations to migration, shifts in species distributions, changes in phenology (timing of biological life cycles), and infectious-disease emergence (21). A powerful reminder to loss of wildlife biodiversity occurred in the Australian mega-fires of 2019–2020 when an estimated 3 billion wild animals, including amphibians, reptiles, birds, and terrestrial mammals, were displaced or killed by over 15,000 bushfires that burned an approximate 97,000 km^2 of area (21–23).

APPROACH TO WEATHER AND CLIMATE-RELATED DISASTERS

Human health and safety is the first priority of disaster management; however, the human impact cannot be the only reason to take action or human risk the only risk to consider. Currently, there are increasing efforts to implement more robust animal health and safety planning and response in disasters. Following are key ways experts and communities can be better positioned to protect animal health from the increasing number of extreme events posed by climate change.

USE OF THE ONE HEALTH AND RELATED SYSTEMS APPROACHES

All disasters involve humans, animals, and the environment, which necessitates the call for a One Health approach with every disaster. The World Health Organization (WHO), the World Organization for Animal Health (WOAH, founded as OIE), the Food and Agriculture Organization of the United Nations (FAO), and the United Nations Environment Programme (UNEP) recently defined One Health as:

> "An integrated, unifying approach that aims to sustainably balance and optimize the health of people, animals and ecosystems. It recognizes the health of humans, domestic and wild animals, plants, and the wider environment (including ecosystems) are closely linked and inter-dependent. The approach mobilizes multiple sectors, disciplines and communities at varying levels of society to work together to foster well-being and tackle threats to health and ecosystems, while addressing the collective need for clean water, energy and air, safe and nutritious food, taking action on climate change, and contributing to sustainable development" (24) (see Figure 9.1).

Preparing for and responding to disasters requires an inclusive and collaborative effort across public and private sectors. These include government, first responders (police and fire), public health, medical and veterinary health and other animal-related professionals, academia, environmental experts, agri/aquaculture, the military, industry, nongovernmental organizations, and the communities themselves. A One Health approach to disasters ensures the consideration and response to immediate and delayed comprehensive impacts of a disaster (human, animal, and environmental) from unique but interconnected perspectives. Experts agree that efforts to assist and protect animals in disasters yield rewards in health and wellness for both animals and humans and prevent secondary environmental effects (25).

The contemporary approach to disasters adopts the concept of disaster risk reduction (DRR). According to the UN Office for Disaster Risk Reduction, DRR aims at "preventing the creation of disaster risk, the reduction of existing risk and the strengthening of economic, social, health and environmental resilience" (4). The Sendai Framework for DRR 2015–2030, supported by the UNDRR, aims to reduce existing risk and strengthen resilience to disasters through an "all-of-society" and "all-of-state institutions" engagement and participation (26). While it is not called a One Health approach, it encompasses

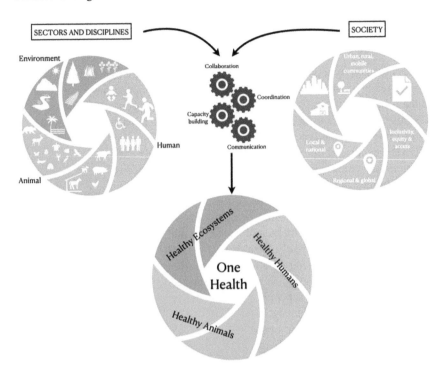

FIGURE 9.1 Representation of One Health. Adapted from: https://www.oie.int/en/tripartite-and-unep-support-ohhleps-definition-of-one-health/.

the multidisciplinary, multi-sectoral inclusive approach to disasters and encourages governments to involve all stakeholders – especially those community members who are embedded and directly affected. A shift from agency to a One Health community-centered focus capable of identifying existing hazards and specific societal vulnerabilities is crucial to improve mitigation and response to disasters.

Harm reduction is another related public health approach that is being applied to, and gaining momentum in helping, One Health problems. Harm reduction is a pragmatic approach to population problems that focuses on reducing the effects of a persistent harm(s) without necessarily eliminating the offending harm. It uses an upstream socio-ecological determinant of health approach to reach beyond the health sector in all health efforts, and it encourages active community participation and empowerment to attain and sustain human and ecological health. Harm reduction supports grassroots efforts to engage with public and private experts and authorities. It involves all individuals and sectors engaged with a problem to participate in working toward a solution, so it is multilevel, multidisciplinary, and empowers people to take control of their health and lives (27). All of the above systemic approaches will serve important purposes toward helping animals in disasters.

Integrating Animals into the Emergency Management System Foundation

While the need to better integrate animals into emergency management is clear, there already exists a set of strong foundational principals that have benefited human health and safety. Emergency management generally uses an "all hazards-" approach to preparedness where potential impacts of events are the focus rather than the type of event. This approach, supported by the WHO and WOAH, recognizes that many of the impacts and their corresponding response solutions are common to diverse types of weather-related events. For example, loss of homes and habitats is common to tornados, flooding, hurricanes, tsunamis, earthquakes, winter storms, and wildfire. Mitigation for each of these is evacuation when possible, while preparedness and response focuses on search and rescue followed by emergency sheltering and mass care. It also allows for addressing specific and likely hazards that exist in certain locations and posing higher risk. Thus, by using an all-hazards approach, emergency management is afforded the opportunity to plan across the spectrum of diverse types of weather-related disaster events for people and animals.

The disaster management cycle is used in most disaster-response situations with the "all hazards" approach to organize, mobilize, prioritize, and manage resources and responsibilities in a crisis (see Figure 9.2). It is typically depicted as four continuous phases: mitigation, preparedness, response, and recovery, which collectively aim for the prevention or reduction in injury, losses of life, and property.

Unique actions occur at each phase; however, they are not always distinct with some overlap of phases. For optimal results in community-health resilience

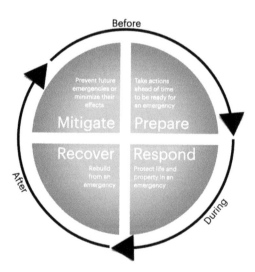

FIGURE 9.2 Disaster Management Cycle. Adapted from: https://www.fairfaxcounty.gov/emergencymanagement/cerg.

for all, One Health partners should work animal-related considerations into each stage of the disaster cycle. A brief explanation of the four phases with animal examples from the U.S. Federal Emergency Management Agency (FEMA) is as follows (28):

- **Mitigation:** Activities that are meant to prevent or reduce (mitigate) hazards and their primary impacts (loss of life and property/habitat damage or loss), while also attempting to reduce delayed effects and other unintended consequences. For animals, this phase may include: securing animal housing to withstand weather damage; reinforcing fencing to prevent animal escapes; farmers and animal caretakers buying insurance for animals in their care.
- **Preparedness:** Activities used to address hazards that cannot be mitigated and includes planning, training, and educating. Using an animal example, this could include: developing home and veterinary preparedness plans (including evacuation); exercising tabletop or full-scale exercises; procuring items to have for animal disaster use.

In these two pre-event phases, it is critical for emergency management to build in robust early-warning messaging for the protection of human and animal health and safety.

- **Response:** Perhaps the most recognized activity, response actions happen immediately following a disaster and encompass activities to save lives and reduce adverse effects from the disaster. For animals, this would include: search and capture; temporary sheltering and emergency veterinary treatment; environmental (habitat) interventions.
- **Recovery:** Activities to restore and/or improve the post-disaster conditions with sustainability and building back a better future as goals. Animal recovery includes: continued veterinary care; rebuilding damaged animal housing; actions to reduce vulnerability to future disasters. An important message is that successful recovery is the hallmark of an adequate preparedness and response system and will result in fewer surrendered/abandoned animals and the decreased need for ongoing housing and medical care.

Incident Command System

To make certain that animal health needs are met in disasters, all managing agencies should consider the use of an established organizational system. One prominent example is the USA's National Incident Management System with its Incident Command System (ICS). Used and recognized internationally, ICS is a "standardized approach to the command, control, and coordination of on-scene incident management that provides a common hierarchy within which personnel from multiple organizations can be effective" (29) (Figure 9.3).

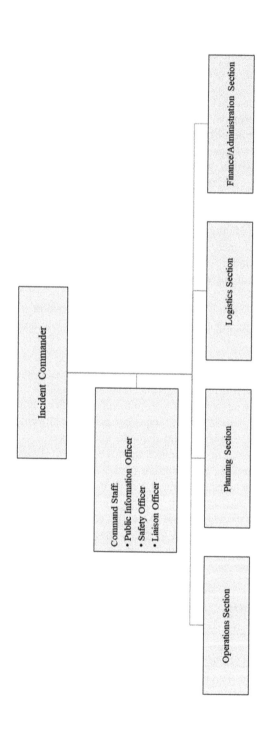

FIGURE 9.3 Standard Incident Command System (ICS) Organizational Structure. Adapted from: https://training.fema.gov/emiweb/is/icsresource/assets/fics%20organizational%20structure%20and%20elements.pdf.

ICS provides an authoritative structure for integrated tactical response that directs information and communication flow, analysis, and implementation of response measures across a wide variety of relevant agencies and partners. This structure addresses the complexity across a spectrum of emergency events that vary in size and intensity to reach common desired objectives. Using this system, different civil jurisdictions and levels of government, nongovernment organizations (NGO) and the public can, through pre-event planning, be safely incorporated into the response to an event and used to the maximum extent possible. ICS has been successfully used in all types of disasters, including natural disasters, terrorist acts, and oil and chemical spills. FEMA has recommended and supported the use of ICS, and it has been used extensively in U.S. animal response during all types of emergencies and disasters (28). In facilitating care to animals using ICS, the care of people is enhanced. Furthermore, ICS can help minimize business and property losses for animal industry, like livestock producers, veterinary practices, humane animal shelters, and many others. In addition, it has been used during infectious disease and invasive species events (30) and repeatedly with oil spills. The Oiled Wildlife Care Network (31) uses ICS to rescue oiled wild animal species in California. International organizations such as Focus Wildlife (32) and International Fund for Animal Welfare (33) also promote and use the ICS framework for their work in global wildlife emergency response. Lastly, it is important to recognize the U.S. ICS is one model of a structural-response management system. Many countries have developed their own response framework that encompass very similar management features for task areas and corresponding roles and responsibilities in disasters. So while the terminology may differ somewhat, the overall functional management and span of control is consistent. For example, Australia operates an Incident Control (versus command) System under its Australasian Inter-Service Incident Management System, which provides seamless integration of activities and resources of multiple agencies to work to resolve any type of emergency (34). It was the system used for human and animal response in the 2019–2020 megafires in Australia.

CONCLUSION

A disaster related to either a weather, climate, or water hazard occurred every day on average over the past 50 years – killing 115 people and causing US$ 202 million in losses daily (2). Societies should certainly consider the number of animals that died in these events and the ripple effects that were caused. It is evident that disasters do not discriminate as we have witnessed them robbing countless species of life, destroying properties and habitats, and diminishing livelihoods and economies. Projections by experts are describing a future where disasters will be even more frequent and damaging. Tireless efforts from many civil, private, and community entities have advanced weather-related animal disaster from near non-existence to impressive capability and capacity. Advancements in animal-disaster response has benefitted from funding, partnership, and mentoring from well-established human-response systems, constant advocacy by NGOs, and dedication of animal professionals. As animal and human health are inextricably connected to each other and our environment, One Health professionals should continue to work collaboratively, with coordinated and clearly communicated approaches to prepare for, respond to, and recover from disasters. This must include protecting the safety, health, and welfare of the Earth's animals, as doing so will help to safeguard valuable planetary resources and build health resilience for all.

REFERENCES

1. Centre for Research on the Epidemiology of Disaster (CRED). Em-Dat | the International Disasters Database [Internet]. [cited 2021 Dec 29]. Available from: https://www.emdat.be/
2. World Meteorological Organization (WMO). WMO Atlas of Mortality and Economic Losses from Weather, Climate and Water Extremes (1970–2019) [Internet]. 2021. Available from: https://library.wmo.int/doc_num.php?explnum_id=10902
3. Intergovernmental Panel on Climate Change (IPCC). Climate Change 2021 the Physical Science Basis Working Group I Contribution to the Sixth Assessment Report of the Intergovernmental Panel on Climate Change [Internet]. 2021. Available from: https://www.ipcc.ch/report/ar6/wg1/downloads/report/IPCC_AR6_WGI_SPM_final.pdf
4. United Nations Office for Disaster Risk Reduction (UNDRR), 2020. Disaster risk reduction [Internet]. [cited 2022 Jan 3]. Available from: https://www.undrr.org/terminology/disaster-risk-reduction
5. Centre for Research on the Epidemiology of Disasters, 2020. Human Cost of Disasters: An overview of the last 20 years 2000-2019 [Internet]. Available from: file:///Users/christa/Downloads/Human%20Cost%20of%20Disasters%202000-2019%20Report%20-%20UN%20Office%20for%20Disaster%20Risk%20Reduction.pdf
6. International Federation of Red Cross and Red Crescent Societies (IFRC). What Is a Disaster? | IFRC [Internet]. [cited 2021 Dec 30]. Available from: https://www.ifrc.org/what-disaster
7. UN Secretary-General António Guterres, 2021. Secretary-General's remarks at Security Council debate on Security in the Context of Terrorism and Climate Change [bilingual, as delivered] [scroll down for all-English and all-French versions] | United Nations Secretary-General [Internet]. [cited 2021 Dec 31]. Available from: https://www.un.org/sg/en/node/261092
8. Asaka J.O., 2021. Climate change - terrorism nexus? A preliminary review/analysis of the literature. *Perspectives on Terrorism*, 15(1), pp. 81–92.
9. Bowles D.C., Butler C.D. and Morisetti N., 2015. Climate change, conflict and health. *Journal of the Royal Society of Medicine*, 108(10), pp. 390–395.
10. United Nations Office for Disaster Risk Reduction (UNDRR). Disaster [Internet]. [cited 2021 Dec 30]. Available from: https://www.undrr.org/terminology/disaster
11. Noji E.K., 2000. The public health consequences of disasters. *Prehospital and Disaster Medicine*, 15(4), pp. 21–31.
12. Paterson D.L., Wright H. and Harris P.N.A., 2018. Health risks of flood disasters. *Clinical Infectious Diseases*, 67(9), pp. 1450–1454.
13. Young S., Balluz L. and Malilay J., 2004. Natural and technologic hazardous material releases during and after natural disasters: a review. *Science of The Total Environment*, 322(1), pp. 3–20.

14. Marshall B.K., Picou J.S. and Bevc C.A., 2005. Ecological disaster as contextual transformation: Environmental values in a renewable resource community. *Environment and Behavior*, *37*(5), pp. 706–728.

15. Albrecht L., Bishop E., Jay B., Lafoux B., Minoves M. and Passaes C., 2021. COVID-19 research: Lessons from non-human primate models. *Vaccines*, *9*(8), p. 886.

16. Gillum R.F. and Obisesan T.O., 2010. Living with companion animals, physical activity and mortality in a U.S. National Cohort. *International Journal of Environmental Research and Public Health*, *7*(6), pp. 2452–2459.

17. Ng Z., Griffin T.C. and Braun L., 2021. The New Status Quo: Enhancing access to human–animal interactions to alleviate social isolation; loneliness in the time of COVID-19. *Animals*, *11*(10), p. 2769.

18. Heath S.E., Kass P.H., Beck A.M. and Glickman L.T., 2001. Human and pet-related risk factors for household evacuation failure during a natural disaster. *American Journal of Epidemiology*, *153*(7), pp. 659–665.

19. Hall M.J., Ng A., Ursano R.J., Holloway H., Fullerton C. and Casper J., 2004. Psychological impact of the animal-human bond in disaster preparedness and response. *Journal of Psychiatric Practice®*, *10*(6), pp. 368–374.

20. Green D., 2019. *Animals in Disasters*. 1st edition. Oxford, United Kingdom; Cambridge, MA: Butterworth-Heinemann, p. 275.

21. Mawdsley J.R., O'malley R. and Ojima D.S., 2009. A review of climate-change adaptation strategies for wildlife management and biodiversity conservation. *Conservation Biology*, *23*(5), pp. 1080–1089.

22. Parrott M.L., Wicker L.V., Lamont A., Banks C., Lang M., Lynch M., et al., 2021. Emergency response to Australia's black summer 2019–2020: The role of a zoo-based conservation organisation in wildlife triage, rescue, and resilience for the future. *Animals*, *11*(6), p. 1515.

23. Ward M., Tulloch A.I.T., Radford J.Q., Williams B.A., Reside A.E., Macdonald S.L., et al., 2020. Impact of 2019-2020 mega-fires on Australian fauna habitat. Nature *Ecology & Evolution*, *4*(10), pp. 1321–1326.

24. Tripartite and UNEP support OHHLEP's definition of "One Health" [Internet], 2021. OIE - World Organisation for Animal Health. [cited 2022 Jan 3]. Available from: https://www.oie.int/en/tripartite-and-unep-support-ohhleps-definition-of-one-health/

25. Gallagher C., Jones B. and Tickel J., 2021. Towards Resilience: The One Health approach in disasters. In: *One Health: The Theory and Practice of Integrated Health Approaches Ed 2*. pp. 310–326.

26. United Nations (UN), 2015. Sendai Framework for Disaster Risk Reduction 2015 - 2030 [Internet]. United Nations. Available from: https://www.preventionweb.net/files/43291_sendaiframeworkfordrren.pdf

27. Gallagher C.A., Keehner J.R., Hervé-Claude L.P. and Stephen C., 2021. Health promotion and harm reduction attributes in One Health literature: A scoping review. *One Health*, *13*, p. 100284.

28. Federal Emergency Management Agency (FEMA). Emergency management in the United States: Livestock in disasters [Internet]. FEMA; Available from: https://training.fema.gov/emiweb/downloads/is111_unit%204.pdf

29. Federal Emergency Management Agency, 2018. ICS Review Document. p. 34.

30. Burgiel S.W., 2020. The incident command system: A framework for rapid response to biological invasion. *Biology Invasions*, *22*(1), pp. 155–165.

31. Cho E., 2018. Response [Internet]. Oiled Wildlife Care Network/School of Veterinary Medicine. [cited 2021 Jun 19]. Available from: https://owcn.vetmed.ucdavis.edu/response

32. Focus Wildlife International, Ltd, 2021. Focus Wildlife [Internet]. Focus Wildlife. [cited 2021 Jun 19]. Available from: https://www.focuswildlife.org

33. International Fund for Animal Welfare, 2021. International Fund for Animal Welfare [Internet]. IFAW. [cited 2021 Jun 19]. Available from: https://www.ifaw.org/eu

34. Australasian Fire and Emergency Service Authorities Council (AFAC), 2011. The Australasian Inter-service Incident Management System: A management system for any emergency [Internet]. Available from: https://training.fema.gov/hiedu/docs/cem/comparative%20em%20-%20session%2021%20-%20handout%2021-1%20aiims%20manual.pdf

10 An Introduction to the Economics of Climate Change and Animal Health

Gregory Graff, Benjamin Nordbrock, and Andrew Seidl

CONTENTS

DOI: 10.1201/9781003149774-10

KEY LEARNING OBJECTIVES

1. Describe what economics is and what range of animal health related issues economics can address.
2. Take an expansive view of the relationship between climate change and human economic activities, beyond just the immediate impacts of climate change.
3. Conceptually map how the economic issues of animal health differ across major categories of animals, including livestock, companion animals, and wildlife.
4. Describe how economic analysis addresses those things of substantial value to us – like ecosystem services generated by a wild animal species or the social and emotional bonds formed with a pet – but for which markets do not (or cannot) exist.

IMPLICATIONS FOR ACTION

1. Animal health professionals, programs, and policies have an enormous opportunity to expand beyond traditional client bases to serve the health of animal populations that may be particularly valuable in addressing climate change. It may require the innovation of nonmarket funding mechanisms to enter such nontraditional areas of practice.
2. Recognize that while adaptation to climate change is inevitable, mitigation is not. Initiative must be taken to craft economically sound incentives to serve animal health in ways that help to abate net greenhouse gas levels, such as by enabling sustainable intensification of livestock production or maintaining keystone animal species within retained forest ecosystems.
3. Economics should be used to provide key criteria in strategic priority setting by animal health research funding agencies, regulators, and companies. Resources and personnel are scarce and should be invested in those initiatives that will make the most valuable contributions in the face of climate change.

INTRODUCTION

The relationship between climate change and animal health has a fundamentally economic dimension, reflecting the extent to which humans value and depend upon animals. Since before the dawn of civilization, animals have been integral to the ecosystems within which humans have lived and upon which humans have depended. Animals have been hunted and harvested for subsistence. Animals have been domesticated and managed for production of food and materials, as well as for labor and transportation. And animals have become human companions in practical tasks, including hunting, herding, protection, and pest control, as well as socially and emotionally. To the extent that climate change affects the health and well-being of animals, it affects the well-being of humans that value and depend upon them. And, conversely, the management of animals and their health present valuable tools in adapting to and influencing the course of climate change.

This chapter seeks to provide a practical introduction to the economics of climate change and animal health. How can we account for the value of animal health amid a changing climate? What are the most important issues on which animal health professionals, business leaders, policy makers, and scholars can focus in response to climate change? This chapter does not assume the reader to be an expert in climate change, nor in economics. Instead, it introduces the animal health professional or policy maker to how economic considerations might inform the broader intervention discussion. A large and ever-growing literature is advancing our understanding of the economics of climate change (1–3). This chapter does not aim to review that literature but rather to draw upon selected sources that help to illustrate and explain key concepts.

We begin by introducing the framework of *animal health economics* and the principal tools of *economic welfare analysis* and *nonmarket valuation*. With these tools, we then explore the three main aspects of the relationship between climate change and human economic activities: climate change *impacts*, *adaptation* to climate change, and *mitigation* of climate change. We then consider how these three aspects of the relationship between climate change and the economics of animal health vary across three broadly defined animal populations: *livestock*, *companion animals*, and *wildlife*. The chapter concludes with advice to animal health professionals, scholars, and policy makers to inform animal health responses to climate change with well-founded economic principles.

HOW TO THINK LIKE AN ECONOMIST

ANIMAL HEALTH ECONOMICS

Economic analysis has, in fact, long informed animal health management decisions, especially when weighing the costs versus the benefits of interventions. The framework of "animal health economics" (AHE) was developed largely to assess the effect of animal health on the profitability of individual livestock operations or entire livestock industry sectors (4–6). While the AHE framework is well suited to questions posed by livestock producers or national animal health authorities, it was not necessarily designed to address the full scope of possible health-related issues among non-livestock animal populations – whether wild animal populations integral to natural ecosystems or companion animals within human households. It was also not designed to address the different types of value that animals can have for different segments of society. Nor was the framework of AHE intended to analyze the influence that animals have on the climate and the value of animal health interventions aimed at mitigating climate change.

However, the broader field of economics does provide both concepts and analytical tools to address the full range of questions forced upon us by climate change. Climate change's effects on animal health and animals' effects on climate change become economic questions when human well-being is ultimately affected. Economics is the science of how humans make decisions under scarcity. Economic decisions are driven by what we value, and economics has developed a variety of methods to measure what we value, and by how much. When individuals are making collective decisions as they interact in markets, direct market-based economic valuation techniques can be used. When changes are not well-reflected in markets, indirect and nonmarket valuation techniques are still available to quantify economic effects.

DIRECT ECONOMIC VALUATION TECHNIQUES: WELFARE ANALYSIS

Valuation is most straightforward in the context of markets. The term *market* is used to describe a complex social behavior that can be observed even in primitive human societies. Markets arise, as Adam Smith famously put it, from the human capacity to *"signify to another, 'this is mine, that is yours: I am willing to give this for that'"* (7). By their very nature, markets elicit value. Key measures can be used to quantify the impact of any sort of change on the welfare of those who choose to be involved as consumers (buyers) or producers (sellers) in a market (Figure 10.1). *Consumer surplus* is defined as the extent to which the gross value that consumers realize from a good or service that they buy in the market exceeds the price that they paid for it. Changes in consumer surplus are also known as *demand-side effects,* which generally result from changes in the population of consumers, consumers' overall income, and consumer tastes or preferences, as well as from changes in government policies or the rules governing the market. On the other hand, *producer surplus* is defined as the amount by which the price producers receive exceeds the costs incurred in providing goods and services to consumers, essentially profit. Changes in producer surplus are also known as *supply-side effects* and result from changes in the number of producers, costs of inputs, the cost of capital, and technology, as well as changes in government policies.

The collective economic *welfare effect* is the sum of changes in consumer and producer surplus and represents the net change in the well-being of all those involved in the market. Welfare effects can be realized today and for years into the future. *Discount rates* reflect the difference in value placed on welfare effects (i.e., net costs or benefits to consumers and producers) realized today versus welfare effects at felt some point in the future. Low discount rates indicate that, collectively, individuals engaged in the market generally place similar values on costs or benefits realized in the future versus costs or benefits realized today. High discount rates imply that individuals place less value on a cost or benefit occurring in the

Where value occurs	**Within markets**	**Outside of markets (market failures)**
	Consumer surplus (demand side effects)	Externalities: positive and negative
	Producer surplus (supply side effects)	Public goods: Ecosystem services and other public goods
		Other market failures

Valuation methods	**Direct valuation methods**	**Indirect or non-market valuation methods**	
	Partial equilibrium analysis of welfare effects within a single market	Revealed preference analyses (utilizing related markets or surrogate markets): Travel costs & hedonic prices	Stated preference analyses (using simulated market): Surveys of contingent valuation and choice experiments
	General equilibrium analysis of the sum of welfare effects within all relevant interconnected markets		

Total Economic Value (TEV)

Types of value	Use value				Non-use value		
	Direct use value		Indirect use value	Option value	Altruism value	Bequest value	Existence value
	Consumptive use value	Non-consumptive use value					

FIGURE 10.1 Depending upon whether economic value is reflected within markets or is outside the scope of markets, economic valuation methods can be used to quantify an entire range of different types of value.

future than if it occurred today. The sum of the stream of discounted welfare effects (benefits and costs) reaching into the future that result from a decision or change or intervention made today is called the *net present value*. As a rule, we choose the course of action with the greatest net present value.

Finally, consideration of the welfare effects within just a single market is called a *partial equilibrium analysis*. Consideration of the sum of welfare effects in all the interconnected markets that are affected by a given change is called a *general equilibrium analysis*. This is the theoretically correct measure of the economic impact of any change. However, due to the data needs and analytical complexity, most empirical studies adopt a partial equilibrium approach to welfare analysis.

There is a range of markets in which valuations are directly placed on animal health. Obviously, markets for animals themselves, such as livestock auctions, sell animals as the product, and within these markets, healthy animals are valued more than unhealthy animals. There are also markets for breeding stock or animal genetics, where the health attributes of an animal or its lineage figure directly into its breeding value and thus end up being reflected in market price. Value placed on animal health can also be observed in markets for veterinary services, pharmaceuticals, and devices, as well as animal feed, supplements, and other products and services that directly influence animal health. Additional secondary or linked markets can provide indications of the value that humans place on animal health, such as products produced from animals or services provided by animals.

INDIRECT ECONOMIC VALUATION TECHNIQUES: NONMARKET APPROACHES

Economics also provides ways to assess the value of things that, although important to humans, remain outside of the scope of our market interactions. Such situations are generally referred to as *market failures* (Figure 10.1), reflecting the conviction that well-functioning markets are the best way we have to account for value. Two important types of market failures are *externalities* and *public goods* – both of which are particularly common in the context of environment al and public health issues (8), including those involving wildlife populations or livestock.

An *externality* occurs when a market fails to account for the full impact, arising from actions and operations within that market, on third parties that are external to the market (8). Externalities (i.e., external impacts) can, in fact, be either positive or negative, but in either case, they effectively leave the market in question misaligned from the socially optimal combination of quantities and prices of goods sold, resulting either in too much of a bad thing or too little of a good thing. If, for example, people begin to "think globally, yet act locally" and consider the broader societal impacts of their energy consumption, perhaps by switching from fossil fuels to renewable energy, they would begin "internalizing the externality" that is currently generated by most energy produced and sold on markets. In markets characterized by externalities, incentives influencing consumers' and producers' decisions are out of alignment with society's overall best interests. They can be adjusted through public policies, thereby improving market outcomes. The extent to which such policy-based incentives change behavior can reveal how much value people place on the external impact. Greenhouse gas emissions, and thus climate change, represent a classic negative externality. Among the external impacts, of course, are those impacting animal health.

A *public good* is a good or service that consumers can access in the normal course of life without having to pay for it, thus undermining the incentive (or even failing to cover the cost) for a business to provide that good or service on the market. Because of this endemic *free rider* problem, a public good or service is also considered a situation of market failure, one in which a governmental, quasi-governmental, or nonprofit mechanism must be relied upon to provide the public good or service, paid for by taxes or sometimes by charging user fees (8). One very important subset of public goods includes a range of climate-related *ecosystem services*, in which natural systems work in the service of humans, and yet no payment is made for those services. The Millennium Ecosystem Assessment (9) described four broad categories of ecosystem services: *supporting, provisioning, regulating,* and *cultural* ecosystem services. Nature provides and climate influences all four types – for example, the *supporting* services of soil formation, nutrient cycling, and primary biological production in natural ecosystems; the *provisioning* of water resources form rainfall and snowpack or wood and fiber from forests; the *regulating* of temperatures and thus propensity for the spread of pathogens; or *cultural* services by maintaining seasonal traditions or recreational opportunities – none of which are paid for by those who benefit.

Given the nature of these market failures, economists have developed a range of indirect nonmarket valuation methods that use techniques such as analyses of indirectly associated market activities, direct surveys, or behavioral experiments to elicit how much value people place upon externalities or public goods. For example, externalities of livestock production include things like odor, flies, and slow-moving machinery, the value of the impacts of which can be assessed by conducting surveys of neighbors. Among public goods, the value of some provisioning ecosystem services can be indirectly measured in related markets, such as measuring the effects on property values (hedonic price effects) in the neighborhood. Since supporting and regulating ecosystem services frequently serve as inputs to production in markets, their values can be elicited based upon estimates of what producers in those markets would be willing to pay for such an input, if they had to pay for it. Some cultural ecosystem services can also be measured through related market approaches, such as calculating the travel costs borne to participate in those cultural activities.

DISTINGUISHING AMONG CLIMATE IMPACT, ADAPTATION, AND MITIGATION

Discussions of climate change generally distinguish among three different aspects of the relationship between climate change and human economic activities (see Figure 10.2): first, there is the *impact* of climate change on human economic

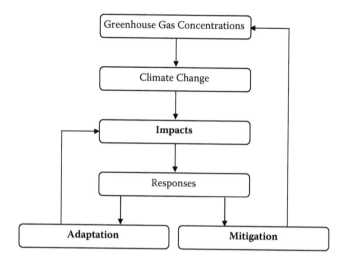

FIGURE 10.2 The interrelationships between climate change, its impacts on human economic activities, and the responses of human economic activities through adaptation and mitigation.

Source: Adapted from (13).

activities; second, in response to those impacts, there is *adaptation* of human economic activities to climate change; and, third, there can be a parallel response in human economic activities, taking measures to *mitigate* climate change, primarily through the abatement of greenhouse gas (GHG) levels in the atmosphere (10–12). These three are intertwined. Each overlaps with and influences the others in potentially complex ways, often with conflicting objectives and tradeoffs.

Throughout this book, mechanisms are detailed by which climate change is observed to impact animal health. Generally, these involve direct physical damages, such as changes in habitat and ecosystem balance, changes in disease prevalence, changes in available nutrition, or changes in reproduction. Such impacts on animal health change the value of animals to humans. In many cases, a loss in value is experienced, such as when an outbreak in avian influenza decimates commercial poultry barns. Yet, there can also be gains in value. For example, losses in livestock production in one country can drive up prices and benefit farmers with healthy livestock in other, unaffected countries. Or changes in climate can create more favorable conditions for some animals in some geographies. It should be cautioned that in even seemingly simple cases, the attribution of losses or gains in value due to changes in animal health resulting from climate change can be complex (14). First, multiple factors interact within the climatic, ecological/biological, and economic systems involved, and causality can be very difficult to identify. Second, all the factors being considered are, in a sense, natural survival challenges already faced by animals (changes in ecosystem, disease burdens, nutrition or reproduction challenges, etc.), and climate is driving shifts in the frequency, within a given location, of highly stochastic events. The Intergovernmental Panel on Climate Change (IPCC) deals with this complexity by describing "key risks" rather than definitive "impacts" (10). Yet, still, overarching trends can be hypothesized and tested.

As impacts are anticipated or begin to be felt, both animal and human systems respond by adapting, either in a fundamentally evolutionary sense through survival of the fittest, or in a more planned and managed sense. The IPCC defines adaptation as "the process of adjustment to actual or expected climate and its effects. In human systems, adaptation seeks to moderate or avoid harm or exploit beneficial opportunities." In the management of animal health, adaptive measures can include changes in location, adjustments in food/feed sources, breeding and the development of new genetics, construction of new infrastructure, and the innovation and adoption of new healthcare technologies and practices. While such adaptive measures can be costly, the economics of adaptation is more complex than simply tallying up the financial costs and attributing them as losses due to climate change. Many of these investments would likely have been undertaken anyway during business-as-usual changes and upgrades, with climate considerations simply affecting when and how such investments are made. Moreover, adaptive investments would naturally be expected to generate positive economic returns.

Finally, animals affect the climate. To the extent that humans place value on ways to mitigate climate change, another response to climate change will come in the form of investments or interventions in animal health as part of strategies for reducing emissions and even sequestering GHGs outright (see, for example, Chapters 3 and 16). For example, concerned pet owners may seek pet food brands and veterinary practices that reduce their GHG footprint (15,16), while larger livestock operations may respond to market-like incentive mechanisms to adjust animal health management in ways that reduce GHGs and thereby provide for economically valuable public goods (17).

ECONOMICS OF CLIMATE CHANGE AND ANIMAL HEALTH ACROSS THREE BROAD CATEGORIES: WILDLIFE, LIVESTOCK, AND COMPANION ANIMALS

The economics of climate change and animal health varies substantially across three broad categories of animals – livestock, companion animals, and wildlife. Animals in these respective categories play very different roles vis-à-vis human society and are thus valued very differently by humans. Concomitantly, the extent of healthcare intervention and management undertaken across these three categories are likely to vary significantly as each is impacted by climate change, adapts to climate change, and is involved in strategies to mitigate climate change. Table 10.1 illustrates this framework of considering the three different aspects of the relationship between climate change and human economic activities (impact, adaptation, mitigation) across these three categories of animal populations (livestock, companion animals, and wildlife) and summarizes some of the key economic issues that arise within each intersection.

Climate Change and Animal Health

TABLE 10.1
A Framework Summarizing the Economic Issues That Arise across the Complex Set of Relationships between Climate Change and Animal Health

	Livestock	Companion Animals	Wildlife
Climate change impacts (key risks)	Livestock productivity and losses mean lower revenues and higher costs to producers, and higher prices to consumers. Some gains to producers may occur in regions or sectors with improved conditions. Given livestock markets, healthcare interventions are calculated to optimize returns, balancing expected benefits against costs.	Health impacts on companion animals are similar to health impacts on humans, based on integration within human households. Healthcare interventions in veterinary practice shift to reflect greater incidence of climate related conditions and emergency responses to extreme events.	Reductions in fishing and hunting and shifts in recreation and tourism. Losses in nonmarket values of wild species and biodiversity, such as existence value of wildlife. Healthcare interventions largely require public support due to lack of private incentives (public goods market failures).
Climate change adaptations	Adaptation strategies include location, food/feed sources, breeding/genetics, infrastructure, technologies. Many adaptations are private decisions to protect against financial losses or to exploit opportunities. Other adaptations are largely public goods and thus require public support.	Companion animals benefit from adaptations made by human households. Adaptations specific to companion animals include preventive veterinary care or innovations in pet comfort accessories.	Wildlife management strategies proposed to assist wild species in adapting to climate, especially for economically or recreationally important species. Healthcare interventions largely require public support due to lack of private incentives (public goods market failures).
Mitigation of climate change	Sustainable livestock management can reduce net GHG emissions. Managing health of livestock in intensified production systems to retain ecosystems critical for GHG abatement. Developing alternative protein sources (plant-, insect-, cell-culture based) less intensive in GHG emissions.	Sustainable pet care management practices to reduce net GHG emissions, such as feed reformulation, and pet waste management.	Managing health of wildlife species may help maintain natural or managed ecosystems and thereby contribute to the abatement of GHGs. Healthcare interventions largely require public support due to lack of private incentives (public goods market failures).

LIVESTOCK

Humans depend upon livestock, as well as farmed aquatic species, for nutrition, including meat and animal products and other useful by-products and materials. These are consumptive uses, and their values can be quantified using welfare analysis in markets for live animals, meat, and other animal products, also considering, of course, the externalities of these markets. For example, in the United States, the retail equivalent value of beef produced in 2020 was about $123 billion (18). Second, people place value also on services provided by livestock species. Within traditional agrarian societies, livestock provide farm labor, transportation, and even a range of financial functions (such as savings, collateral, and insurance). In high-income countries today, livestock species are enjoyed in agritourism, sports, and even therapeutic programs. Companies also look to the farm or ranch for marketing symbols (e.g., Mustang, Ram, Red Bull, etc.). These are all non-consumptive use values of livestock. Third, agrobiodiversity of livestock has value for maintaining or introducing important genetic traits in breeding programs, such as traits for disease resistance, productivity, or product quality. Thus, there is an option value in the diversity of genetic stocks, for example, by preserving heritage breeds. Finally, humans place an existence or bequest value on livestock breeds and species, given their historical and cultural significance.

Impacts of Climate Change on Livestock

Direct health-related losses in the value of livestock that can be attributed to climate change include higher mortality, lower reproduction rates, slower growth, lower weights, and lower production of milk and eggs, caused by physical factors such as higher temperatures, reduced availability or quality of feed, grazing, or water sources, or greater disease pressures, as well as the frequency and severity of extreme events such as storms, floods, and wildfires (19–27). In any single market, livestock production losses can result in decreased market supply, higher prices and, as a result, economic losses shared by both livestock producers and consumers in the form of smaller producer surplus and consumer surplus. Increased variability in weather-related losses also increases financial risks, reducing production below optimal levels, and increasing the occurrence of bankruptcies among livestock producers.

Yet, changing weather patterns result in both economic damages and economic benefits (28). Conditions may improve in some locations currently unsuitable for livestock production, expanding economic opportunities in those areas. Or changes in climate may change relative prices across sectors of livestock production, shifting comparative advantage from one sector to another. For example, climate change may improve conditions for some types of freshwater aquaculture, providing alternatives to wild capture fisheries (26). Animal health interventions can be made in direct response to the impacts of climate change to reduce losses and speed recovery. In such cases, the tools of animal health economics can inform the balance of expected benefits against the costs of care and compliance with veterinary and public health regulatory requirements.

Adaptations by Livestock to Climate Change

As climate impacts occur, animal populations, livestock producers, markets, and consumers respond in ways that are inevitably adaptive (29). In fact, calculations of losses from climate impacts on business-as-usual or status quo scenarios are naïve in the sense that these systems are highly dynamic and constantly adapting to changes in a variety of ways, including the following (10):

- new enterprises or business models
- new capital investments (e.g., infrastructure, shelter)
- new technologies (e.g., animal genetics)
- new information systems (e.g., weather modelling and reporting)
- new financial institutions or instruments (e.g., insurance)
- new norms and regulations (e.g., building codes, trade agreements)
- new emergency management procedures.

In livestock health management, a wide range of potential adaptation strategies have been proposed and analyzed (12,17,23,24,30). Adaptation can include the innovation and adoption of new technologies and standards of practice for veterinary care, veterinary pharmaceuticals, animal vaccines, or feed supplements. Investments can be made in the selection of locally adapted breeds and in the development of genetically stress tolerant and disease resistant breeds. The geographic distribution of livestock and aquaculture can shift as animals are physically moved, as producers in some locations downsize or exit the industry, and as producers in other locations increase their operations or enter the industry (26,31). Other adaptations include investments in construction, ranging from simple shade or wind shelters to full containment in indoor climate-controlled facilities, reconfigurations of supply chains to minimize risk and increase resilience, adoption of new handling procedures and food-safety standards in light of increased pathogen loads, and development and utilization of financial risk management and insurance.

The economics of adaptation in livestock are driven largely by private decisions to protect against potential losses or to exploit potential opportunities. However, a range of adaptive measures requires collective action, particularly in the areas of research and development of new technologies, informational resources, food safety standards, and public health infectious disease management strategies.

Livestock and Mitigation of Climate Change

Livestock agriculture uses roughly 30% of global land area and accounts for an estimated 12% of human-related GHG emissions, meaning there is significant potential for livestock to contribute to reductions in emissions, mostly through improved productivity and associated land savings (32). A range of sustainable livestock-management practices are supported by policies such as the U.S. Department of Agriculture's (USDA's) Environmental Quality Incentives Program (EQIP). Increasingly, livestock producers are also finding such practices can generate carbon credits that can then be sold on voluntary private carbon markets.

Animal health management is integral to such changes, including changes in grazing practices, feed-ration compositions, feed supplements to increase weight gain and reduce time on feed, reductions of methane emissions from ruminant animals, manure management to limit methane and nitrate emissions, and genetic improvements or shifts in breed composition (17).

At a global scale, two trends are likely to be most impactful in terms of achieving emissions reductions from livestock, both of which will involve extensive animal health management. The first trend is the prevention of land-use change (including deforestation) that results from sustainable intensification of livestock production systems. Intensification is estimated to result in avoiding the conversion of over 150 million hectares and thereby saving over 700 million metric tons of CO_2 emissions per year between 2000 and 2030 (32). In fact, it is estimated that mitigation through preventing land-use changes are 5 to 10 times more efficient than direct emissions reductions from current livestock operations (32). Of course, intensification of production, whether from primarily grass-based to mixed-crop and livestock systems or into confined production systems, involve a range of animal health management, nutrition, and innovation challenges.

The second trend, which has seen surprising momentum in recent years, is the development of alternative protein sources that effectively substitute for animal products. These include, broadly, (a) plant-based substitutes for dairy, egg, and meat products, (b) fermentation-based substitutes of single molecules, including proteins, amino acids, vitamins, etc., and (c) cell-culture based substitutes for meat tissues. Proliferation of plant-based, fermented, or cultured substitutes for livestock products are expected to lead to reductions in demand for livestock products to the extent that these alternative protein production systems are less intensive in GHG emissions on a per-ton or per-calorie basis. They would result in a net reduction in GHG emissions by substituting for livestock production (33,34). Moreover, for these techniques to produce close facsimiles of animal-based products will require extensive knowledge of animal biology, genetics, and pathology.

COMPANION ANIMALS

There is an estimated global population of almost 400 million cats and almost 500 million dogs kept in human households, plus an additional 480 million free-roaming cats and 700 million to 1 billion free-roaming dogs (35). The value of companion animals is inherently difficult to quantify due to the personal utility that humans gain from the companionship. Companion animals that serve a specific purpose, such as shepherding, guarding, or hunting, could be valued in terms of the monetary value of the service that animal provides. The utility value of a cat that catches mice in the house could be calculated based on the costs of hiring a pest control service or the increase in the assessed value of the home due to being free of pests and pest damage. However, companion animals' predation and wildlife habitat intrusion may be considered as social cost rather than a social benefit when the species involved are not considered pests (e.g., rabbits, squirrels, songbirds).

Companion animals kept mainly for companionship are valued in the personal satisfaction and emotional utility they bring to their humans. That relationship is a nonmarket function, meaning that humans do not pay a market price for the daily companionship or emotional well-being. Still, a healthy pet likely brings greater emotional, social, and health benefits to their humans than a sick or ailing pet could, but quantifying the value of an animal's state of health is hindered by the highly subjective nature of the comparison: How much would a person pay to prevent the complete loss of a pet? How much would they pay to keep a pet healthy? These may seem to be impossible questions.

It is possible, however imperfectly, to impute the value of animal health from markets for veterinary care. For example, if a dog owner pays $1,000 for a nonemergency surgery that keeps the dog able to run and walk with them, then the value of that aspect of the dog's health is worth at least $1,000 to the human owner; otherwise, the owner would forego the surgery. Some critics become frustrated with economics, arguing that it is not right to put a price on the health of a beloved companion. Yet, the exercise is not intended to trivialize health or the importance of the human-animal bond; rather, it is to analyze the tradeoffs among diverse and sometimes competing human objectives against a common metric.

Impacts of Climate Change on Companion Animals

We expect impacts of climate change on the health of companion animals to be similar to impacts on health of humans, based upon their degree of integration into human households, especially in urban areas of high- and middle-income countries. Impacts on health of companion animals in rural areas and lower-income countries may be more similar to impacts on livestock. Impacts on health can be of two general types (36). The first are changes in frequency or severity of existing health problems influenced by weather or climate. The second are new health problems resulting from new factors introduced or exacerbated by climate. Such factors include heat, air quality, water quality, food safety, prevalence of pathogens, and physical damages, displacements, or migration due to extreme events. Additional concerns include changes in caretaking patterns by humans such as decreased exercise or increased abandonment (35).

It is not straightforward to quantify the economic value of impacts on companion animal health. A first approach would be to consider increased demand for, or expenditures on, veterinary care, pharmaceuticals, vaccines, pet food and supplements, and other related healthcare products. However, in such an analysis, accurate attribution is not feasible since it requires separating the baseline of what expenditures would have been in the absence of climate change from the actual expenditures observed in the market. Still, even greater nonmarket values are placed on animals that are lost or human-animal bonds that are weakened due to health problems caused by climate change.

Adaptations to Climate Change by Companion Animals

Similarly, we can expect adaptations for companion animals to result from measures taken by humans out of concern for their own health. Increases in air conditioning, air filtration, water filtration and purification, control of vectors of infectious diseases (e.g., mosquitos, ticks), and even relocation, in what is known as environmental migration (37) undertaken by humans for their own sakes, are likely to help their companion animals adapt as well. Other adaptive measures may be specific to animals, including preventive veterinary care or innovations in pet comfort accessories.

Companion Animals and Mitigation of Climate Change

Given the sizable populations of companion animals, their feed and care contribute a significant environmental "paw print" (15,38,39), with pet food production estimated to account for about 1–3% of the net carbon emissions of the entire agricultural sector (15). Control of population and selection of breeds could reduce the net GHG emissions, as could reformulation of pet foods to use ingredients with a lower GHG burden. Management of animal waste, similar to concerns in the livestock industry, would help reduce nitrate and methane emissions from leaching and decomposition (40). Sustainable veterinary practices are also a consideration in the mitigation of climate change from companion animals. Pet owners may be willing to pay more for a veterinarian that is knowledgeable and actively conscious of reducing net GHGs (16,41).

The value of mitigative measures in the highly diffused markets for companion animal health and wellness are less likely to be captured by those providing the mitigative products and services. As a result, there is likely less economic incentive to engage in such measures. Some larger pet food manufacturers may be better positioned to collaborate with their livestock, and crop ingredient suppliers to engage in sustainable production practices supported by government programs or generating carbon credits sold on voluntary carbon markets. Other opportunities may arise for innovative solutions that reward the actions of individual pet owners or veterinary practices.

WILDLIFE

Humans depend upon and benefit from wildlife in many ways. First, humans have long obtained important nutrition from wild terrestrial, fresh water, and marine animal species. Economically, this nutritional value is considered the *consumptive use* value of wildlife. Quantitative economic welfare estimates of climate impacts on consumptive use values can be made based upon markets for game, bush meat, or wild-caught fish, or indirectly based upon changes in travel cost expenditures or tourism spending incurred to participate in recreational hunting or fishing.

Second, people enjoy and therefore place a value on viewing or photographing wildlife. Companies look to nature for marketing symbols (e.g., Puma, Jaguar, Gecko) and for design inspiration, or biomimicry (again, Gecko). In the economics literature, these are just some examples among a wide range of *nonconsumptive direct use* and *indirect use* values that humans place upon wildlife that are not part of any market and are therefore evaluated using methods developed for assessing the value of public goods (see Figure 10.1).

Thirdly, biodiversity and wild animals are valued for their potential for enabling researchers to identify and synthesize new medicines and health treatments, a process known as bioprospecting. This use is described as an *option value* of biodiversity and wildlife in the economics literature. Value estimates of this sort are based on bioprospecting theories (42–45) and formalized in biodiversity access and benefit-sharing protocols (46).

Finally, people place value on protecting species, or returning endangered species to health, even if they never interact with or use those species themselves. Instead, they value the peace of mind that comes with knowing that the species is alive and well in the world and playing its role within natural ecosystems. This is called the *existence* value. Or, they may place a value on saving the species for other people of the current or future generations. These are called *altruism* or *bequest* values, respectively. Existence, altruism, or bequest value can be estimated using nonmarket valuation methods (Figure 10.1).

For example, a recently published average value per day of consumptive uses of wildlife include: fishing at $58.21 and hunting at $57.90 per day, for the fishers and hunters engaged in these activities (47). Nonconsumptive use values of wildlife include snorkeling at $37.41, scuba diving at $39.94, wildlife viewing at $52.28, and birdwatching at $36.54 (47). However, nonconsumptive use value can be multiplied many times for the similar experiences, whereas consumptive use is a one-off: the difference between the value of a fish in the sea versus a fish on the plate. Published existence values for terrestrial habitat are $36–136 per household per year, $12–69 per household year for coastal habitats, and $11–129 per household-yr for wetland habitats (48). Existence values for habitat are derived from their potential to provide a safe and productive location for flora and fauna to thrive for future generations of human benefit.

Impacts of Climate Change on Wildlife

Direct effects of climate change on the health of wildlife can impact each of the above types of value. For example, marine acidification, warming seas, and shifting currents that are brought on, in part, by climate change add to already existing stresses on global fish stocks and thus declines in valuable and important sources of nutrition, particularly for poor and marginalized people dependent upon substance fishing for their livelihoods (49). The result is a loss in consumptive use value. Climate change can be expected to affect the nature of terrestrial ecosystems, changing habitats and habitat corridors for terrestrial wildlife species. Changes in food availability, seasonal patterns of migration, breeding and survival of young, and disease prevalence can have direct and indirect effects on species health, population numbers, and even drive them to extinction (50). This can result is the loss of the nonconsumptive use value, as well as existence, altruism, and bequest values of the species.

To the extent that veterinary healthcare interventions improve the survival likelihood of wildlife species, the value saved can be approximated by the percentage improvement in survival likelihood multiplied by estimates of the value of the species in question. While such estimates are fraught with over-simplifications and assumptions, examples include Northern Spotted Owls at $116 per person year (p-yr), grizzly bears at $76/p-yr, whooping cranes at $58/p-yr, red-cockaded woodpeckers at $22/p-yr, bald eagles at $40/p-yr, bighorn sheep at $35/p-yr, illustrating the potential value of veterinary health interventions in the survival of such species (51). Then, multiplying by the millions of people valuing each of these individual species today and into the future, the results are in the billions of dollars. Since there is little or no profit incentive to motivate or support such veterinary healthcare interventions, their lack of provision is essentially a public goods market failure, and thus, if they are provided at all, it will be through public policies or philanthropic initiatives.

Adaptations by Wildlife to Climate Change

Ecosystems naturally adapt to change, and there is ample ecological evidence of individual species already adapting to climate change, such as changing morphological characteristics (52) or shifting habitat range (53). Other species, especially those that are highly specialized or have limited habitat ranges, may lack adaptive capability (54). A range of wildlife management strategies have been proposed to assist wild species in adapting to climate change, such as protected areas, invasive species management, and assisted migrations. Far less has been recommended in terms of direct animal health interventions, including survival strategies, reproduction assistance, or disease management (55). In general, strategies for maintaining ecological complexity and ecosystem resilience are central to current thinking about wildlife adaptation to climate change (56), rather than direct healthcare treatments. To the extent that direct interventions are warranted, they will most likely focus on adaptation of economically or recreationally important harvested wild species, such as wild salmon, deer, and elk, or wild species that pose a specific identifiable risk, such as a reservoir for pathogens that may readily spread to closely related livestock species. Even where there might be clear consensus about the efficacy and value of a specific animal health intervention to assist wildlife adaptation to climate change, the economics are much like those discussed above for animal health interventions to help wildlife recover from direct impacts: without a clear profit incentive, any such wildlife health intervention to assist adaptation is effectively a public good (or service) and thus dependent on public policy or philanthropy for its provision.

Wildlife and Mitigation of Climate Change

Natural ecosystems – including forests, grasslands, and marine ecosystems – are optimal for absorbing atmospheric GHGs relative to alternative types of land use. So much so that global accounting of GHG emissions typically divides emissions into two categories: (a) direct emissions and (b) emissions retained in the atmosphere due to lost natural sequestration capacity resulting from land use changes such as deforestation. Conversely, one of the most effective ways to reduce GHGs and mitigate climate change is to restore lands and marine habitats to their natural state. Given practical limitations on the feasibility of complete restoration, often a compromise outcome is a managed mixed-use ecosystem.

The health of wild animal populations may both help and be helped by habitat restoration efforts. Several reforestation and landscape restoration efforts, such as those supported by the Global Environmental Facility (GEF), the Global Climate Fund (GCF), and the United Nations Reducing Emissions from Deforestation and forest Degradation plus (REDD+), have dual goals of climate change mitigation and wildlife habitat improvement. Wildlife habitat improvement helps to assure wildlife health improvement. Direct wildlife health-management initiatives – including captive breeding, reintroductions, and health interventions – especially for apex predators or other key species, can be key components in restoring ecosystems' balance and increasing their capacities for carbon sequestration and storage. Examples include reducing overpopulation of endemic species or management of invasive species to restore ecosystem balance.

Management of wildlife health for purposes of climate mitigation largely lacks private profit incentives, and thus, faces the classic public goods dilemma for financial support. National protected areas are, in some countries, financially supported by fuel taxes paid by automobile drivers and air travelers (e.g., Costa Rica, Belize). Other protected areas are supported by supplying carbon credits to carbon markets and emissions offset programs created by governments to meet their commitments for emissions reductions. Additional "climate finance for ecosystem services" schemes can be envisioned to include support of wildlife health with an aim of restoring and maintaining wild and mixed-use ecosystems to draw down GHGs and mitigate climate change.

CONCLUSIONS

Climate change presents a broad set of challenges to the health of animals. There is great concern over potential for economic losses and disruptions among livestock, companion animals, and wildlife populations, which in turn affect the well-being of humans that depend upon and value them in ways summarized in Table 10.1. But the challenges of climate change also present new economic opportunities. Climate-related risks create increased demand for goods and services that help animals and humans adapt, as well as additional incentives for investments in research and development and infrastructure that will improve productivity, reduce negative externalities, and provide the public goods and services upon which healthy animal and human populations depend.

Looking to the future, an economic lens is critical for developing plans and making investments to protect animal health. Economic criteria are central to establishing priorities. What new animal health goods and services are likely to be demanded, and to what extent? Where should incentives be targeted, given that resources are limited? When does it make more sense to allocate resources to adapt to climate change or to mitigate climate change? When two opportunities are compared, economic analysis helps inform the choice between them, even if one is characterized as a commercial market value and the other is a nonmarket value. Economics provides a systematic approach to establishing a common metric of value across widely divergent contexts.

Still, we must recognize limitations that restrict our ability to use economics as a tool for climate change actions. Economic value is only one criterion competing in the arena of collective decision making. Social acceptability and political feasibility impose strong constraints as well. In the end, economics works together with science, law, and other influences, including political pressure groups and global events, to shape the actions that we take as a society, including our care for the health of animals that we value and depend upon.

REFERENCES

1. Raworth, K., 2017. *Doughnut Economics: Seven Ways to Think Like a 21st-century Economist.* London: Random House.
2. Stern, N., 2007. *The Economics of Climate Change: The Stern Review.* Cambridge, UK: Cambridge University Press.
3. Stern, N., 2008. The economics of climate change. *American Economic Review,* 98(2), pp. 1–37.
4. Dijkhuizen, A.A. and Morris, R.S., 1997. *Animal Health Economics. Principle and Applications.* Sidney: University of Sidney.
5. Otte, M. and Chilonda, P., 2000. Animal health economics: An introduction. *Animal Production and Healthy Division (AGA).* Rome, Italy: FAO, 12 p.
6. Rushton, J., 2017. Improving the use of economics in animal health–Challenges in research, policy and education. *Preventive Veterinary Medicine, 137,* pp. 130–139.
7. Smith, A., 1776. *An Inquiry into the Nature and Causes of the Wealth of Nations.* London: Methuen & Co., Ltd.
8. Cornes, R. and Sandler, T., 1996. *The Theory of Externalities, Public Goods, and Club Goods.* Cambridge, UK: Cambridge University Press.
9. Millennium Ecosystem Assessment, 2005. *Ecosystems and Human Well-being: Synthesis.* Washington, DC: World Resources Institute, Island Press.
10. IPCC, 2014. Summary for Policymakers. In: Climate Change 2014: Impacts, Adaptation, and Vulnerability. Part A: Globaland Sectoral Aspects. Contribution of Working Group II to the Fifth Assessment Report of the Intergovernmental Panel on Climate Change. [Field, C.B., V.R. Barros, D.J. Dokken, K.J. Mach, M.D. Mastrandrea, T.E. Bilir, M. Chatterjee, K.L. Ebi, Y.O. Estrada, R.C. Genova, B . Girma, E.S. Kissel, A.N. Levy, S. MacCracken, P.R. Mastrandrea, and L.L. White (eds.)]. Cambridge, United Kingdom and New York, NY, USA: Cambridge University Press. pp. 1–32. https://www.ipcc.ch/site/assets/uploads/2018/02/ ar5_wgII_spm_en.pdf
11. McCarl, B.A. and Schneider, U.A., 2001. *Greenhouse Gas Mitigation in US Agriculture and Forestry.* American Association for the Advancement of Science, pp. 2481–2482.
12. McCarl, B.A., Yu, C-H and Attavanich, W., 2021. *Climate Change Impacts and Strategies for Mitigation and Adaptation in Agriculture.* Multidisciplinary Digital Publishing Institute, 545 p.
13. Locatelli, B., 2011. *Synergies between Adaptation and Mitigation in a Nutshell.* Center for International Forestry Research.
14. Cramer, W., Yohe, G., Auffhammer, M., Huggel, C., Molau, U., da Silva Dias, M.A.F., et al., 2014. Detection and attribution of observed impacts. *Climate Change 2014: Impacts, Adaptation, and Vulnerability Part A: Global and*

Sectoral Aspects. Cambridge, UK, and New York, NY, USA: Cambridge University Press, pp. 979–1037.

15. Alexander, P., Berri, A., Moran, D., Reay, D. and Rounsevell, M.D., 2020. The global environmental paw print of pet food. *Global Environmental Change, 65*, p. 102153.

16. Deluty, S.B., Scott, D.M., Waugh, S.C., Martin, V.K., McCaw, K.A., Rupert, J.R., et al. 2021 Client choice may provide an economic incentive for veterinary practices to invest in sustainable infrastructure and climate change education. *Frontiers in Veterinary Science*, p. 1184.

17. Zhang, Y.W., McCarl, B.A. and Jones, J.P.H., 2017. An overview of mitigation and adaptation needs and strategies for the livestock sector. *Climate, 5*(4), p. 95.

18. USDA, 2021. Cattle & Beef Statistics & Information. In: Service E.R., ed. Washington, DC.

19. De La Rocque, S., Rioux, J-A and Slingenbergh, J., 2008. Climate change: Effects on animal disease systems and implications for surveillance and control. *Revue scientifique et technique (International Office of Epizootics), 27*(2), pp. 339–354.

20. Gaughan, J., Lacetera, N., Valtorta, S.E., Khalifa, H.H., Hahn, L. and Mader, T., 2009. Response of domestic animals to climate challenges. *Biometeorology for Adaptation to Climate Variability and Change*. Springer, pp. 131–170.

21. Mills, J.N., Gage, K.L. and Khan, A.S., 2010. Potential influence of climate change on vector-borne and zoonotic diseases: A review and proposed research plan. *Environmental Health Perspectives, 118*(11), pp. 1507–1514.

22. Özkan, Ş, Vitali, A., Lacetera, N., Amon, B., Bannink, A., Bartley, D.J., et al., 2016. Challenges and priorities for modelling livestock health and pathogens in the context of climate change. *Environmental Research, 151*, pp. 130–144.

23. Seo, S.N. and McCarl, B., 2011. Managing livestock species under climate change in Australia. *Animals, 1*(4), pp. 343–365.

24. Tirado, M.C., Clarke, R., Jaykus, L., McQuatters-Gollop, A. and Frank, J., 2010. Climate change and food safety: A review. *Food Research International, 43*(7), pp. 1745–1765.

25. Wall, E., Wreford, A., Topp, K. and Moran, D., 2010. Biological and economic consequences heat stress due to a changing climate on UK livestock. *Advances in Animal Biosciences, 1*(1), p. 53.

26. Weatherdon, L.V., Magnan, A.K., Rogers, A.D., Sumaila, U.R. and Cheung, W.W.L., 2016. Observed and projected impacts of climate change on marine fisheries, aquaculture, coastal tourism, and human health: An update. *Frontiers in Marine Science, 3*.

27. Ekine-Dzivenu, C., Mrode, R., Oyieng, E., Komwihangilo, D., Lyatuu, E., Msuta, G., et al., 2020. Evaluating the impact of heat stress as measured by temperature-humidity index (THI) on test-day milk yield of small holder dairy cattle in a sub-Sahara African climate. *Livestock Science, 242*, p. 104314.

28. Auffhammer, M., 2018. Quantifying economic damages from climate change. *Journal of Economic Perspectives, 32*(4), pp. 33–52.

29. Tol, R.S., 2005. Adaptation and mitigation: Trade-offs in substance and methods. *Environmental Science & Policy, 8*(6), pp. 572–578.

30. McCarl, B.A., Thayer, A.W. and Jones, J.P., 2016. The challenge of climate change adaptation for agriculture: An economically oriented review. *Journal of Agricultural and Applied Economics, 48*(4), pp. 321–344.

31. Wang, M. and McCarl, B.A., 2021. Impacts of climate change on livestock location in the US: A statistical analysis. *Land, 10*(11), p. 1260.

32. Havlík, P., Valin, H., Herrero, M., Obersteiner, M., Schmid, E., Rufino, M.C., et al., 2014. Climate change mitigation through livestock system transitions. *Proceedings of the National Academy of Sciences, 111*(10), pp. 3709–3714.
33. Alexander, P., Brown, C., Arneth, A., Dias, C., Finnigan, J., Moran, D., et al., 2017. Could consumption of insects, cultured meat or imitation meat reduce global agricultural land use? *Global Food Security, 15*, pp. 22–32.
34. Lynch, J. and Pierrehumbert, R., 2019. Climate impacts of cultured meat and beef cattle. *Frontiers in Sustainable Food Systems*, p. 5.
35. Protopopova, A., Ly, L.H., Eagan, B.H. and Brown, K.M., 2021. Climate change and companion animals: Identifying links and opportunities for mitigation and adaptation strategies. *Integrative and Comparative Biology, 61*(1), pp. 166–181.
36. USGCRP, 2016. *The Impacts of Climate Change on Human Health in the United States: A Scientific Assessment.* Washington, DC: U.S. Global Change Research Program. 312 p.
37. Martin, S., 2010. Climate change, migration, and governance. *Global Governance, 16*(3), pp. 397–414.
38. Su, B., Martens, P. and Enders-Slegers, M-J., 2018. A neglected predictor of environmental damage: The ecological paw print and carbon emissions of food consumption by companion dogs and cats in China. *Journal of Cleaner Production, 194*, pp. 1–11.
39. Martens, P., Su, B. and Deblomme, S., 2019. The ecological paw print of companion dogs and cats. *BioScience, 69*(6), pp. 467–474.
40. Stevens, D. and Hussmann, A., 2017. *Wildlife Poop Versus Dog Poop Explained.* Leave No Trace. online article https://lnt.org/wildlife-poop-versus-dog-poop-explained/
41. Stephen, C., Carron, M. and Stemshorn, B., 2019. Climate change and veterinary medicine: Action is needed to retain social relevance. *The Canadian Veterinary Journal, 60*(12), p. 1356.
42. Artuso, A., 2002. Bioprospecting, benefit sharing, and biotechnological capacity building. *World Development, 30*(8), pp. 1355–1368.
43. Barrett, C.B. and Lybbert, T.J., 2000. Is bioprospecting a viable strategy for conserving tropical ecosystems? *Ecological Economics, 34*, pp. 293–300.
44. Costello, C. and Ward, M., 2006. Search, bioprospecting and biodiversity conservation. *Journal of Environmental Economics and Management, 52*(3), pp. 615–626.
45. Rausser, G.C. and Small, A.A., 2000. Valuing research leads: Bioprospecting and the conservation of genetic resources. *Journal of Political Economy, 108*(1), pp. 173–206.
46. Rosendal, G.K., 2006. Balancing access and benefit sharing and legal protection of innovations from bioprospecting: Impacts on conservation of biodiversity. *The Journal of Environment & Development, 15*(4), pp. 428–447.
47. Kaval, P. and Loomis, J., 2003. Updated outdoor recreation use values with emphasis on National Park recreation. *Final Report, Cooperative Agreement*, pp. 1200–1299.
48. Nunes, P.A. and van den Bergh, J.C., 2001. Economic valuation of biodiversity: Sense or nonsense? *Ecological Economics, 39*(2), pp. 203–222.
49. Daw, T., Adger, W.N., Brown, K. and Badjeck, M-C., 2009. Climate change and capture fisheries: Potential impacts, adaptation and mitigation. In: Cochrane, K., De Young, C., Soto, D. and Bahri, T. eds., *Climate Change Implications for Fisheries and Aquaculture: Overview of Current Scientific Knowledge.* Rome: Food and Agricultural Organization.

50. Hofmeister, E., Rogall, G.M., Wesenberg, K., Abbott, R., Work, T., Schuler, K., et al. 2010. Climate change and wildlife health: Direct and indirect effects. *US Geological Survey Fact Sheet*, 3017. https://pubs.usgs.gov/fs/2010/3017/pdf/fs2010-3017.pdf

51. Loomis, J.B. and White, D.S., 1996. Economic benefits of rare and endangered species: Summary and meta-analysis. *Ecological Economics*, *18*(3), pp. 197–206.

52. Ryding, S., Klaassen, M., Tattersall, G.J., Gardner, J.L. and Symonds, M.R., 2022. Shape-shifting: Changing animal morphologies as a response to climatic warming. *Trends in Ecology & Evolution*, *37*(1), p. 106.

53. Lenoir, J. and Svenning, J.C., 2015. Climate-related range shifts – A global multidimensional synthesis and new research directions. *Ecography*, *38*(1), pp. 15–28.

54. Hoffmann, A.A. and Sgrò, C.M., 2011. Climate change and evolutionary adaptation. *Nature*, *470*(7335), pp. 479–485.

55. LeDee, O.E., Handler, S.D., Hoving, C.L., Swanston, C.W., Zuckerberg, B., 2021. Preparing wildlife for climate change: How far have we come? *The Journal of Wildlife Management*, *85*(1), pp. 7–16.

56. Markham, A. and Malcolm, J., 1996. Biodiversity and wildlife: Adaptation to climate change. In: Smith, J.B., Bhatti, N., Menzhulin, G.V., Benioff, R., Campos, M., Jallow, B., et al., eds., *Adapting to Climate Change: An International Perspective*. New York, NY: Springer New York, pp. 384–401.

11 The International Response to Animal Health and Climate Change

Maud Carron, Tianna Brand, Sophie Muset,
Keith Hamilton, Chadia Wannous,
François Diaz, Guillaume Belot,
Stéphane de la Rocque, Julio Pinto,
Delia Grace Randolph, and Ahmed. H. El Idrissi

CONTENTS

DOI: 10.1201/9781003149774-11

KEY LEARNING OBJECTIVES

1. Describe the current fields of action of international animal health organizations relevant for climate change response and their limitations
2. Understand how global One Health agendas support the integration of animal health into climate change response

---

IMPLICATIONS FOR ACTION

1. Greater collaboration is needed between climate, environmental, conservation, and "classic" animal health organizations to address all determinants of animal health in support of climate change response
2. Integrating international climate-related data and climate-vulnerability indices into animal disease surveillance, early warning and preparedness tools will be a priority to support the mitigation of climate change impacts.
3. International support is needed for Veterinary Services to endorse a leadership role in responding to the impacts of climate change on animal health

---

INTRODUCTION: THE NEED FOR AN INTERNATIONAL RESPONSE TO ANIMAL HEALTH AND CLIMATE CHANGE

Anthropogenically-driven warming of the global climate threatens natural and social systems (1,2), leading to inevitable health impacts (3). Livestock supports the livelihoods of 750 million of the poorest worldwide (1), the same groups being also the most vulnerable to climate change (4). Disease emergence at the human-animal-ecosystem interface in recent decades illustrates the complex dependencies between livestock production methods, changes in ecosystems and wildlife population dynamics, zoonotic diseases transmission pathways, and their fragile equilibrium in a climate change context (5).

Climate change impacts animal health in different ways, altering broad health determinants such as habitats, water and feed accessibility, but also animal disease epidemiology (5–7). In particular, vector-borne diseases such as Rift valley fever (RVF), Bluetongue, West Nile, African horse sickness, and trypanosomiasis appear to be influenced by climatic changes (8). The recent increased frequency of RVF outbreaks in Kenya, a disease usually following 5- to 15-year cycles associated with flooding, have highlighted the need to improve surveillance during inter-epidemic periods in an already resource-scarce environment (9). Unusual severe rainfalls have been associated with anthrax outbreaks in European regions previously free of the disease for decades. Milder winters are shifting wildlife species' ranges, bringing novel pathogens into naïve populations and increasing the disease burden of some wild cervid and bovid species in the northern hemisphere (5). Changing disease dynamics leave Veterinary Services (VS) unprepared for early detection and response and in greater need of support for timely collection of data on evolving risk factors (10). Climate variations are shifting global needs in terms of animal health management and support (11), calling for an international response to these challenges, and for international animal health organizations to adapt their strategies and interventions to support national VS (10).

The present chapter presents how leading international organizations in animal health mainstream their climate change response into their strategic orientations and activities. Their role in further integrating climate change within Veterinary Services' governance and specific Veterinary Services functions is reviewed, as well as existing gaps and future avenues for a more robust response to climate change.

VETERINARY SERVICES (VS)

"VS" most commonly refers to the governmental body(ies) responsible for the development and implementation of national animal health and welfare policies. Their roles usually include (but are not limited to) official disease surveillance and control, disease reporting, ensuring that abattoirs/facilities producing animal products, food of animal origin and by-products meet sanitary standards, ensuring prudent use of veterinary medicinal products, including antimicrobial agents, and the certification of live animals/animal products for trade purposes. National governmental VS are the primary point of contact for international animal health organisations.

However, in its broader sense, and following the OIE (World Organisation for Animal Health, now WOAH) definition, "VS" means the governmental and nongovernmental organisations that implement animal health, welfare measures and other international OIE standards in the territory. This highlights the key and complementary role of the private sector in managing animal health nationally, in particular for early disease detection and treatment of diseases in animals.

MAINSTREAMING CLIMATE CHANGE RESPONSE INTO GLOBAL ANIMAL HEALTH AGENDAS AND VETERINARY SERVICES GOVERNANCE

GLOBAL ANIMAL HEALTH GOVERNANCE IN SUPPORT OF CLIMATE CHANGE RESPONSE

The World Organisation for Animal Health (OIE) (12) is recognised by the World Trade Organisation as the reference organization for the development of international animal health standards, aimed at ensuring safe trade in animal and animal products. The standards also serve as the basis for prevention and control of animal diseases globally (13). The Food and Agriculture Organization of the United Nations's (FAO) broad portfolio (14) includes supporting food security by raising agricultural production and improving efficiency in the use of natural resources. A major part of FAO's work on livestock relates to promoting sustainable animal production, including improved animal health. Both OIE and FAO disseminate scientific information, support international disease reporting, and contribute to the development of global strategies for the control and eradication of animal diseases with high impact on food security, livelihoods, and public health.

The OIE 7th Strategic Plan (2021–2025) illustrates a shift toward a more systemic approach through its commitment to further "exercise its voice in global discussions on [climate change, (…) animal welfare or societal expectations for more environmentally friendly animal production]." In particular, the plan states that "Veterinary Services must be better prepared to respond to these complex, multiple challenges"(13). FAO's significant involvement in climate change response related to agricultural practices is not new and culminated in 2017 with the FAO Strategy on Climate Change (15). However, the latter only briefly referred to the impacts of climate change on animal health. FAO's strategic framework 2022–2031 (16) brings climate change again to the centre of FAO's work and identifies making agri-food systems more resilient to climate hazards a major challenge to achieving food security. A recent FAO call for action entitled "Animal Health and Climate Change" (1) underlines how the roles of animal health, climate change, and sustainable development still need to be further explored to achieve both climate and food security objectives.

Broader determinants of animal health, such as ecosystem and biodiversity protection and restoration, both highly impacted by climate change, are the focus of other international players, including the United Nations Environment Programme (UNEP)[1] (17) and the International Union for Conservation of Nature (IUCN) (18). UNEP and IUCN's work further contribute to animal health and welfare in its broader sense by guiding the global environmental agenda and promoting the conservation of wildlife species and habitats, with specific work programs dedicated to climate action (19,20). The following section presents how greater collaboration among international organizations to better incorporate these complementary domains of expertise is being initiated.

PARTNERING FOR ONE HEALTH

The value of the One Health concept for addressing health threats at the human-animal-ecosystem interface has gained significant traction over the past two decades with the emergence of zoonotic diseases with pandemic potential such as Ebola, Middle East Respiratory Syndrome, Highly Pathogenic Avian Influenza, and COVID-19 (21). These global health challenges have created momentum for international animal health organizations to further strengthen their One Health strategic orientation, endorse a transdisciplinary approach, and raise awareness on the need for greater inclusion of the environmental dimension, including climate change, in global animal health management activities (5).

FAO, OIE, and the World Health Organization (WHO) have been working together for years to address risks at the human-animal-ecosystem interface (22). In 2021, this tripartite coalition was joined by UNEP to collectively drive change toward a system-based approach to support healthy people, animals, and ecosystems, while addressing the underlying factors of disease emergence, spread, and persistence, and the complex socio-economic and environmental determinants of health. By integrating environmental and climate change considerations into a more comprehensive understanding of disease dynamics, One Health can reveal its entire potential, thereby improving disease prevention, preparedness, and control in a holistic manner.

This international One Health approach has branched out to address national needs via a series of country-level tools, with new or expanded One Health components (23). Synergies between public and animal health capacity assessment tools (e.g., on the public health side, the WHO's State Party Annual Reporting and Joint External Evaluations that assess compliance with the International Health Regulations (IHR), and on the animal health side, the OIE Performance of Veterinary Services (PVS) Evaluations that assess compliance with OIE international standards) have grown, multisectoral coordination capacities becoming a key indicator of health systems strengthening (24,25). Efforts to integrate environmental considerations into existing tools, such as IHR-PVS national bridging workshops that support joint planning between human and animal health sectors (26) or the Tripartite Zoonoses Guide and its operational tools (on multisectoral coordination mechanisms, joint-risk assessments, and surveillance and information sharing) (27) are increasing. These various initiatives directly support multisectoral governance of animal health functions, a key requirement to build Veterinary Services (VS) capacity in climate change response (4).

With COVID-19, tripartite partners have taken actions to better include wildlife health matters into their activities. The OIE wildlife health-management framework (28), endorsed during the 88th General Session in 2021, is a good example of a new commitment to further address animal health at its intersection with conservation and environmental issues.

Finally, the tripartite is also collaborating on education and expanding the integration of One Health, the environment, and climate change in its related activities. The new WHO Academy is partnering with OIE for the development of a learning framework for the One Health workforce, which will support the continuous acquisition of new skills required to address upcoming challenges, including climate change. This represents a key opportunity to bring change in how veterinarians, public health, and environmental health professionals perceive their role in the One Health approach, and to better integrate preparedness and leadership competencies to address global climate challenges in health curricula.

ANIMAL HEALTH NEEDS TO BE A CENTRAL COMPONENT IN ADDRESSING CLIMATE CHANGE

The One Health approach in animal health-management activities represents a fertile ground for future, more climate-oriented activities, but it remains a humble progress considering the response required to address climate change. While recognition of the importance of climate change is growing among the animal health community, Veterinary Services' preparedness and involvement remain limited. The 87th OIE General Session presented a study on "How External Factors (e.g., climate change, conflicts, socioeconomics, trading patterns) will impact Veterinary Services, and the Adaptations required" (29). Results (illustrated in Figure 11.1) showed that "extreme weather" was among external factors considered by OIE members as highly important, but for which preparedness was low. While "livestock pandemics," "emerging diseases" and "human pandemics" were ranked as top priorities, "biodiversity loss" and "improvement of the livestock contribution to global greenhouse gas (GHG)" were ranked lower. These findings suggest either a disconnect between how VS perceive their role in climate change response, or a lack of capacity of the VS to get involved in this field (11).

Indeed, current VS functions tend to focus on preventing and managing animal infectious diseases, in line with a definition of animal health focused on the absence of infectious pathogens. A multisectoral approach is recognized as key

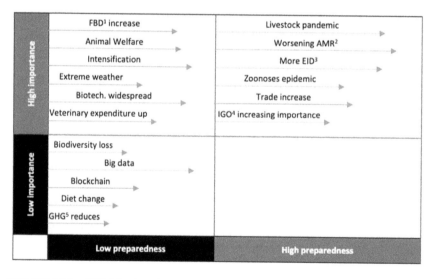

FIGURE 11.1 OIE Member Country assessment of the importance of and their preparedness for external factors. The figure shows a limited number of key external factors as an example. [1]FBD: Foodborne disease, [2]AMR: Antimicrobial resistance, [3]EID: Emerging infectious diseases, [4]IGO: Inter-governmental organisations,organizations, [5]GHG: Greenhouse gases.

Source: World Organisation for Animal Health (OIE). Reproduced with permission.

to support One Health (5), yet VS do not consider themselves as privileged actors in conservation and other environmental efforts that impact animal health at large (30).

This idea is mirrored by environmental agencies' initiatives that seldom consider animal health, even in their agricultural-related activities. For example, UNEP's and IUCN's latest work programs only mention "animal" twice each, mainly in relation to species' extinction (19,20). Synergies between animal health activities in the domains of livestock, wildlife, conservation, and sustainable agriculture still need to be harnessed further by international agencies to better support VS in filling a leadership gap at the intersection of animal, ecosystem, and human health, in society-wide climate change plans and action (2).

A recurring recommendation in the literature for climate change response, including in the animal health realm, is to strengthen multisectoral partnerships (1,2,4,19) at all levels. Beyond the commitment of international animal health organizations to tackle climate change more directly, initiatives specifically aimed at mitigating and adapting to climate change often lack animal health considerations. For example, an analysis of nationally determined contributions (NDCs under the Paris Agreement) of agricultural adaptation and mitigation in 2016 showed that only 61 countries included livestock in their plans (4). Robustness in mechanisms to support a two-way dialogue between animal health and climate organizations is needed at multiple levels and should be supported by foresight exercises and engagement with the private sector.

MAINSTREAMING CLIMATE CHANGE INTO INTERNATIONAL SUPPORT TO NATIONAL VETERINARY SERVICES

The livestock sector is responsible for 14.5% of GHG emissions, driving further global warming, while also playing a key role in food security and representing significant opportunities for climate change adaptation in sustainable agriculture (4,31,32). FAO has a leading role in developing guidance and tools for sustainable and climate-smart livestock production (e.g., Global Livestock Environmental Assessment Model – GLEAM (33) a climate-smart Agriculture Sourcebook (34,35)) and produced a series of key reference assessments on the environmental impact of livestock and potential mitigation options (31,32,36–38). These reports mention the impacts of climate change on animal health in terms of changes in disease patterns and new public health risks. They refer to healthy animals as a positive factor in support of climate change mitigation: improving animal health enhances resource use efficiency by reducing mortality, improving productivity and fertility, and requiring fewer animals to meet demands for animal products (1,39).

In this sense, FAO animal health activities and OIE's core work support countries worldwide in removing an inefficiency, in other words, poor animal health, from the livestock sector.[2] Leading international animal health organizations support healthier animals and the strengthening of VS through information and guidance dissemination, standards setting, as well as capacity-building activities, in which climate challenges are already being, or can be further integrated, as presented below.

Strengthening the Capacities of Veterinary Services and Improving Animal Health to Support Climate Change Response

Animal health, and its management by national VS, is chronically under-resourced, despite making a vital contribution to public health, food security, and poverty alleviation (40). Persisting gaps in VS capacities and difficulties to control animal diseases highlight how "classic" OIE and FAO activities on disease control and VS strengthening remain relevant for global climate change action. National capacity-building activities that directly benefit VS on the ground, such as the OIE Performance of Veterinary Services (PVS) Pathway (24) could in the future be targeted in priority to countries most vulnerable to climate change impacts and/or further adjusted to incorporate climate change considerations.

Both OIE and FAO already support countries in emergency preparedness, through the development of guidelines, deployment of experts to support members' response to animal health crises, and regional analyses of gaps in emergency management. This support could further include animal health-related emergencies triggered by climate change and consider the animal health and welfare impacts of adverse weather events. National VS need to be included in cross-government disaster-management frameworks to build response capacity (41). To allow greater customization of existing risk man-agement and disease control strategies designed by national VS to a climate-change context, well-adapted surveillance data is needed. To address this gap, the following activities should be leading priorities.

Strengthening Animal Health Intelligence

OIE and FAO are routinely involved in guiding animal disease surveillance systems and have developed online tools to collect and share animal health data, such as the OIE World Animal Health Information System (WAHIS for live-stock and wildlife) and the FAO Global Animal Disease Information System (EMPRES-i). FAO has been working for years in developing and disseminating early-warning messages about the risk of occurrence of climate sensitive dis-eases through the Emergency Prevention System (EMPRES) and associated tools such as the FAO/OIE/WHO – Global Early Warning System (GLEWS+) that map and track disease outbreak to develop risk-management strategies. The GLEWS+, is a joint platform that integrates data from different sources and conducts disease intelligence and epidemiological analysis to provide warning messages and a basis for more accurate risk assessments to be conducted by the international scientific community, with the ultimate goal of contributing to the forecasting of disease patterns (42). FAO and technical partners have been monitoring climatic conditions to better forecast the risk of RVF vector ampli-fication in East Africa for the past several years using modeling tools (43).

The OIE is exploring how to best support its Members in identifying, tracking, and prioritizing climate-sensitive diseases, as well as strengthening its activities in wildlife disease surveillance. Accordingly, OIE international standards and global disease-notification requirements will need to remain up to date and fit for purpose with respect to disease prevention, mitigation, and surveillance in the context of climate change.

ADDRESSING DATA GAPS AND DATA INTEGRATION NEEDS TO STRENGTHEN EVIDENCE

Despite considerable investments globally and the large body of work undertaken by the Intergovernmental Panel on Climate Change (IPCC) and other research entities (44), (45) to better understand climate processes, the basis of evidence remains weak on the risks for animal health and how these could be mitigated. To make existing information more relevant to animal health risk management, climate trends need to be understood with greater spatial resolution for priority regions and with greater integration of socio-climatic factors. This clarity would allow international animal health organizations to provide improved decision support tools to policy makers and the private sector on the management of diseases.

Exploring ways to integrate international climate-related data and climate-vulnerability indices (e.g., IUCN's Guidelines for Assessing Species' Vulnerability to Climate Change (46)) into animal disease surveillance, early-warning, and preparedness tools will be a priority for supporting the mitigation of climate change impacts. For example, IUCN's "Contribution to Nature Initiative" targets the development of a platform to overlay data from conservation, restoration actions, biodiversity, potential for species extinction (in particular IUCN's Red List of Threatened Species), as well as nature-based solutions to climate change, including in terms of carbon sequestration (20). This platform could present synergies with some of FAO's carbon footprint assessment tools (33) and other international animal disease surveillance tools and datasets. To support data integration, a global observatory focused on climate fluctuations and animal disease/health determinants could also be envisaged.

CONCLUSIONS

While much attention has been given to climate change implications for human health and other agricultural activities, animal health has been relatively neglected. Furthermore, animal production systems are targeted as contributors to climate change, while at the same time, being adversely impacted by it. Recent international efforts are gradually recognizing the importance of nurturing links between animal health management and climate change response. The strategic orientations of OIE and FAO support this vision, and efforts to integrate environmental considerations into animal health activities are expanding.

Given the urgent need to respond and adapt to climate change, a One Health approach with a strong environmental dimension would be extremely beneficial for mainstreaming climate change in global animal health agendas and governance of VS. To achieve this goal, new partnerships and a continuous dialogue are needed between climate and animal health players to leverage existing data and capture opportunities for synergies. Further efforts are needed to address the gaps in cross-sectoral coordination, governance processes, and legal frameworks. More fundamentally, a key divide between wildlife and livestock-orientated activities, and/or animal production and conservation activities, must be reduced to bolster the impact of current climate change response.

NOTES

1 The Convention on Biological Diversity (CBD) and the Convention on International Trade in Endangered Species of Wild Fauna and Flora (CITES) are under UNEP.
2 This does not translate into an overall positive or negative impact of the sector in terms of addressing the roots of climate change. The complex cost-benefit equation of the sector, both on the environmental side (i.e., livestock as a source of GHG) as well as on the food security and livelihoods side (i.e., livestock as a key source of livelihoods and food) (35), is not discussed further, as it goes beyond the scope of this chapter.

REFERENCES

1. Food and Agriculture Organization of the United Nations (FAO), 2020. Animal health and climate change [Internet]. [cited 2021 Oct 4]. Available from: http://www.fao.org/3/ca8946en/CA8946EN.pdf
2. Stephen, C. and Soos, C., 2021. The implications of climate change for Veterinary Services. *Rev Sci Tech l'OIE*, *40*(2), pp. 421–430.
3. Intergovernmental Panel for Climate Change (IPCC), 2014. Climate change 2014: Synthesis Report. Contribution of Working Groups I, II, III to the Fifth Assessment Report of the Intergovernmental Panel on Climate Change. Pachauri, R.K. and Meyer, L.A., eds. Geneva.
4. Wannous, C., 2020 Aug 1. Climate change and other risk drivers of animal health and zoonotic disease emergencies: The need for a multidisciplinary and multi-sectoral approach to disaster risk management. *Rev Sci Tech*, *39*(2), pp. 461–470.
5. Black, P.F. and Butler, C.D., 2014. One Health in a world with climate change. *OIE Rev Sci Tech*, *33*(2), pp. 465–473.
6. Baylis, M., Caminade, C., Turner, J. and Jones, A.E., 2017 Aug 1. The role of climate change in a developing threat: The case of bluetonque in Europe. *OIE Rev Sci Tech*, *36*(2), pp. 467–478.
7. Maksimovic, Z., Cornwell, M.S., Semren, O. and Rifatbegovic, M., 2017 Dec 1. The apparent role of climate change in a recent anthrax outbreak in cattle. *Rev Sci Tech l'OIE*, *36*(3), pp. 959–963.
8. De La Rocque, S., Rioux, J.A. and Slingenbergh, J., 2008. Climate change: Effects on animal disease systems and implications for surveillance and control. *OIE Rev Sci Tech*, *27*(2), pp. 339–354.
9. World Health Organization (WHO), 2021. Rift Valley fever – Kenya (12 February 2021). reliefweb [Internet]. Available from: https://reliefweb.int/report/kenya/rift-valley-fever-kenya-12-february-2021

10. Martin, V., Chevalier, V., Ceccato, P., Anyamba, A., De Simone, L., Lubroth, J., et al., 2008. The impact of climate change on the epidemiology and control of Rift Valley fever. *Rev sci tech Off int Epiz*, 27(2), pp. 413–426.
11. Eloit, M., 2020. Climate change: An added complexity to addressing health risks [Internet]. *F@rmLetter of the World Farmers' Organisation*. [cited 2021 Oct 30]. pp. 30–32. Available from: https://www.wfo-oma.org/frmletter-3_2020/climate-change-an-added-complexity-to-addressing-health-risks/
12. World Organisation for Animal Health (OIE), 2021. World Organisation for Animal Health [Internet]. [cited 2021 Oct 30]. Available from: https://www.oie.int/en/home/
13. World Organisation for Animal Health (OIE), 2021. OIE Seventh Strategic Plan for the period 2021–2025 [Internet]. [cited 2021 Oct 4]. Available from: https://www.oie.int/app/uploads/2021/08/a-88sg-14.pdf
14. Food and Agriculture Organization of the United Nations (FAO), 2021. Food and Agriculture Organization of the United Nations [Internet]. [cited 2021 Oct 30]. Available from: http://www.fao.org/home/en/
15. Food and Agriculture Organization of the United Nations (FAO). FAO Strategy on Climate Change [Internet]. [cited 2021 Oct 4]. Available from: http://www2.ecolex.org/server2neu.php/libcat/docs/LI/MON-093088.pdf
16. Food and Agriculture Organization of the United Nations (FAO). Strategic Framework 2022-31 [Internet]. [cited 2021 Oct 4]. Available from: http://www.fao.org/3/ne577en/ne577en.pdf
17. United Nations Environment Programme (UNEP), 2021. United Nations Environment Programme [Internet]. [cited 2021 Oct 30]. Available from: https://www.unep.org/
18. International Union for Conservation of Nature (IUCN). International Union for Conservation of Nature [Internet]. [cited 2021 Oct 30]. Available from: https://www.iucn.org/
19. United Nations Environment Programme (UNEP), 2018. For pepole and planet: The United Nations Environment Programme strategy for 2022–2025 to tackle climate change, loss of nature and pollution [Internet]. Available from: https://papersmart.unon.org/resolution/uploads/k1900699.pdf
20. International Union for Conservation of Nature (IUCN), 2021. Programme 2030 – One Nature, One future [Internet]. [cited 2021 Oct 4]. Available from: https://portals.iucn.org/library/node/49292
21. The Authors, 2020. Emerging zoonoses: A one health challenge. EClinicalMedicine [Internet], *19*. Available from: https://www.thelancet.com/journals/eclinm/article/PIIS2589-5370(20)30044-4/fulltext#:~:text=Emergingzoonoses%3A A one health challenge December%2C 2019%2C,virus 2 %28SARS-CoV-2%29%2C crossing species to infect humans
22. Food and Agriculture Organization of the United Nations (FAO), World Organisation for Animal Health (OIE), World Health Organization (WHO), 2017. The Tripartite's Commitment – Providing multi-sectoral, collaborative leadership in addressing health challenges [Internet]. Available from: www.oie.int/2010tripartitenote
23. Pelican, K., Salyer, S.J., Barton Behravesh, C., Belot, G., Carron, M., Caya, F., et al., 2019. Synergising tools for capacity assessment and One Health operationalisation. *Rev Sci Tech*, 38(1), pp. 71–89.
24. Stratton, J., Tagliaro, E., Weaver, J., Sherman, D.M., Carron, M., Di Giacinto, A., et al., 2019. Performance of Veterinary Services Pathway evolution and One Health aspects. *Rev Sci Tech.*, 38(1), pp. 291–302.

25. De La Rocque, S., Caya, F., El Idrissi, A.H., Mumford, L., Belot, G., Carron, M., et al., 2019. One Health operations: A critical component in the International Health Regulations Monitoring and Evaluation Framework. *Rev Sci Tech.*, *38*(1), pp. 303–314.

26. Belot, G., Caya, F., Errecaborde, K.M., Traore, T., Lafia, B., Skrypnyk, A., et al., 2021. IHR-PVS National Bridging Workshops, a tool to operationalize the collaboration between human and animal health while advancing sector-specific goals in countries. *PLoS One*, *16*(6 June), pp. 1–16.

27. Food and Agriculture Organization of the United Nations (FAO), World Organisation for Animal Health (OIE), World Health Organization (WHO), 2019, A tripartite guide to addressing zoonotic diseases in countries taking a multi-sectoral, one health approach [Internet]. [cited 2021 Oct 6]. Available from: file:///C:/Users/m.carron/Downloads/9789241514934-eng (1).pdf

28. World Organisation for Animal Health (OIE), 2021. OIE Widlife Health Framework – Protecting Widlife Health to achieve One Health [Internet]. Paris; [cited 2021 Oct 6]. Available from: https://www.oie.int/fileadmin/Home/ eng/Internationa_Standard_Setting/docs/pdf/WGWildlife/A_Wildlifehealth_ conceptnote.pdf

29. Grace, D., Caminiti, A., Torres, G., Messori, S., Bett, B.K., Lee Hu, S., et al., 2019 May. How external factors (e.g. Climate change, conflicts, socio-economics, trading patterns) will impact Veterinary Services and the adaptations required [Internet]. Available from: https://doc.oie.int/dyn/portal/index.xhtml?page=alo& aloId=39938

30. Stephen, C., Carron, M. and Stemshorn, B., 2019. Climate change and veterinary medicine: Action is needed to retain social relevance. *Can Vet J*, *60*(12), pp. 1356–1358.

31. Gerber, P.J., 2013. Food and Agriculture Organization of the United Nations. Tackling climate change through livestock: A global assessment of emissions and mitigation opportunities [Internet]. [cited 2021 Oct 4]. 115 p. Available from: http://www.fao.org/3/i3437e/i3437e.pdf

32. Food and Agriculture Organization of the United Nations (FAO), 2013. *Tackling Climate Change through Livestock: A Global Assessment of Emissions and Mitigation Opportunities*. Rome.

33. Food and Agriculture Organization of the United Nations (FAO), 2019. Five practical actions towards low-carbon livestock. Rome.

34. Food and Agriculture Organization of the United Nations (FAO), 2013. Climate-smart agriculture sourcebook [Internet]. [cited 2021 Oct 4]. 557 p. Available from: http://www.fao.org/3/i3325e/i3325e.pdf

35. Food and Agriculture Organization of the United Nations (FAO), 2017. Leap at a glance 2016–2017 [Internet]. Available from: http://www.fao.org/3/i7804e/ i7804e.pdf

36. Food and Agriculture Organization of the United Nations (FAO), 2006. Livestock's Long Shadow – Environmental issues and options [Internet]. [cited 2021 Oct 4]. Available from: http://www.fao.org/3/a0701e/a0701e00.htm

37. Steinfeld, H., Mooney, H.A., Schneider, F. and Neville, L.E., 2010. *Livestock in a Changing Landscape – Drivers, Consequences, and Responses*. Island Press.

38. Gerber, P., Mooney, H.A., Dijkman, J., Tarawali, S., De Haan, C., 2010. Livestock in a Changing Landscape, Experiences and Regional Perspectives, 2, Washington, DC: Island Press. Available from: www.islandpress.org/bookstore/ details5173.html?prod_id=1950

39. Food and Agriculture Organization of the United Nations (FAO), 2017. Livestock solutions for climate change [Internet]. Available from: http://www.fao.org/gleam/results/en/

40. World Organisation for Animal Health (OIE), 2019. Strengthening Veterinary Services through the OIE PVS Pathway – The Case for Investment in Veterinary Services [Internet]. [cited 2021 Oct 6]. Available from: https://www.oie.int/app/uploads/2021/03/20190513-business-case-v10-ld.pdf

41. World Organisation for Animal Health (OIE), 2016. Guidelines on disaster management and risk reduction in relation to animal health and welfare and veterinary public health (Guidelines for National Veterinary Services) [Internet]. Available from: https://www.oie.int/app/uploads/2021/03/disastermanagement-ang.pdf

42. Food and Agriculture Organization of the United Nations (FAO), World Organisation for Animal Health (OIE), World Health Organization (WHO). 2013. GLEWS+. The Joint FAO–OIE–WHO Global Early Warning System for health threats and emerging risks at the human–animal–ecosystems interface [Internet]. Available from: http://www.glews.net/?page_id=5

43. Jeffrey, M., 2018. Rift Valley Fever Surveillance [Internet]. FAO Animal Production and Health Manual No. 21. Rome;. Available from: http://www.fao.org/3/i8475en/I8475EN.pdf

44. International Livestock Research Institute (ILRI), 2019. Introduction to International Livestock Research Institute's agenda for climate change adaptation in livestock systems [Internet]. *Climate Change Adaptation in Livestock Systems Brief 1: Introduction to ILRI's agenda*. Nairobi. Available from: https://cgspace.cgiar.org/handle/10568/101605

45. Intergovernmental Panel for Climate Change (IPCC), 2021. Intergovernmental Panel for Climate Change [Internet]. [cited 2021 Oct 30]. Available from: https://www.ipcc.ch/

46. Foden, W., Garcia, R.A. and Platts, P.J., 2016. Selecting and evaluating CCVA approaches and methods [Internet]. Available from: https://www.researchgate.net/publication/307420414

12 Preparing For the Unanticipated

Craig Stephen

CONTENTS

KEY LEARNING OBJECTIVES

1. Have a rudimentary understanding of how systems attributes and operational or organizational factors will lead to surprising health outcomes due to climate change.
2. Link a taxonomy of surprise to general actions to reduce surprise or be better prepared to resist its effects.
3. Describe why promoting health in advance of surprising threats and harms is a necessary component of an animal health and climate change agenda.

DOI: 10.1201/9781003149774-12

IMPLICATIONS FOR ACTION

1. Animal health programs that plan only for known and anticipated impacts of climate change will be necessary but inadequate for climate change preparedness.
2. Investing in building healthy, resilient populations by protecting and promoting the determinants of health is a multi-solving solution that can better prepare animals for a wide variety of known and surprising events.
3. Surprise preparedness will require multidisciplinary partnerships to address the drivers of vulnerability to changing health circumstances.

CLIMATE CHANGE WILL BRING SURPRISES

There is something comfortable about planning animal health programs by seeing cause-effect relationships as stable, linear, and predictable. But "things that have never happened before, happen all the time" (1). Health and ecological sciences have long recognized that health and resilience are products of complex, dynamic social and ecological interactions (see Chapter 1) and that surprises are normal in such systems (2). Animal health managers and researchers must anticipate change but should not expect to predict it (3).

Anthropogenic climate change is piling on top of existing natural variability creating a situation wherein past conditions are less reliable indictors of future conditions. As a result, climate change is leading to surprises (4). Climate change studies have witnessed surprising events, surprising outcomes, surprising responses, and surprising sources of greenhouse gases. Climate-related surprises can come as abrupt climate system changes and/or as steady changes in climate that trigger abrupt changes in other physical, biological, and human systems. Surprises provide society or ecosystems little time to adapt or resist the change, making them of special concern.

Surprise will generally come from lack of sufficient information or knowledge, operational factors that limit our openness to surprise, and the basic dynamics of complex adaptive systems (5). The drivers of climate change adaptation will be mediated by changing hazard-climate-animal interactions and modified by global trends such as landscape change, animal overexploitation, globalization, agricultural intensification, and urbanization. Variations in local biotic interactions will compound the challenge of anticipating how populations will respond to interventions or climate change. Animal health management responses to climate change will be subject to inherent variability of socio-ecological systems that will be further complicated by trade-offs between multiple objectives, values, and interest and further limited by uncertainties and ambiguities about what causes the problems and the effects of interventions.

The future of health in the face of climate change will be replete with "foreseeable unexpected events." They are foreseeable in that our experiences and research will reveal the possibility of new health impacts of climate change. They are unexpected because we can rarely predict their timing, location, and impacts with accuracy.

ORIGINS OF SURPRISE

CLIMATE–HEALTH RELATIONSHIPS ARE COMPLEX

Climate change and health are influenced by multiple, interacting, complex systems. Economic systems drive greenhouse gas production, physiological systems influence individual animal health, socio-ecological systems affect population health, governance systems affect how we make decisions and work together, social systems affect how we value and treat animals, regulatory systems determine how society responds, and ecosystems provide support on which animals depend. Health outcomes are multi-scale, complex phenomena affected by relationships and interactions at cellular, individual, population, social, and ecosystem scales and by the influences and feedbacks across and between these various scales, which change over time. Each of these systems have fuzzy boundaries that affect each other in complex and sometimes chaotic ways. This lack of clarity creates uncertainty about how the cascade of biological and social outcomes will interact to affect each other and climate change (6). The messy interactions between climate change and health limit our ability to anticipate the course of climate change, foresee its effects, discover how to best manage a health problem, and avoid unintended side effects of our actions.

EMERGENCE AND SURPRISE ARE EXPECTED IN COMPLEX SYSTEMS

When multiple systems affect and interact with each other, as seen in climate change and its relationships to health, surprising events are inevitable due to the complex and chaotic interactions within and between those systems. The concepts of systems, complexity, and chaos appear frequently in the climate change and ecology literature, and with increasing frequency in the health literature. The many parts in a complex system interact in a nonlinear fashion such that a change in one part does not have a fixed effect on the whole system; rather, it depends on the current state of that part and other parts in the system (7). The parts, and the organized structures they create, change dynamically, interdependently, and often unpredictably over time. Small changes such as internal stochastic effects or external perturbations can be amplified to have a drastic impact on the system, such as causing a sudden transition into another regime of stability (8). The extreme sensitivity to initial conditions can make system change appear irregular and even random, even though the system is evolving deterministically according to some simple rules. We may be able to retrospectively explain how these surprise events emerged in a complex system, but we are rarely able to predict them in advance.

TIPPING POINTS

A tipping point is the point at which an abrupt, irreversible transition occurs in a complex system wherein the system changes from one state to an alternate state. Small perturbations to a system's state or its parameters can trigger a cycle of positive feedback, causing the system to dramatically transition from one steady state to another. Shifts between disease elimination and outbreak regimes, for example, can occur through tipping points, such as abrupt changes in vaccine uptake in response to a social media posting (9). There has been a tendency to assume that tipping points in the Earth's system were of low probability and little understood, but mounting evidence suggests that tipping points in climate change are "more likely than was thought, have high impacts and are interconnected across different biophysical systems, potentially committing the world to long-term irreversible changes" (10). When the climate tips from one state to another, past predictions of the future of our climate need to be recalculated. But, as a climate regime tips into a new state, that new state may make it more susceptible to transition, which then necessitates a new evaluation of what the future may bring. The loss of arctic ice shields would, for example, fundamentally change coastal ecosystems due to sea level rise. The loss of permafrost would release more greenhouse gases, shifting the world into a new rate of climate change. Such changes will put the world into a new state for which our past experiences may be insufficient to see the new future.

Systems diagnostics can act as early-warning signals for critical transitions at tipping points (11). They have been successfully applied to warn of a wide variety of phenomena, including species extinction or population regime shifts, seizures or heart attacks, and climate changes (8). A key advantage of early-warning systems for critical transitions is they are based on general features of nonlinear dynamic systems so they do not require precise knowledge of the feedback loops that may trigger the transition. One such behavior is critical slowing down, which describes the fact that, near a tipping point, a system will recover more slowly from perturbations. However, early-warning systems for tipping points will need to be customized. For example, in disease emergence, the number of infected individuals grows continuously albeit explosively, versus more commonly studied critical transitions where the transition between two states is discontinuous (12). These types of early-warning systems are still in their infancy. High-quality epidemiological data will be crucial to reduce statistical uncertainty to a point where early-warning systems of climate change effects on surprising health outcomes are reliable enough to motivate proactive behaviour (2).

OPERATIONAL AND SOCIAL ORIGINS OF SURPRISES

Individual and organizational attributes can predispose us to being surprised (2). We might be surprised because of the way we see phenomena and how we believe the world works. The more dissimilar a situation is from what we expect, the more surprising we find the effect. But not everyone is surprised by everything. What is surprising to someone who must respond to changing climate (e.g. a politician or business leader) may not be surprising to the person who recognized the need to respond (e.g. a climatologist). Insensitivity to new information and perceived power dynamics can lead to failure to recognize and act on early-warning signals. People can become desensitized to warning messages that are too frequent, too many, or too disconnected with their lived experiences or when not accompanied with solutions.

Organizations can be unreceptive to warning signals outside of their usual scope of practice when cross-sectoral communication breaks down, and when bureaucratic conflicts and inadequate protocols preclude information sharing (2). Priorities and overcrowded agendas may discourage collaborations that extend beyond immediate interests, thereby reducing opportunities for novel information to be brought into view and alter our expectations.

Operational surprises arise when the links between policy, intelligence, warning, and response fail (13). Many surprises are rarely a "bolt out of the blue," but are events whose warning signals were not tracked, recognized, or appropriately interpreted. An organization may be unaware of an event or consequence until it becomes severe or affects a population of special interest. This is frequently the case for emerging diseases. These surprises are knowable in retrospect, but they elude detection because of lack of surveillance or interest in a place or population.

TAXONOMY OF SURPRISE

There are four broad types of surprise (14–16). First, surprises may occur because we lack the capacity, methods, or insights to detect or anticipate a surprising event, or available methods lack sensitivity to detect an effect until its impacts are severe. Some surprises are eminently understandable with hindsight, but organizations may lack the foresight, surveillance, knowledge systems, resources, or will to anticipate such events in time to minimize or prevent their impacts and spread. An example might be a fish die-off due to low water oxygen secondary to warming waters in a noncommercial species in a remote body of water that is not subject to surveillance. Second are surprises that come from failure to recognize an actionable signal or respond to that signal despite ample warning. An example of this type of surprise could be when a public health agency fails to see the detection of a novel zoonotic pathogen in wildlife in the absence of human illness as an early warning sign. Responses to these first two categories of surprise have focused on connecting specialized pools of knowledge and improving access to the larger network of information and expertise.

Unanticipated consequences of socio-ecological interactions are the third category of surprise. These surprises are conceivable in retrospect once additional investigation reveals connections that drove an emerging phenomenon that was not previously anticipated. For example, changes in agricultural yield and crop pests under climate change are likely, but the environmental and societal problems resulting from these changes are almost completely unknown (17). Changing crop-pest interactions could, for example, affect pesticide use and the secondary exposure of animals to these compounds. The fourth type of surprise is new, previously inconceivable events. These latter two surprise types arise due to un-certain, ambiguous, or unanticipated circumstances. Strategies for these surprises focus on building population health and resilience against the unforeseen. Figure 12.1 expands on the typology of surprise developed by Stephen et al. (18), based on concepts from climatology, surprise science, emerging disease research, and military characterizations of surprise.

EXAMPLES OF HEALTH SURPRISES

In general, we can say that climate change will have direct and indirect impacts on animal determinants of health in multiple, interacting ways, across a range of scales. These impacts will amplify existing risks, shift known risks to new places, times and/or species, and create new unanticipated risks. Many examples of sur-prising health events are described throughout this book. Below are three general types of surprise to illustrate, but not totally account for, some of the anticipated themes of potentially surprising climate change effects on animal health.

1. Emerging diseases

Emerging infectious diseases, such as COVID-19, SARS, HIV, swine influenza, and chronic wasting disease, have caught us by surprise, even though the drivers of emergence are broadly known and sophisticated surveillance systems are in place (19,20). Climate change forms the background context for changes to occur in susceptibility and infectiousness, and opportunities for the transmission of human and animal disease (21). Climatological variation and ecological changes are key drivers of the diversity and distribution of pathogens and parasites (22).

2. Extreme and unexpected weather events

Extreme weather events will cause droughts, fires, and floods for which we will be ill prepared. Predictions of future weather are inherently uncertain. They grow more unreliable the farther out in time that we want to predict the weather and in the absence of local information. Climate change is expected to shift the frequency, intensity, and geographical distribution of weather variability and ex-tremes (23). These realities restrict the forewarning of extreme weather events. Extreme events often bring unexpected situations and impacts (24). Anomalously hot or cold weather can cause direct mortality, can change how animals interact with each other and their environment (thus changing infectious diseases dynamics or predator-prey relations), and can physically harm animals (e.g. fires and floods). As examples, late freeze events could endanger larval or juvenile animals, or flash floods and fire could displace livestock from critical food sources.

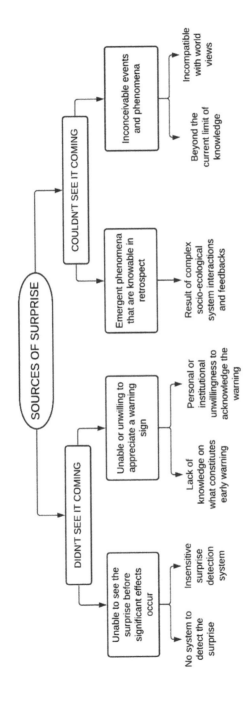

FIGURE 12.1 A typology of the sources of surprise.

3. Changes in ecological relationships.

One of largest uncertainties in forecasting climate change effects is in understanding how it will affect species interactions (25). "Inconclusive, unexpected,
or counter-intuitive results should be embraced in order to understand apparent
disconnects between theory, prediction, and observation" (26) when planning for
how ecological changes will affect the quality, access to, or availability of the
determinants of health. Changes in the dates of recurrent natural events (such as
animal migration or reproduction) in relation to seasonal climatic change will
affect food web interactions that will influence transmission and exposure to
pathogens, parasites, and pollutants, as well as access to the needs for daily
living. For example, strong spatial variation has been documented for transmission of Brucella in wild elk (*Cervus canadensis*) at daily and annual scales.
This variation is strongly governed by variation in host movement in response to
spring phenology (27). As another example, late seasonal breeding timing of
wood frogs (*Rana sylvatica*), when coupled with colder breeding seasons, has
been seen to influence offspring response to wetland contamination (28). Shifting
climate regimes will alter biotic communities in surprising ways, leading to new
and unanticipated opportunities for pathogens, parasites, and pollutants to move
between species (29). Interactions of management factors, climate change effects, and natural histories will result in unanticipated impacts on conservation
and food security, as exemplified by changes in phenology affecting fish migration and reproduction (30).

PREPARING FOR CLIMATE CHANGE INFLUENCED ANIMAL HEALTH SURPRISES

Despite promises for scientific and technological means to improve our ability to
predict impeding threats, there is presently not enough predictive power to accurately forecast health outcomes resulting from the environmental and social
changes anticipated in the 21st century (31). Predictions of future health effects
cannot be made with confidence (apart from broad generalizations) because of
the prevailing uncertainties in how the climate will respond to future greenhouse
gas levels and uncertainties over how societies and ecosystems will respond.

   Complexity and uncertainty are not excuses for inaction. There will always be
inherent uncertainty and unpredictability in the dynamics and behavior of animal
health, but management decisions must be made. When we do not prepare for
surprise or wait too long to implement measures to deal with it, the social and
ecological costs can be remarkably high. Although predicting future animal
health impacts of climate change may rarely be possible, being attentive to what
might happen helps us to identify vulnerabilities in advance of harms and consider opportunities to make populations more resilient and thus better able to
ward off future surprises.

   A surprise-oriented health-management system will need to be multifaceted
to: (i) identify and address socio-economic, ecological, and health circumstances

that are conducive to disease or threat emergence and increase vulnerability to their effects; (ii) build capacity for individuals and populations to cope with multiple interacting threats and stressors to build resilience against the additional pressure of an unexpected health threat in advance of harms; (iii) develop capacity to adaptively respond to surges in unexpected problems; and (iv) integrate surprise activities into routine health management (32)

The taxonomy of surprise outlined in figure 12.1 provides a convenient way to think about how to manage for surprise. For surprises that elude detection, are not recognized, or are not acted upon, we need to improve our situational awareness so that managers can understand if the circumstance are conducive to a new problem. Situational awareness can be enhanced by better information sharing to detect early-warning signals and by including signals of vulnerability (such as evidence of exposure and sensitivity to threats or compromises in ability to cope with new harms) in population health-monitoring programs. For unanticipated and inconceivable surprises, the best strategy is to invest in making populations healthy so that they can resist a wide suite of new threats and be resilient enough to recover. Disaster management, climate change preparation, and pandemic planning are increasingly recognizing that, to deal with a surprising and unpredictable future, communities, populations, and ecosystems must have the assets needed to cope with what may come (2). A surprise-based approach is concerned with identifying and sustaining the shared protective factors that support health.

ANIMALS AS SENTINELS FOR EMERGING CLIMATE CHANGE HARMS

The primary role of animal sentinels is to produce early-warning signals that empower individuals, communities, and/or organizations to respond in a timely and appropriate manner to avoid, reduce, or mitigate harm. Warning occurs when a change in risk or vulnerability status is revealed, and that change is rapidly communicated to those able to respond (33). Early-warning systems for climate change adaptation, preparedness, and response should, therefore, produce signals that help to detect and respond to a range of factors that can drive risk and vulnerability and inspire actions.

Given the projections for changing distributions and burdens of pathogens and pollutants in the face of climate change, the role of animals as climate change bio-indicators is anticipated to increase (34). Animal sentinels can document changes in host ranges, host responses to pathogens, and the relationships between hosts (35). There is a long history of wildlife serving as bio-sentinels for the effects and distribution of environmental pollutants and pathogens (36,37). Examples of wildlife serving as climate change sentinels are (i) the spread of Lyme disease vectors by birds that have begun to shift their migratory timing and locations (38); (ii) warmer Arctic summer temperatures shortening the lifecycle of a muskoxen lung nematode larvae (39); and (iii) heat stress negatively influencing rates of mortality from parasitic and infectious disease in moose in the northern United States (40).

There are no standard, evidence-based guidelines for the use of animal sentinel data in general (35) nor specifically for climate change. This reflects: (i) the challenge of deciding the proportional contribution of climate change to health outcomes, as many of the outcomes of concern are influenced by other macro and micro-factors; (ii) the dearth of assessments of the utility of climate change sentinels or early warning surveillance; and (iii) insufficient time for surveillance systems to have been operating to determine if wildlife health changes precede human health effects of climate change (34).

Hazen et al. (41) proposed two types of sentinels: elucidating sentinels that indicate past or ongoing changes that are otherwise unobserved, and leading sentinels that foreshadow future change. Elucidating sentinels provide an observable link between physical processes and biological responses and can be used to monitor processes that are difficult to observe directly. Leading sentinels have a lower threshold for responding to changes than other species or system components, or they are exposed to changes earlier than others. The attributes of the ideal sentinel varies with the objective of a sentinel surveillance system, but in general, a good sentinel species; (i) is sufficiently abundant or resistant to the impacts of sample collection, and surveillance activities do not have significant impacts on the population; (ii) are easy to identify and find; (iii) are in the same exposure pathways of the populations of interest; (iv) have measurable responses to the changes of interest, and (v) have a large body of information on the species pathophysiology and ecology (42).

SUMMARY

Animal health managers need to be prepared for surprise. The grand health challenges of the Anthropocene are made up of multiple, simultaneous assets, deficits, and problems that interact to pull us closer to or further from critical tipping points. Unconnected approaches to managing the health of people, animals, or environments are failing to meet today's complex health challenges and are proving to be unsustainable (43). Responses that react to these challenges after harms occur and in disciplinary silos have been unable to get ahead of the unprecedented social and environmental changes of the 21st century. Transformative changes are needed to accelerate and amplify innovations to protect health in the face of climate change and the surprises it will cause. A transformative climate change agenda will need more than better disease control. It will also need mechanisms for building a system of interdependent and mutually supportive actions to promote health and resilience of people, animals, plants, and ecosystems today and in the future. Without transformation to address root causes of health and resilience, we will continue to battle new crises as they emerge.

Adopting a surprise-oriented approach will require concurrent shifts in perspective that support a systems-based approach to protecting assets to be resilient to surprise while building the necessary situational awareness and intelligence through partnerships for early warning. As it will not be possible to be ready for all unexpected events, collaborative, multi-solving solutions that look across a spectrum of potential threats to find root causes of vulnerability and resilience may be an effective strategy for climate change preparedness.

REFERENCES

1. Sagan, S.D., 1993. *The Limits of Safety Organizations, Accidents, and Nuclear Weapons.* Princeton University Press. Cited in Stikeleather, J., & Masys, A. J. (2020). Global Health Security Innovation. In *Global Health Security* (pp. 387–425). Springer, Cham.
2. Stephen, C., 2020. Rethinking pandemic preparedness in the Anthropocene. *Healthcare Management Forum, 33*(4), pp. 153–157.
3. Kareiva, P. and Fuller, E., 2016. Beyond resilience: How to better prepare for the profound disruption of the Anthropocene. *Global Policy, 7*, pp. 107–118.
4. Pershing, A.J., Record, N.R., Franklin, B.S., Kennedy, B.T., McClenachan, L., Mills, K.E., Scott, J.D., Thomas, A.C. and Wolff, N.H., 2019. Challenges to natural and human communities from surprising ocean temperatures. *Proceedings of the National Academy of Sciences, 116*(37), pp. 18378–18383.
5. Gross, M., 2019. The paradox of the unexpected: Normal surprises and living with nonknowledge. *Environment: Science and Policy for Sustainable Development, 61*(3), pp. 20–25.
6. Darmstadter, J. and Toman, M.A., 2015. Nonlinearities and surprises in climate change: An introduction and overview: Joel Darmstadter and Michael A. Toman. In *Assessing Surprises and Nonlinearities in Greenhouse Warming*. Routledge, pp. 13–22.
7. Strogatz, S.H., 2018. *Nonlinear Dynamics and Chaos with Student Solutions Manual: With Applications to Physics, Biology, Chemistry, and Engineering.* CRC Press.
8. Stephen, D., Stephen, C., Carmo, L.P. and Berezowski, J., 2020. Complex systems thinking in health. In *Animals, Health, and Society*. CRC Press, pp. 207–222.
9. Phillips, B., Anand, M. and Bauch, C.T., 2020. Spatial early warning signals of social and epidemiological tipping points in a coupled behaviour-disease network. *Scientific Reports, 10*(1), pp. 1–12.
10. Lenton, T.M., Rockström, J., Gaffney, O., Rahmstorf, S., Richardson, K., Steffen, W. and Schellnhuber, H.J., 2019. Climate tipping points—Too risky to bet against. *Nature, 575*, pp. 592–595
11. Scheffer, M., Carpenter, S.R., Lenton, T.M., Bascompte, J., Brock, W., Dakos, V., Van de Koppel, J., Van de Leemput, I.A., Levin, S.A., Van Nes, E.H. and Pascual, M., 2012. Anticipating critical transitions. *Science, 338*(6105), pp. 344–348.
12. Drake, J.M., Brett, T.S., Chen, S., Epureanu, B.I., Ferrari, M.J., Marty, É., Miller, P.B., O'dea, E.B., O'regan, S.M., Park, A.W. and Rohani, P., 2019. The statistics of epidemic transitions. *PLoS Computational Biology, 15*(5), p. e1006917.
13. Parker, C.F. and Stern, E.K., 2002. Blindsided? September 11 and the origins of strategic surprise. *Political Psychology, 23*(3), pp. 601–630.
14. Betts, R.K., 1980. Surprise despite warning: Why sudden attacks succeed. *Political Science Quarterly, 95*(4): 551–572.
15. Kates, R.W. and Clark, W.C., 1996. Environmental surprise: Expecting the unexpected?. *Environment: Science and Policy for Sustainable Development, 38*(2), pp. 6–34.
16. Cunha, M.P.E., Clegg, S.R. and Kamoche, K., 2006. Surprises in management and organization: Concept, sources and a typology. *British Journal of Management, 17*(4), pp. 317–329.

17. Taylor, R.A.J., Herms, D.A., Cardina, J. and Moore, R.H., 2018. Climate change and pest management: Unanticipated consequences of trophic dislocation. *Agronomy*, 8(1), p. 7.

18. Stephen, C., Berezowski, J. and Misra, V., 2015. Surprise is a neglected aspect of emerging infectious disease. *Ecohealth*, 12(2), 208–211.

19. Wolfe, N.D., Dunavan, C.P. and Diamond, J., 2007. Origins of major human infectious diseases. *Nature*, 447(7142), pp. 279–283.

20. Jones, K.E., Patel, N.G., Levy, M.A., Storeygard, A., Balk, D., Gittleman, J.L. and Daszak, P., 2008. Global trends in emerging infectious diseases. *Nature*, 451(7181), pp. 990–993.

21. Heffernan, C., 2018. Climate change and multiple emerging infectious diseases. *Veterinary Journal* (London, England: 1997), 234, p. 43.

22. Hoberg, E.P. and Brooks, D.R., 2015. Evolution in action: Climate change, biodiversity dynamics and emerging infectious disease. *Philosophical Transactions of the Royal Society B: Biological Sciences*, 370(1665), pp. 20130553.

23. Rummukainen, M., 2012. Changes in climate and weather extremes in the 21st century. *Wiley Interdisciplinary Reviews: Climate Change*, 3(2), pp. 115–129.

24. Dilling, L., Morss, R. and Wilhelmi, O., 2017. Learning to expect surprise: Hurricanes Harvey, Irma, Maria, and beyond. *Journal of Extreme Events*, 4(03), pp. 1771001.

25. Winder, M. and Schindler, D.E., 2004. Climate change uncouples trophic interactions in an aquatic ecosystem. *Ecology*, 85(8), pp. 2100–2106.

26. Parmesan, C. and Hanley, M.E., 2015. Plants and climate change: Complexities and surprises. *Annals of Botany*, 116(6), pp. 849–864.

27. Cross, P.C., Maichak, E.J., Rogerson, J.D., Irvine, K.M., Jones, J.D., Heisey, D.M., Edwards, W.H. and Scurlock, B.M., 2015. Estimating the phenology of elk brucellosis transmission with hierarchical models of cause-specific and baseline hazards. *The Journal of Wildlife Management*, 79(5), pp. 739–748.

28. Buss, N., Swierk, L. and Hua, J., 2020. Phenological shifts in amphibian breeding influences offspring size and response to a common wetland contaminant. *Biological Sciences Student Scholarship*. 4. https://orb.binghamton.edu/bio_students/4. Last accessed Feb 2, 2022

29. Williams, J.W. and Jackson, S.T., 2007. Novel climates, no-analog communities, and ecological surprises. *Frontiers in Ecology and the Environment*, 5(9), pp. 475–482.

30. Peer, A.C. and Miller, T.J., 2014. Climate change, migration phenology, and fisheries management interact with unanticipated consequences. *North American Journal of Fisheries Management*, 34(1), pp. 94–110.

31. Whitmee, S., Haines, A., Beyrer, C., Boltz, F., Capon, A.G., de Souza Dias, B.F., Ezeh, A., Frumkin, H., Gong, P., Head, P. and Horton, R., 2015. Safeguarding human health in the Anthropocene epoch: Report of The Rockefeller Foundation–Lancet Commission on planetary health. *The Lancet*, 386(10007), pp. 1973–2028.

32. Stephen, C. and Soos, C., 2021. The implications of climate change for Veterinary Services. *Revue scientifique et technique* (International Office of Epizootics), 40(2), pp. 421–430.

33. Yamin, F., Rahman, A. and Huq, S., 2005. Vulnerability, adaptation and climate disasters: A conceptual overview. *IDS Bulletin*, 36(4), pp. 1–14. 10.1111/j.1759-5436.2005.tb00231.x

34. Stephen, C. and Duncan, C., 2017. Can wildlife surveillance contribute to public health preparedness for climate change? A Canadian perspective. *Climatic Change*, 141(2), pp. 259–271.

35. Halliday, J.E., Meredith, A.L., Knobel, D.L., Shaw, D.J., Bronsvoort, B.M.D.C. and Cleaveland, S., 2007. A framework for evaluating animals as sentinels for infectious disease surveillance. *Journal of the Royal Society Interface*, *4*(16), pp. 973–984.

36. Reif, J.S., 2011. Animal sentinels for environmental and public health. *Public Health Reports*, *126*(1 suppl), pp. 50–57.

37. Kuiken, T., Leighton, F.A., Fouchier, R.A., LeDuc, J.W., Peiris, J.S.M., Schudel, A., Stöhr, K. and Osterhaus, A.D.M.E., 2005. Pathogen surveillance in animals. *Science*, 309(5741), pp. 1680–1681.

38. Ogden, N.H., Lindsay, L.R., Hanincová, K., Barker, I.K., Bigras-Poulin, M., Charron, D.F., Heagy, A., Francis, C.M., O'Callaghan, C.J., Schwartz, I. and Thompson, R.A., 2008. Role of migratory birds in introduction and range expansion of *Ixodes scapularis* ticks and of *Borrelia burgdorferi* and *Anaplasma phagocytophilum* in Canada. *Applied and Environmental Microbiology*, *74*(6), pp. 1780–1790.

39. Kutz, S.J., Hoberg, E.P., Polley, L. and Jenkins, E.J., 2005. Global warming is changing the dynamics of Arctic host–parasite systems. *Proceedings of the Royal Society B: Biological Sciences*, 272(1581), pp. 2571–2576.

40. Murray, D.L., Cox, E.W., Ballard, W.B., Whitlaw, H.A., Lenarz, M.S., Custer, T.W., Barnett, T. and Fuller, T.K., 2006. Pathogens, nutritional deficiency, and climate influences on a declining moose population. *Wildlife Monographs*, *166*(1), pp. 1–30.

41. Hazen, E.L., Abrahms, B., Brodie, S., Carroll, G., Jacox, M.G., Savoca, M.S., Scales, K.L., Sydeman, W.J. and Bograd, S.J., 2019. Marine top predators as climate and ecosystem sentinels. *Frontiers in Ecology and the Environment*, *17*(10), pp. 565–574.

42. Beeby, A., 2001. What do sentinels stand for?. *Environmental Pollution*, *112*(2), pp. 285–298.

43. Kock, R., Queenan, K., Garnier, J., Nielsen, L. R., Buttigieg, S., De Meneghi, D., Holmberg, M., Zinsstag, J., Rüegg, S., and Häsler, B., 2018. Health solutions: Theoretical foundations of the shift from sectoral to integrated systems. In Rüegg, S.R., Häsler, B., Zinsstag, J., *Integrated Approaches to Health: A Handbook for the Evaluation of One Health*. Wageningen Academic Publishers, pp. 22–37.

13 Climate Change and Animal Health: The Role of Surveillance Systems

Katie Steneroden, Colleen Duncan, and Craig Stephen

CONTENTS

KEY LEARNING OBJECTIVES

- Understand that surveillance is an action-oriented tool for the timely collection of information on changing health and disease risks.
- Differentiate between current traditional single-agent surveillance systems and "climate ready" animal health surveillance systems that are adaptable to unexpected events and future surprises.
- Recognize that the assessment of animal vulnerability and resiliency are critical components of future surveillance systems.

DOI: 10.1201/9781003149774-13

> **IMPLICATIONS FOR ACTION**
> - Animal health surveillance systems must expand beyond disease to include animal health and climate factors.
> - Animal health surveillance must focus on vulnerable people, places, and animals.
> - Robust and informative systems for climate change adaptation will result from cross-sectoral collaborations between animal and human health teams at local and international levels.

INTRODUCTION

Surveillance, in the broadest sense, means paying close watch to something. Surveillance has been a critical element of animal health systems for more than 10,000 years, when humans first domesticated livestock (1). Human life and health depended on it. Early livestock keepers kept a close watch over their animals' social and natural environment, using indigenous practices, handed down over generations, to address health and disease problems (2). Animal surveillance systems have been evolving ever since that time. Calls for "climate change ready" animal health surveillance are growing (ex. (3–6) see also Chapter 4, 7 and 11). Climate change poses unique challenges for current animal surveillance systems. Despite the recognized utility for these networks to monitor and respond to climate-associated health threats, current systems must evolve to meet present and future needs.

SURVEILLANCE OVERVIEW

Surveillance systems are the primary source of timely information about changing health and disease risk used for health decision making (7). More detailed definitions are as numerous as the epidemiologists who study it, but generally, surveillance is seen as the systematic, continuous collection and evaluation of pertinent data (such as disease, mortality, or test information) that is promptly assessed and shared with those who need to know to launch an effective response. Animal health surveillance should be designed to improve health by providing timely and reliable signals that inform decisions and actions to protect, manage, and maintain health.

The goals of animal surveillance follow from, and are in step with, the goals to support and conserve animal health and welfare, as well as public health. Further goals are to inform risk analysis, management practices, and biosecurity measures and provide data for trading partners. (8) In practice, these goals have most often translated into demonstrating the presence, absence, emergence, impacts, and distribution of animal diseases and risk factors, including infectious, zoonotic, toxicologic, neoplastic, and production-limiting diseases. Surveillance activities range from small and local (e.g., farmers monitoring herd somatic cell counts on their farms) to international (e.g., the Global Early Warning System (GLEWS) for health threats emerging at the human–animal–ecosystems interface). The capacities and resources needed for surveillance must be scaled to the purpose of the surveillance system and its scope of observation.

Surveillance systems are purpose oriented. Their essential purpose is to create information and knowledge for health decision making, leading to action. The specific purposes vary with the situation of concern. For example, a system intended to establish the absence of a disease in a wildlife population to support claims of disease freedom will use different sampling protocols and case definitions than will a surveillance program intended to find the first incursion of a pathogen into a marine ecosystem or one to detect environmental thresholds of excess heat harms to dairy cattle.

Surveillance systems designers need to know the objectives and rationale of the system, who will use the data, and how they will use it (9). The answers to these questions help set the scope of the system, informing issues such as who should participate and which methods are most appropriate. Designers need to be aware of the epidemiological situation, what decisions need information and understanding, and what data needs to be collected, analyzed, assessed, and communicated to help make those decisions. Explicit objectives provide a mechanism for designing and evaluating a surveillance system. Interested readers can find a wide variety of further information on the principles and practice of animal health surveillance elsewhere (ex (7,9–11)).

CONSIDERATIONS FOR THE DESIGN AND IMPLEMENTATION OF "CLIMATE CHANGE READY" ANIMAL HEALTH SURVEILLANCE SYSTEMS

Given the complexity of direct and indirect animal health impacts associated with climate change, surveillance systems must evolve to meet current needs and be ready for future demands. This evolution will ensure the ongoing provision of reliable data to inform appropriate climate adaptation and mitigation actions. Climate-associated animal impacts occur through multiple pathways, affecting all body systems, and exposure outcomes are influenced by both physical and social environments (refer to Chapters 4, 5 and 8). The multiple, interacting downstream health effects of climate change complicate the selection of specific indicators and thresholds surveillance systems will use to signal action (12). Therefore, it is not possible to prescribe an all-embracing surveillance system that tracks climate change impacts on all aspects of health for all species.

Some surveillance programs will need to be purpose-built to detect, assess, and report on specific direct effects of climate change for specific species and places. All other surveillance systems will need to consider how climate change impacts their capacity to collect surveillance data and to turn it into meaningful action information. Climactic drivers have not historically been accounted for when planning these systems. Table 13.1 describes some of the challenges to anticipate when designing and implementing climate-ready animal health surveillance systems and potential solutions.

TABLE 13.1

Animal Surveillance Challenges, Their Link to Climate Factors, and Potential Solutions. This Table Was Informed by (6,13–15)

Challenging issues	Climate change aspects	Solutions
Understanding climate-health relationships requires data from multiple sources, often collected on different data platforms, with different data requirements, and on different spatial and temporal scales.	Growing expectations for surveillance to generate data to detect and confirm climate effects on animal health will suffer until the capacity for integrated data collection and analysis is furthered.	Cross and intersectoral collaborations to develop data standards need to include animal health needs and data considerations. Animal health sectors need to understand the sources and nature of data taken from other sectors and used for surveillance analysis.
Important animal diseases are usually clustered by geographical locations either due to animal population density or production systems.	Geographical and production locations and animal spatial distributions are influenced by specific climate conditions and are sensitive to changes in these factors.	Adjustment for the influence of climate on the distribution of populations and hazards/risks must be considered in sampling design and analysis of the collected data.
Dynamic changes in animal populations, particularly among food-producing and free-ranging animals, can change disease patterns.	Seasonal variations or shift in climatic patterns influence animal movement by time and location.	Dynamically reallocate surveillance capacities as climate change alters the locations and incidence of known problems.
Trade and animal movement impact incidence and prevalence data from surveillance systems.	Climate changes and their frequency are closely linked to animal movement and food trade.	Include weather data (and trends) when making inferences from surveillance data.
Dependency on livestock owners for the collection of animal health and risk factor data.	Inconsistency or bias in climate data can result when variations in climate observations occur.	Validate measurements of reliable weather and climate observations when assessing data obtained from owner-reported surveillance systems.
The epidemiological unit for an animal health surveillance system is usually a cluster or group of animals.	Usually, an epidemiological unit shares the same climate conditions. Management or human intervention may alter the natural conditions of sub-populations, such as artificial cooling environments, for a portion of the population.	Match the scales and places for climate data collection to the scales and places for animal health data when analysing the link between climate factors and animal health status.
Animal health surveillance systems are usually focused on	The impact of climate change on overall social wellbeing in	Surveillance data to assess climate change impacts on

TABLE 13.1 (Continued)
Animal Surveillance Challenges, Their Link to Climate Factors, and Potential Solutions. This Table Was Informed by (6,13–15)

Challenging issues	Climate change aspects	Solutions
specific diseases or groups of diseases with economic importance.	addition to ecosystem functions cannot be ignored.	animal health needs to also include data on changing social and ecological impacts.
Animal health surveillance systems usually focus on infectious diseases or production diseases with limited inclusion of other adverse health events.	The impact of climate change on the overall health status of animals cannot be ignored, including animal welfare issues that are influenced by climate change.	Broaden the definition of adverse animal health status to include animal welfare issues and positive and negative environmental impacts of raising animals.
Animal health surveillance systems must be ready for surprises and unanticipated health impacts.	Climate change will affect epidemiological relationships, including exposure pathways and species sensitivities, in unexpected ways due to the complex dynamics of climate-animal-ecosystem-society-systems.	Surveillance systems focused on determinants of risk and vulnerability, rather than exclusively on etiologic agents and adverse health impacts, will be better suited to "surprise surveillance."

SURVEILLANCE IN A RAPIDLY CHANGING WORLD

ADAPTIVE SURVEILLANCE

The inherent uncertainties, one-to-many relationships between climate change and health impacts, and long-time frames between climate adaptation or mitigation actions and changed health risk means that surveillance systems will need to be learning systems. Surveillance programs will need to develop the capacity to adaptively respond to surges in known problems in new amounts, new species, or new places and respond to unexpected problems. In return, surveillance systems will be expected to provide animal health managers with ongoing monitoring and evaluation data to support their programs' adaptive management.

There will always be uncertainty and unpredictability in the prevention and management of climate change-related health impacts, so learning should be incorporated into the operation of existing surveillance systems whenever possible (16). This means surveillance programs will need the internal capacity to characterize the full extent of a problem(s) they are tracking, anticipate the possible consequences of the choices they make on surveillance targets, analyze approaches and thresholds for action, evaluate impacts of changes in the surveillance system, and incorporate lessons learned into future operational decisions (17). Learning through management can be achieved by (i) trying different surveillance approaches concurrently, evaluating one against another; (ii) trying approaches in a step-wise fashion where one surveillance system is tried and if it fails, a different system is launched and monitored, or (iii) the same system is modified as monitoring and evaluation of its performance are assessed while surveillance is underway (16).

SURVEILLANCE FOR SURPRISE

We cannot be sure what the future will look like, but we can be sure that it will look different from today. Surveillance systems operating under the forecasted climate change scenarios will be challenged to deal with new normals, "foreseeable unexpected events," and surprises. Diseases and disorders with which we are familiar will change in impact, distribution, and frequency (i.e., new normals). Some unexpected events will be foreseeable in that our experiences and research will reveal the possibility of new health impacts of climate change, but they will be unexpected because we can rarely predict their timing, location, and impacts with accuracy. There will also be true surprises because there is presently not enough predictive power to accurately forecast health outcomes resulting from the interaction of climate change with other environmental and social changes anticipated in the 21st century (18). The need for preparing for surprise has been previously acknowledged with respect to emerging diseases (19). Chapter 12 explores ways we can prepare for climate surprises.

FROM SURVEILLANCE TO INTELLIGENCE

Health intelligence is the process of creating knowledge products resulting from collecting and analyzing data, experience, and other learning to make them understandable and usable for future decision making (15). Knowledge products result from putting surveillance analyses into context by connecting surveillance outputs with other forms of information to create meaning and answer the question "why does this matter"? The goal is to protect and promote health by early actions in advance of harm. This approach contrasts with the usual animal surveillance goal of detecting harms early to quickly minimize their effects. Given time lags between climate change, adaptation and mitigation actions, and health impacts, only waiting for adverse impacts is a "too late" strategy for climate change surveillance.

The health intelligence process begins by observing, assessing, synthesizing, and communicating existing data and information using a network for accessing data. The resulting information outputs and signals are assessed to support strong messages that inform policy making to protect population health (20). The goal is to support strategic actions that avoid and mitigate animal health threats. Two purposes support this goal; the first is to transparently and collectively provide agile and ongoing information to identify targets, priorities, and purposes for early action. The second is to identify actions adapted to local conditions to address strategic priorities. Health intelligence for climate change would need to track changes in threats and adverse outcomes and track changes in vulnerability and the resilience of populations of concern.

Vulnerability Surveillance

Vulnerability describes the characteristics and circumstances that make an individual, population, or system more likely or prone to be harmed by a hazard. Vulnerability assessments have been used in various settings (ex. business and environmental monitoring) to better understand capacity-building needs and identify actions to reduce vulnerability. Warnings occur when a change in vulnerability status is revealed, and that change is rapidly communicated to those able to respond (21). Vulnerability is a logical topic of climate change surveillance interest. However, different vulnerability assessment tools can generate different findings under the same circumstances because different tools may use different indicators, thresholds, and decision criteria. Variations in definitions and in the approaches to vulnerability assessment used, plus a lack of evaluation studies, result in the inability to specify a single approach that can be applied in all contexts for all possible threats. There has been little attention to these questions and issues in veterinary surveillance science.

Vulnerability is a combined outcome of being sufficiently exposed to a hazardous agent, being sensitive to its effects, having limited capacity to adapt to a new situation by being able to resist impacts and be resilient to them (22). Vulnerability changes may result from changes in an interacting set of positive and negative factors or circumstances rather than a discrete hazard. Therefore, the targets for vulnerability surveillance can include the presence of harmful substances or situations (ex. pathogens or extreme weather events), situations that can make animals more likely to be exposed to those harms (ex. changes in animal trade or animal movement patterns), or factors that will affect animals' ability to resist or cope with the effects of a harm (ex. changes in animal health resources or the loss of critical animal habitat). Vulnerability surveillance builds on the ideas of risk factor surveillance (surveillance of changes in the determinants of disease rather than disease outcomes or etiologic agents) and risk-based surveillance (using information about the probability of hazard occurrence and the magnitude of the biological and economic consequence of hazards to plan, design, and interpret the results obtained from surveillance systems).

Knowledge gaps and local circumstances will affect the most influential contributors to the vulnerability of a situation, population, or place. This precludes the prescription of a specific, standard set of recommendations on the best variables to monitor for assessing animal health vulnerability and their associated action thresholds. Selection of the most useful variables to be assessed and tracked will be context-specific and require a combination of data, experience, consultation, and judgment.

Climate change health impacts are increasingly defined by the vulnerability of populations. Many of the factors influencing population vulnerability are linked to the determinants of health (ex. access to veterinary care, inhospitable weather, changes in food security – see Chapter 5 for further examples). Maintaining animal health and welfare in the face of climate change is about more than managing animal husbandry and disease control. It is also about ensuring the compatibility of the animal's environment with its evolved needs such that animal can continue to live well in the face of climate change (23). Health, vulnerability, and resilience are all affected by access to the determinants of health (24,25). The determinants of health, therefore, can be a useful concept to consider when proposing sources of information for vulnerability surveillance intended to signal actions in advance of harms.

Building healthy animal populations is a multifaceted undertaking that requires transdisciplinary teams that can foster the social and environmental conditions conducive to keeping animals healthy (26). While the exact composition of transdisciplinary surveillance teams will vary with issues and circumstance, they should include those who can see changes in the variables under surveillance, record, analysis and assess those changes, communicate their significance to decision makers, and implement required responses. Healthy populations that are better equipped to cope with stressors and changes arising from climate change are a major component of climate change adaptation. Surveillance to ensure the protection of the determinants of health and the resulting climate change resilience will require a re-imagination of animal health surveillance to also track elements of animals' social and physical environments.

COLLABORATION

Because no single sector has full control over the various factors that determine animal health, climate-ready animal health surveillance will require multi-sectoral coordination, collaboration, and engagement. While this ideal is actively being promoted in the shadow of the COVID-19 pandemic, there is, as yet, no proven path to realizing an integrated animal health surveillance system. The importance of collaboration in addressing the animal health threats arising from climate change is a recurring theme throughout this book. Specific to surveillance, there are both structural and social collaboration needs. Data standards, sharing protocols, and use agreements are foundational elements of shared surveillance systems. Integration of data from a multitude of animal, environmental, and human health systems, with associated climate information, requires expertise in both data management and computing. Interpersonal collaborations are dependent on capacity for team building, leadership, shared governance, goal setting and more.

STRENGTHENING EXISTING CAPACITIES

Better data and knowledge about health capacity and threats in a local area contribute to better resource allocation to support climate change adaptation actions. Greater surveillance, diagnostic and investigative capacity, and enhanced data sharing can help prioritize harms to manage, improve the burden of illness estimates, and enhance situational awareness (27). The designers of future surveillance systems must be particularly mindful of the disproportionate impacts of climate change in lower income countries (28). Human health is tightly connected to animal health everywhere, but nowhere are the adverse effects felt more strongly and deeply than for those living in poverty. Climate-based animal health surveillance systems of the future must be designed with the world's poorest communities front and center.

Sadly, climate change will disproportionally impact the most vulnerable animals, people, and places (28,29). Developing and validating viable, sustainable, and transportable models for animal health surveillance underpins all attempts to combat climate change threats strategically and equitably. Advocacy for equivalent surveillance capacity worldwide must continue. Pragmatic surveillance operators will need to understand how to adapt what we know works in one setting to different social, political, financial, and ecological settings to deliver and sustain global surveillance capacity (see Chapter 11).

CONCLUSIONS

The essential elements of climate change management for effective adaptation to the future will be unpredictable and emergent rather than predictable and planned (30) because of the unprecedented rate of social and environmental change and the complexity of interactions between co-occurring global threats. Animal health surveillance systems will have the dual tasks of providing information to best prevent and manage existing problems and protect current health assets while also helping anticipate what may come with climate change in time for actions to mitigate, adapt to, or avoid their effects.

Animal health surveillance systems cannot isolate their climate change response from their responses to other threats. Animals are being impacted by multiple anthropogenically-driven global and local crises simultaneously, such as emerging diseases, increasing pollution, and pressures on their food, water, and habitats. Many of these concurrent threats are linked not only in their causes but also in their solutions. Growing experience points to the need for surveillance of the "causes-of-the-causes" shared between climate change, the extinction crisis, habitat degradation, and other mega-threats to create multi-solving solutions that address the root causes of growing pressures on animal health. The need for surveillance innovations has never been greater.

REFERENCES

1. Frantz, L.A.F., Bradley, D.G., Larson, G. and Orlando, L., 2020. Animal domestication in the era of ancient genomics. *Nature Reviews Genetics*, 21(8), pp. 449–460.
2. Wanzala, W., Zessin, K.H., Kyule, N.M., Baumann, M.P.O., Mathias, E. and Hassanali, A., 2005. Ethnoveterinary medicine: A critical review of its evolution,

perception, understanding and the way forward. *Livestock Research for Rural Development*, *17*. p. 119.

3. Stephen, C. and Soos, C., 2021. The implications of climate change for Veterinary Services. *Revue scientifique et technique (International Office of Epizootics)*, *40*(2), pp. 421–430.

4. Halabi, S.F., 2020. Adaptation of animal and human health surveillance systems for vector-borne diseases accompanying climate change. *Journal of Law, Medicine & Ethics*, *48*(4), pp. 694–704.

5. Stephen, C. and Duncan, C., 2017. Can wildlife surveillance contribute to public health preparedness for climate change? A Canadian perspective. *Climatic Change*, *141*(2), pp. 259–271.

6. Heffernan, C., Salman, M. and York, L., 2012. Livestock infectious disease and climate change: A review of selected literature. *Animal Science Reviews, 2013*, pp. 19–44.

7. Berezowski, J.S.C. and Carmo, L.P. 2020. Building health surveillance for decision support at the animal, human, environment nexus. In: Stephen, C., ed., *Animals, Health, and Society: Health Promotion, Harm Reduction, and Health Equity in a One Health World*. CRC Press, pp. 113–134.

8. (OIE) WOfAH, 2021. Chapter 1.4 Animal Health Surveillance. In: *Terrestrial Animal Health Code [Internet]*. Paris: OIE. 29th Ed. Available from: https://www.oie.int/en/what-we-do/standards/codes-and-manuals/terrestrial-code-online-access/?id=169&L=1&htmfile=sommaire.htmer.

9. Salman, M.D., 2008. *Animal Disease Surveillance and Survey Systems: Methods and Applications*. John Wiley & Sons.

10. Thrusfield, M., 2018. *Veterinary Epidemiology*. John Wiley & Sons.

11. Martin, S.W., Meek, A.H. and Willeberg, P., 1987. *Veterinary Epidemiology: Principles and Methods*. Ames IA: Iowa State University Press.

12. Ebi, K.L., Boyer, C., Bowen, K.J., Frumkin, H. and Hess, J., 2018. Monitoring and evaluation indicators for climate change-related health impacts, risks, adaptation, and resilience. *International Journal of Environmental Research and Public Health*, *15*(9). p. 1943.

13. Robertson, C., Nelson, T.A., MacNab, Y.C. and Lawson, A.B., 2010. Review of methods for space-time disease surveillance. *Spatial and Spatio-Temporal Epidemiology*, *1*(2–3), pp. 105–116.

14. Robertson, C., Yee, L., Metelka, J. and Stephen, C., 2016. Spatial data issues in geographical zoonoses research. *Canadian Geographer / Le Géographe canadien*, *60*. pp. 300–319.

15. Stephen, C.B.J., 2022. Wildlife health surveillance and intelligence. Challenges and opportunities. In: Stephen, C., ed., *Wildlfie Popualtion Health*. Springer Nature.

16. Allen, C.R., Fontaine, J.J., Pope, K.L., and Garmestani, A.S., 2011. Adaptive management for a turbulent future. *Journal of Environmental Management*, *92*(5), pp. 1339–1345.

17. Ebi, K., 2011. Climate change and health risks: Assessing and responding to them through 'adaptive management'. *Health Affairs (Millwood)*, *30*(5), pp. 924–930.

18. Whitmee, S., Haines, A., Beyrer, C., Boltz, F., Capon, A.G., de Souza Dias, B.F., et al., 2015. Safeguarding human health in the Anthropocene epoch: Report of The Rockefeller Foundation–*Lancet* Commission on planetary health. *The Lancet*, *386*(10007), pp. 1973–2028.

19. Stephen, C., Berezowski, J. and Misra, V., 2015. Surprise is a neglected aspect of emerging infectious disease. *EcoHealth*, *12*(2), pp. 208–211.

20. Hemmings, J. and Wilkinson, J., 2003. What is a public health observatory? *Journal of Epidemiology and Community Health*, *57*(5), pp. 324–326.

21. Yamin, F., Rahman, A. and Huq, S., 2005. Vulnerability, adaptation and climate disasters: A conceptual overview. *IDS Bulletin*, *36*(4), pp. 1–14.

22. Stephen, C. and Berezowski, J., 2022. Wildlife health surveillance and intelligence. Challenges and opportunities. In: Stephen, C., ed., *Wildlife Population Health*. Springer Nature.

23. Fraser, D., Duncan, I.J., Edwards, S.A., Grandin, T., Gregory, N.G., Guyonnet, V., et al., 2013. General principles for the welfare of animals in production systems: The underlying science and its application. *Veterinary Journal*, *198*(1), pp. 19–27.

24. Karpati, A., Galea, S., Awerbuch, T. and Levins, R., 2002. Variability and vulnerability at the ecological level: Implications for understanding the social determinants of health. *American Journal of Public Health*, *92*(11), pp. 1768–1772.

25. Gowan, M.E., Kirk, R.C. and Sloan, J.A., 2014. Building resiliency: A cross-sectional study examining relationships among health-related quality of life, well-being, and disaster preparedness. *Health Qual Life Outcomes*, *12*, p. 85.

26. de Goede, D.M., Gremmen, B. and Blom-Zandstra, M., 2013. Robust agriculture: Balancing between vulnerability and stability. *NJAS - Wageningen Journal of Life Sciences*, *64–65*, pp. 1–7.

27. Belay, E.D., Kile, J.C., Hall, A.J., Barton-Behravesh, C., Parsons, M.B., Salyer, S.J., et al., 2017. Zoonotic disease programs for enhancing global health security. *Emerging Infectious Diseases*, *23*, pp. S65–S70.

28. IPCC, 2014. Summary for policymakers. In: Field, C.B., Barros, V.R., Dokken, D.J., Mach, K.J., Mastrandrea, M.D., Bilir, T.E., Chatterjee, M., Ebi, K.L., Estrada, Y.O., Genova, R.C., Girma, B., Kissel, E.S., Levy, A.N., MacCracken, S., Mastrandrea, P.R. and White, L.L., eds., *Climate Change 2014: Impacts, Adaptation, and Vulnerability Part A: Global and Sectoral Aspects Contribution of Working Group II to the Fifth Assessment Report of the Intergovernmental Panel on Climate Change. ambridge*, United Kingdom and New York, NY, USA: Cambridge University Press, pp. 1–32.

29. Thomas, K., Hardy, R.D., Lazrus, H., Mendez, M., Orlove, B., Rivera-Collazo, I., et al., 2019. Explaining differential vulnerability to climate change: A social science review. *WIREs Climate Change*, *10*(2), pp. e565.

30. Hanlon, P. and Carlisle, S. 2008. Thesis: Do we face a third revolution in human history? If so, how will public health respond? *Journal of Public Health (Oxf)*, *30*(4), pp. 355–361.

14 Climate Change Leadership: Team Building, Change Agents, Planning, Strategy

Tim K. Takaro

CONTENTS

KEY LEARNING OBJECTIVES

1. Each of the three examples listed here have a wealth of online information and examples of good visuals for teaching.
2. Your community (or classroom) has leaders like those described here who have a story to tell and can connect with students. Entice them into telling this story in the classroom.
3. Teach a knowledge-to-action framework so that your students can see the value of evidence contributing to policy.

DOI: 10.1201/9781003149774-14

IMPLICATIONS FOR ACTION

1. Build an open, trusting community around the goals of your campaign.
2. Early on, establish achievable goals, preferably goals that each individual can see themselves in and want to achieve.
3. Celebrate your victories, and invite others to celebrate with you.

Addressing complex and far-reaching issues like health effects of climate change on animals is a classic "wicked" problem in policy and planning due in part to the dynamic character of the science and physical impacts of climate change, the shifting socio-political landscape, and the lack of precedent and experience with solutions to such a complex problem. Climate change presents problems and threats to life that will need to be solved again and again for generations to come. "Wicked" problems require special skills for leaders, some of which will be addressed in this chapter. Such an extraordinary challenge requires multifaceted, diverse leadership by individuals who can establish a vision, generate the energy to grow and sustain the vision, mobilize collective action around the vision, collaborate with others, and value diversity in the expressions of leadership. For many, becoming this type of leader requires closely examining our own role in society's power structure and learning new skills and theories of change to foster the cooperation needed in this moment.

Effective climate action can feel like a distant dream at times, when policies lag far behind what the climate crisis demands. In countries like Canada that have a strong fossil-energy lobby, we see government ministries associated with resource extraction "captured" by the industries their job is to regulate. (1) Government subsidies to these industries is costly ($18 billion in 2020) (2), and it allows for maintaining the status quo, thereby stifling innovation, particularly in advancing energy technology. Politicians promise more and more ambitious targets for emissions reductions, which rarely, if ever, materialize. More courageous political leadership is needed to envision a future with renewable energy and robust resilience to the ecological changes that are already "locked in" to the climate future. The current lack of clear leadership is why civil society must step up with clear demands for the future planet we need.

The leadership for such civic action is among us now, but it may be unrecognized in the quiet presence of a schoolteacher or nurse, or an Indigenous rights activist or Black Lives Matter organizer, student or union steward, or a mother who decides she is done with being quiet and goes to her first civic action. These are the ranks of the future climate action leadership: people who understand the value of evidence and who value planetary wellness; people who respect themselves and others and can earn the respect of diverse populations, people who want to make a positive difference in their lives and the lives of others; people with integrity and moral courage who will boldly call out injustice and dishonesty when they see it and who support the growth of others; wise people who have a broad understanding of the human condition and can balance the perspectives of multiple stakeholders; and people who show personal humility for the sake of the greater good, and can place protecting the planet before their personal priorities.

The specific goals of effective climate leadership will shift with time as achievements are integrated along with setbacks, producing continual innovation and growth. Overall, the long-term goal is to ensure that future generations of animals and humans inherit a planet upon which they can continue to thrive. Shorter-term goals include emissions reductions and transition to renewable energy, building community-level resilience to the climate futures we can anticipate, along with education and knowledge translation about the relationships between humans and animals; all of these show the interdependence in the web of life and how that web is fragile and threatened by a rapidly changing environment. Climate change is already having a dramatic impact on biodiversity, habitat loss, and species extinctions. These dynamics, coupled with demographic pressures of humans building on animal habitat, in turn impact the spread of zoonotic diseases and spillover events that lead to disease in humans (see Chapters 6 & 7). Understanding these connections could help in preventing the next pandemic and promote healthier animal populations. To achieve these goals, we need to define the change that climate leadership should be seeking. We need to understand what is required for that overall goal of ensuring that future generations have a healthy planet. In the following section, we will explore three recent and very distinct examples of climate leadership in Canada that have come at these questions from very different perspectives. The first, Protect the Planet Stop TMX, is a grassroots citizen organization working to stop the Trans Mountain Expansion pipeline from the Alberta oilsands to Vancouver, BC. The second is the Wet'suwet'en peoples' struggle to protect their territory from a methane gas pipeline in Northern British Columbia (BC), and the third, the David Suzuki Foundation's national Bluedot campaign for municipality-based environmental rights.

PROTECT THE PLANET STOP TMX

This leadership example is more personal and involves another fight at the frontlines of climate change. As a physician-scientist specialized in occupational and environmental health, the impacts of climate change on health have been a major focus of my career. I have published dozens of papers, been involved in studies of climate impacts on waterborne disease, and participated in government reviews and policy making. I've treated asthma, asbestos disease, and all manner of environmentally caused maladies, all in the comfort of the academy or clinic. So, at 63 years of age, it was almost as much of a surprise to me as it was to my colleagues when I found myself literally at the end of my rope in August 2020, high in the trees just downstream from my university in British Columbia, Canada (Figure 14.1).

FIGURE 14.1   Dr. Tim Takaro continued to work from his aboreal "office" on a portaledge 20 meters up in the trees in Burnaby, BC blocking construction of the Trans Mountain pipeline. 3 Aug 20. The movement called Protect the Planet Stop TMX (stoptmx.ca) grew strong, and the blockade in the trees went on until 26 Nov 21. The project budget has ballooned, and it is years behind schedule.

With support of local First Nations and other groups of activists opposed to the pipeline, I organized an ever-growing group called Protect the Planet Stop TMX (PPST) to block the contentious Trans Mountain Pipeline expansion project (TMX) beginning at a strategic spot in my hometown. TMX would enable expansion of the Alberta Tarsands with a seven-fold planned increase in oil tanker traffic through First and Second Narrows in Vancouver. The construction plans called for the trees along the Brunette River conservation area to be felled during a six-week window. We were determined to keep this felling from happening by occupying those trees.

I had done all the usual, comfortable academic things to draw attention to the project that was slated to tunnel underneath my university. As part of the project-review process, I had written two extensive reports about the negative health impacts of the project. I later learned that the review process was fraudulent, and the climate impacts were not allowed to be reviewed. With my public health colleagues, I brought the issue to the Provincial Health Officers Council, which passed a resolution requesting an independent, cumulative, and comprehensive health impacts assessment (HIA) due to the inadequacies of the contractor's HIA. I testified at hearings and visited government ministries. None of these efforts seemed to make a difference. There was no response about mitigating the health impacts I described, and after cases brought by First Nations about their rights and title were refused by the Supreme Court of Canada,

I knew that nonviolent civil disobedience was the only option left for blocking this deadly project.

When I began my protest by tree-sitting, I did not expect to be a leader in a coalition of groups striving to stop the TMX pipeline, but our timing and location in Metropolitan Vancouver and the compelling nature of our arguments caught fire. The COVID-19 response opened the public's eyes to a very different equitable and green future, and a climate emergency was declared in Parliament. It became more and more clear that the Canadian government must stop building large fossil-energy projects and instead invest in sustainable, renewable energy infrastructure. Citizens from all walks of life began to gather in a camp below the tree-sit, and online, to support this work. Colleagues from my university, Simon Fraser University, joined in, along with students, pastors, nurses, teachers, doctors, construction workers, organizers, and lawyers, including Indigenous peoples and settlers. Each brought their gifts: artistry, building, media, music, graphic design, writing, environmental science and fish biology, surveillance electronics wizardry, and tree climbing. Many had the gift of leadership, and our structure was designed to support these leaders with task-oriented affiliation groups that they co-lead. The governance structure had a coordinating committee and a commitment to equity, nonviolence, and love for all living things. This structure enabled continued growth of PPST and blossoming of new leaders and new affinity groups that expanded the range and diversity of our work. Dozens of people have now taken to the trees, and many more gathered to support them with home-cooked meals, press events, and expansion into churches and bird-watching and other community groups. Together, we feel strong and empowered despite the immense powers of government arrayed against us. We continue to delay the project, pushing it further and further into economic and political non-viability. COVID-19 has shown us that we can respond to a health emergency with incredible strength by citizens and government. We are transforming our world at work, in school, and in our economic priorities. We've known that we must transform our energy system for decades, and now is the time to bounce back better from COVID-19 by phasing out fossil-fuel energy sources and fostering new ideas and new leaders.

WET'SUWET'EN

The ongoing expansion of the oil and gas industry in Canada is a threat to planetary health and Indigenous rights. Movements to resist expansion of this industry and transition more rapidly to an energy system based upon renewables are emerging around the world. The efforts of the Wet'suwet'en in Northwestern BC, led by hereditary chiefs of that nation, are a strong example of this movement. Five clans make up the Wet'suwet'en Nation – Gilseyhu (Big Frog), Laksilyu (Small Frog), Gidimt'en (Wolf/Bear), Likhts'amisyu (Fireweed), and Tsayu (Beaver). In the traditional Wet'suwet'en governance system, the hereditary chiefs from each clan are title holders of the land, and each clan has the right to control access over a territory (3). The local politics are complex, with elected chiefs on the Indian Act reserves sometimes in conflict with hereditary leadership due to benefit agreements made by elected councils with extractive industries. As determined in the landmark court case *Delgamuukw vs. BC.*, UBC law professor Gordon Christie notes, "There's really no question in Canadian law that the Wet'suwet'en hereditary chiefs are the ones who represent the houses that have the say over what happens on the territory, not the tiny reserves." (3)

The state and industry have worked together in colonial fashion to exacerbate the divisions between Indigenous peoples. They use the court system to enforce injunctions against Indigenous land defenders who are protecting food and water resources on their traditional territory. Out of this jurisdictional conflict and struggle, strong leadership has emerged. One standout leader, Freda Huson, was recently recognized internationally with the Swedish 2021 Right Livelihood Award (aka, the Alternative Nobel Prize). Huson (aka Chief Howihkat) led construction of the Unist'ot'en Healing Centre near the Wedzin Kwa (Morice) River, a very strategic spot on traditional Wet'suwet'en territory. This trauma-recovery center and land re-occupation project has become a focal point for blockades against the Coastal GasLink pipeline. The pipeline is a key link in the province's efforts to expand fracking in BC and is designed to supply a new LNG terminal under construction on the coast.

Huson was arrested in February 2020 in the blockades that precipitated solidarity protests and rail blockades across Canada, but at 57 years of age, she has been fighting for her people much longer. In 2009, she moved to the forest camp at Uni'sto'ten to protect her peoples' sacred land and point a better way for the planet's future. This is a very critical spot on Wet'suwet'en territory, where the steep topography relaxes slightly in a tight spot that pipelines and utilities also find strategic. Building the healing center was an idea from Huson's niece, Karla Tait, a clinical psychologist and director of clinical services at the center. The presence of the several buildings that now make up the camp has served as a nidus for attracting land defenders from around the world and as a physical reminder that new pipelines, presently the Coastal Gas Link (CGL) pipeline, are not wanted on their territory. "We are stewards of the land. We don't own … [it]; we're entrusted to take care of it, so the land will take care of us. But if we destroy the land, we destroy ourselves" (4). With dignity and ceremony, she and other land defenders continue to fight and slow construction of CGL on their lands. In 2019, Huson's leadership at the barricades along the construction route galvanized railroad blockades and other actions by Indigenous peoples across the nation. With these tactics, and using the frequently shoddy permitting and approval processes followed by CGL, Huson and her followers with the hereditary chiefs on Wet'suwet'en territory have put CGL one to two years behind schedule. Every delay is a small victory since the world is engaged in an inexorable energy transition, and delays of new fossil energy infrastructure make their cancellation more likely every day.

BLUE DOT

The Blue Dot campaign was launched in 2014, promoting the idea that everyone in Canada has the right to a healthy environment. Over 175 municipalities across Canada have passed resolutions in support, and the effort has blossomed into a movement to amend the Canadian Charter of Rights and Freedoms to include the right to a healthy environment, joining 110+ other countries enshrining this right in their constitutions (5). The effort has been backed by the David Suzuki Foundation and Ecojustice to inspire the next generation of environmental leaders. The campaign appears to be working since millions of Canadians have engaged in this effort at education and advocacy for the environment, and many new leaders have emerged, including Rupert and Franny Yakelashek, described below.

The effort was designed to foster leadership beginning at the municipal level. This vision has been accomplished by reaching first for goals that were clearly achievable and appealing to "people like you." Achievable goals are important to empower people to act. In the case of Blue Dot, an easy win was getting over 110,000 people to sign a petition supporting the right to a clean environment. Then, with a little more effort and certainly more leadership, getting a municipality on board. This goal entailed learning about civic processes, getting on council agendas, arguing for the right to a clean environment, and shepherding a resolution through council. With confidence gained at this level, efforts at higher levels of government were launched using similar tools such as the Minister of Parliament pledge for environmental rights. Ultimately, the goal of amending the Charter of Rights and Freedoms was announced, but only after enough smaller goals had been reached. The continued relevance of the work is reinforced with timely, related follow-up campaigns, such as a call to strengthen the Canadian Environmental Protection Act (Bill C-28) in April 2021 and a parliamentary petition delivered to the House of Commons in December 2020 with 8,437 signatures. In this incremental way, the movement has grown and sustained interest in the ultimate goal, which at one time appeared too daunting. But now, with dozens of active citizen groups and 25,000 volunteers pushing the call forward, it seems within reach.

These groups formed at the municipal organizing stage of the campaign and consist of many new leaders. David Suzuki recognized one in a piece he wrote in 2015 (6). His story was about 10-year-old Rupert Yakelashek who held a packed room of politicians, campaigners, and municipal administrators in rapt attention with his passion for the environment. Along with his younger sister Franny, he has organized rallies, written countless letters to local leaders, and spoken at city council meetings. The next day, Rupert's audience, the Union of BC Municipalities, passed a resolution calling on the BC government to enact an environmental bill of rights. Rupert and Franny are still working, and they have inspired many others with their simple, clear message: "We need to protect the rich land that gives us life. Adults should think about kids when they make decisions that will affect future generations."

Young climate leaders like Rupert Yakelashek – or Greta Thunberg on the international stage – are the hope of the planet. They are not the only ones, however. Many bear witness to the habitat loss, extinctions, and loss of bio-diversity at an ever-increasing rate, and they can connect these impacts to the climate emergency and the role of society in controlling emissions and protecting nature with all its creatures. All of us have consumer practices that are not the rapacious ones of our parents' generation, and there is a passionate and righteous urgency to the call for meaningful policy change. All of us can be "influencers." Planet protectors are now vocal in every nation and on the international stage. You, too, are a potential climate leader for the future, which is, of course, now.

CONCLUSION

Though we face a wicked problem in the climate emergency, and we are not all professional organizers, climate scientists, project engineers, lawyers, or com-munication specialists, we all have the capacity to learn from others, unite against a common foe, and be agents for change. Each of us approaches direct action from different starting points. One doesn't have to lock themselves to a gate or sleep in a tree to participate in protecting the planet. The three examples outlined above show how a bite-sized chunk of a bigger, even "wicked," problem can be addressed with the focus on local individuals and local action in the case of the Wet'suwet'en and Protect the Planet Stop TMX; or, in the case of Blue Dot, a well-organized and financed national solution that at its core is also distributed into achievable municipal level actions in support of a bigger vision. Each one feeds off the empowerment of individuals who seek more equity for others and feel the strength in numbers striving for a common goal. They all recognize the intersectoral nature of the problem they confront and manage this problem with integrated planning and the strength of diversity in their orga-nizing. Expertise in climate change mitigation, environmental economics and law, or human health impacts of climate change are certainly desirable but are not required for action and change. The experts who wish to see their evidence used in a good way will support meaningful action and contribute to the cause. It is the observed effectiveness of the action (or the hope of effectiveness) that will attract them and fuel the movement and its evidence base.

The most hopeful aspect of the global movement to move society forward to a sustainable future, one where animals, humans, and the natural environment all thrive, is that the leadership required for this transformation is transferable and infectious. Anyone can be inoculated with the drive to make a difference and learn the skills needed to reach their first attainable goals. Small successes give hope in an often-depressing time. Movements thrive off this hope. The emerging leaders are then further energized to stretch the groups they build and achieve goals that once seemed impossible. The relentless drive to policy solutions, and societal and personal changes that emerge from this work, is how the world will save itself from climate catastrophe in the coming decade.

REFERENCES

1. Taft, K., 2018. *Oils Deep State*. Toronto: James Lorimer & Co.
2. Thurton, D., 2021. Canada spent $18B on financial supports for the fossil fuel industry last year. *CBC News*. Available at: https://www.cbc.ca/news/politics/fossil-fuel-subsidy-canada-1.5987392
3. Seatter, E and Turner, J., 2021. Untangling the 'rule of law' in the Coastal GasLink pipeline standoff. Richocet, 5 Feb 20. Available at: https://ricochet.media/en/2904/untangling-the-rule-of-law-in-the-coastal-gaslink-pipeline-standoff.
4. Morin, B., 2021. Freda Huson: An indigenous 'warrior' for the next generation. *Al Jazeera* 29 Mar 21. Available at: https://www.aljazeera.com/features/2021/3/29/freda-huson-an-indigenous-warrior-for-the-next-generation.
5. David Suzuki Foundation, 2021. Internet website. Available at: https://bluedot.ca/about/ accessed 20 Oct 21.
6. Suzuki, D., 2021. Growing environmental leadership fills me with gratitude. *David Suzuki Foundation*. Available at https://bluedot.ca/stories/david-suzuki-growing-environmental-leadership-fills-me-with-gratitude/

15 Hope for Health in the Anthropocene

*Chris G. Buse, Maxwell J. Smith,
and Diego S. Silva*

CONTENTS

KEY LEARNING OBJECTIVES:

- Public health is increasingly implicated in transdisciplinary research and practice that realizes connections between the health of humans, nonhuman animals, and ecosystems in the Anthropocene. These actions can promote hope and a sense of possibility.
- The recognition of this need for integrative practice can engender a sense of hope and optimism in the future, provided practices promote appropriate intersectional and equity-informed interventions.

DOI: 10.1201/9781003149774-15

- Theories of interspecific justice are increasingly required to think through the moral and practical trade-offs in emphasizing the health of humans over that of other species.

IMPLICATIONS FOR ACTION:

- Hope can be a helpful aspect of promoting meaningful and sustained action when intervening in complex systems.
- Public health actors may be unconsidered or unlikely allies for those working in conservation science. Enhancing collaborations across sectors can promote transdisciplinary engagement that leverages unique insights of multiple fields to bolster sustainability in the broadest sense.
- Interventions seeking to redress environmental harms that influence the health of both humans and non-human animals should highlight solutions that are integrative, attentive to intersectional identities, and attempt to promote interspecific justice.

INTRODUCTION: GROUNDING HOPE IN THE ANTHROPOCENE

"Hoping is out of style, so look happy, it's the end of the world."

– Matthew Good Band

Having hope in the 21st century can be downright difficult. Every day, we are inundated with a steady stream of negative news about the state of the environment and its implications for humans and nonhuman animals alike. From climate change being referred to as "code red for humanity" (1) to COVID-19 cases and deaths overwhelming healthcare systems around the world, we are increasingly realizing that many emergent existential threats to multiple species are rooted in humanity's deeply unhealthy relationship with other life and the Earth systems with which we share our planet.

In this chapter, we explore the concept of hope in the context of human and nonhuman animal health in the Anthropocene – a geological epoch which acknowledges that human beings are now the principal driving force behind environmental change (2). Through the deployment of a public health lens that aims to be attentive to health in the broadest sense (i.e., environments, nonhuman animals, humans), this chapter considers the relationship between hope and (in) action, and the types of practical actions that can not only bolster hope, but also attend to the reciprocal relationships between ecosystems, human, and nonhuman animal health. The focus on public health actions in this chapter is intentional, insofar as public health is a transdisciplinary sector that is increasingly wading into the nexus of overlapping human, animal, and ecological health. The natural orientation toward integrative and inclusive research and practice may therefore lend lessons to other sectors seeking to work beyond the narrowly defined boundaries of their own disciplines.

This chapter begins by exploring the relationship between hope and optimism, and presenting cases for hope being both detrimental and helpful to driving action on anthropogenically-driven environmental change. The goal of the chapter is to probe the idea of whether hope does indeed "know no bounds" in the face of potentially catastrophic environmental changes. In doing so, the chapter unpacks key lessons for how integrative, intersectoral, and intersectional public health approaches can clarify connections between all living things and drive sustainable actions that are both health promoting and health protecting for multiple species. This, in turn, raises critical consideration of the need for a robust theory of interspecific justice to account for tradeoffs between interventions that seek to remedy impacts on multiple species – rather than maintaining the status quo of siloed assessment approaches to unique species, an approach that risks privileging some over others without adequate consideration of the implications of doing so.

THE RELATIONSHIP BETWEEN HOPE AND OPTIMISM

"The situation ain't all that new. Optimism's my best defense."

– Rod Stewart

Hope is easily confused with optimism, but the two concepts are distinct and nuanced. Hope refers to a confident feeling that something will unfold in the future. Optimism is the quality of being full of hope, with a strong emphasis on having a positive outlook and emphasizing the good that will potentially unfold in the future. Thus, "hope is distinct from optimism by being an emotion, representing more important but less likely outcomes and by affording less personal control" (3). In short, hope assumes a degree of uncertainty in an outcome, lower control over influencing that outcome, but still holding a wanting or desire for that outcome to be positive. Optimism ascribes a higher likelihood of the desired event to occur, and greater belief in a positive outcome. As Orr (4) puts it:

> optimism is the recognition that the odds are in your favour; hope is the faith that things will work out whatever the odds. Hope is a verb with its sleeves rolled up. Hopeful people are actively engaged in defying or changing the odds. Optimism leans back, puts its feet up, and wears a confident look knowing the deck is stacked.

There are other important nuanced differences between these concepts. Consider a recent study that examined the attitudinal beliefs of both sports fans and voters. When asked about a team's or candidate's prospects for the future, supporters tended to respond in unique ways depending on whether they were feeling hopeful or optimistic about the chances of a specific team or politician, which were also directly related to their past track record of success (5). For teams that had traditionally struggled, what the authors refer to as "bottom-tier football teams," fans were incredibly invested in the hopeful outcome of their team winning, where hope accelerated exponentially with mere possibility of a desirable outcome, whereas the relationship between optimism and likelihood was linear. Similar results were found with voters. This study demonstrates that "hope is distinct from optimism and positive expectation; hope is tapped into when odds are low yet individuals are highly invested in the outcome" (5).

In summary, while conceptually and empirically distinct, hope tends to assume a lower likelihood of a desired outcome, and despite lower degrees of autonomy or control over the outcome, a greater personal investment in the issue. Optimism, by contrast, assumes a higher degree of control over the outcome, and therefore, a higher degree of likelihood in achieving that outcome.

Semantics aside, these distinctions raise important questions for ecosystem, nonhuman animal, and human health. A key challenge of the Anthropocene is that humans believe they have some degree of autonomy over their own health through the lifestyle decisions that we make, and it could be argued there is a universal desire to achieve a good outcome (i.e., good health). Surely this means that many of us would find it easy to be optimistic about our health in the future. Yet, that same recognition of control has proliferated into the paradigm of management and conservation; since we are a driving force on the planet's geophysical processes, we ought also to have control over the health of ecosystems and nonhuman animals.

However, when you consider the factors that are beyond an individual's control – especially given the growth in understanding of the determinants of health and the role that macro-level environmental, social, economic, political, and cultural contexts play in ill-health outcomes – is hope a more relevant concept? We may be able to be optimistic about our ability to change our diet or exercise because of the autonomy we have over the decision, but what about our collective ability to wean ourselves off of fossil fuels or successfully remediate a large landscape that has been host to multiple land uses and extractive industries? Is hope something worth holding, and what role might it play in shifting the contextual conditions by which environments are influencing environmental, nonhuman animal and human health in the 21st century?

THE CASE FOR DESPAIR: FROM LIMITS TO GROWTH TO LIMITS TO HOPE?

"All I hear is doom and gloom."

– The Rolling Stones

As the lead author was contributing to this chapter from his home on the unceded, traditional territory of the Syilx Okanagan Nation in the Okanagan Valley of British Columbia, the province of British Columbia was simultaneously emerging from the hottest summer on record and the worst fire season the province has ever seen. During the early months of the summer of 2021, a "heat dome" descended over much of western North America, leading to 569 heat-attributable deaths (6) – likely a conservative estimate – in British Columbia alone, all under the additional cumulative burden of the global pandemic driven by SARS-CoV-2. To put these numbers into perspective, the number of people who succumbed to extreme heat over the course of only 10 days was a little less than half the number of people who had died from COVID-19 in BC in the preceding 16 months. Among them were our most vulnerable: the elderly, people with pre-existing chronic respiratory and circulatory conditions, and those living alone.

Not only did the extreme heat impact human populations, but it may have killed as many as a billion shellfish in the Salish Sea (7) and impacted numerous others species from fish (8) to birds (9). Elsewhere, catastrophic wildfires have raged across the western United States, Russia and Australia, and European floods in the summer of 2021 were estimated to cost more than $10 billion Euros and result in 242 deaths, with most occurring in Germany (10). Similar events unfolded across large the Asian continent, displacing and negatively impacting hundreds of thousands of people. We are also in the midst of a human-induced mass extinction event (11), oceans are taking on more carbon dioxide and acidifying at alarming rates (12), and the list goes on … It's no wonder that the use of the terms "eco-anxiety," "climate anxiety" and "solastalgia" are increasingly common in our vocabularies (13–15).

The research is abundantly clear: our planet is in trouble, and it's largely our own doing. Moreover, environmental change is precipitating rapid social change, with considerable implications for human health now and into the future (16). Layer on rising rates of income inequality within and between generations, increasing patterns of resource exploitation, growing social unrest, the spread of misinformation that is undermining democratic institutions, and the threats of geopolitical instability, it can be downright impossible for some to be hopeful in the 21st century. To quote novelist and humorist Kurt Vonnegut, "so it goes."

What is perhaps even more depressing is that these troubling trends are not exactly "new" information. Our collective hope has been consistently challenged about the state of the biosphere since at least the 1970s when a group out of MIT wrote the first edition of *The Limits to Growth*. A recent analysis of their modelling efforts suggests the predictions made in the 1970s proved to be startling accurate (17) – that we are on track to exceed the Earth's carrying capacity driven by the exploitation of natural resources and increasing pollution. Whether you brand these trends with the language of "limits to growth" (18), the "great acceleration" (19), or the "Anthropocene" (2), the results are still the same: an increasingly unequal world, teetering on the verge of ecological and corresponding social collapse.

Given the decades of recent research on biosphere integrity, hundreds of years of old thinking from Thomas Malthus on the limits of living within the bounds of a finite world, and millennia-old wisdom of many Indigenous cultures on the need for symbiosis between humans and our environments, we must accept that societal integrity is integrally tied to the health of ecosystems and their ability to provide so-called "ecosystem services," or the things we take from nature, free of cost. These regulating, provisioning, supporting, and cultural services are the foundation of all life on this planet. We continue to ignore them at our own peril.

One source of hope might be the technocratic fantasy that we will simply invent our way out of these problems, to foster a vision that uses the best available technological advances to innovate our way into good health for all. Capture enough carbon, and climate change might cease to be an issue. While technological quick fixes might help to instill hope, we also know that the pace of technological innovation lags behind the scale of our ecological problems: "even if our technology became 5-times more efficient in terms of resource use and pollution reduction, the human impact over [a] child's lifetime would more than quadruple" (20). As a result, most analyses suggest the solutions to the known ecological challenges facing multiple species on our planet are not technical, but rather reflect a failure to harness social, cultural, and political ideologies, economic and legal systems, and ethical commitments to implementing solutions (21). This dire situation requires those with power to engage in rapid actions to support sustainability initiatives and avoid the worst of climate change, something we are projected to have only nine short years to intervene so we can avoid the most deleterious impacts (22).

In terms of human health, even when acknowledging the fairly morose and depressing backdrop, we are told we have a great deal to be optimistic about. Rates of infant mortality are the lowest they have ever been, we are living longer than we ever have before as a species, and medical innovations, including the rapid, albeit inequitable, global deployment of COVID-19 vaccines, are certainly cause for celebration. These markers indicate a certain degree of control over our collective fate, and highlighting these successes is sure to inspire further optimism about our ability to solve complex problems. However, these gains have come at significant, and in some cases, irreversible expense to the biosphere – the foundation for all of life's ability to flourish (23). It's these changes that are now beginning to negatively impact not only our health, but also the fate of our species and many others on our shared planet.

This fact raises another particularly stark challenge to hope: that the dominant societal discourse frames many of these environmental challenges as threats to the human species, with little consideration for impacts to ecosystems and nonhuman animals.

Globally, 23% of all deaths and 26% of deaths among children aged five years and younger are due to modifiable environmental exposures, pressures, or changes (24), and are projected to increase in the future. This reality is a scathing indictment of the glaring omission of environmental determinants in the WHO's 2008 framework on the social determinants of health, let alone how those very social determinants are often a product of, or may be modified by, environmental conditions. Despite widespread acknowledgement that the social, cultural, and even political contexts in which we live, work, and play can influence or modify our health status, the environment has, until recently, been relegated to an entirely separate system, one which is often acknowledged as a threat or something to be protected from. This idea is clearly evidenced by the fact that many public health agencies around the world have environmental "health protection" offices responsible for dutifully monitoring air and water quality, or conducting restaurant inspections, often with little consideration for the broader biophysical processes that sustain and allow life to flourish in the first place (25).

When considering animal health, the picture may be even bleaker. Since 1970, it is estimated that bird, amphibian, mammal, fish, and reptile populations declined by 68% on average, and that biodiversity is being lost at an alarming rate (26). This decline has arguably precipitated shifts in conversations from broad, system-wide conservation goals to the need to grapple with so-called "triage conservation." Here, *triage* refers to the human decision-making process of deciding whether to allocate resources to conservation efforts for specific species based on cost-benefit accounting. This process can, in turn, legitimize extinction by defunding conservation programs that may run counter to dominant economic interests (27). It raises considerable ethical and moral considerations in terms of balancing trade-offs in impacts within and between species (28).

Put simply, this disconnect epitomizes a systemic problem whereby humans imagine ourselves to be greater than, rather than as part of, nature. Chandler (29) argues that hope is challenged in the Anthropocene because of an inability for such "re-enchantment" with nature to be possible. The pessimistic take would therefore suggest that hope is actually just a form of collective denial, a form of grieving over the loss of a hopeful future in the Anthropocene (30), and that the limits to hope are becoming apparent, particularly if hope is merely an escape from the reality of our current situation. This idea should be terrifying: that our collective denial is so widespread and all-encompassing that it usurps our ability to act. So, has hope "gone out of style"? If so, what then can foster and sustain a feeling of hope that is rooted in action when we have so inexplicably stacked the planetary odds against ourselves?

THE CASE FOR HOPE: TOWARD INTERSECTORAL, INTEGRATIVE, AND INTERSECTIONAL ENVIRONMENTAL HEALTH PRACTICE

"Hey world, you know you got to put up a fight."

– Michael Franti

It's easy to make a case for grief and despair, to attribute ecological disaster as a problem beyond our control and willfully turn a blind eye to addressing the problem. It's also easy to identify a litany of problems or critically dissect a proposed intervention to flag all the drawbacks of attempting to solve them, which academic literature on the nature of our ecological challenges does all too well. We are long on problems, but short on solutions.

For hope to be compelling, it must be more than naive optimism (31); it requires action, even in the face of long odds. In other words, we need to move beyond "tough talk" and toward implementing "tough solutions" (32). For example, at the climate meetings in Paris 2018, world leaders set an aspirational target of limiting warming to 1.5 degrees Celsius relative to pre-industrial averages. Yet, according to Climate Tracker (see: https://climateactiontracker. org/countries/), only one country who signed the Paris Agreement is currently implementing strategies and policies compatible with emissions reductions, seven are described as "almost sufficient," and 29 as insufficient, highly insufficient, or critically insufficient. Tough talk, indeed. Moreover, even if it were possible to keep warming to 1.5 degrees, the likelihood for significant disruptions and risks to human health due to shifting precipitation and temperature remain highly likely (33). While perhaps not the point, the commitment to try and limit to 1.5 degrees signaled hope, but that kind of hope is meaningless if it is not backed up with appropriate actions to achieve the goal.

So what actions are necessary to protect and promote human, nonhuman animal, and ecosystem health? In this section, we argue that public health is well-situated to take the very types of actions, in collaboration with other actors, to lead on these issues through integrative, intersectoral, and intersectional actions. To engender a sense of hope, we highlight recent successes on these fronts and demonstrate the potential that robust and concrete action can make in not only tackling the health impacts of environmental change, but also using this time as a foundational opportunity to promote social and gender equity, build a sustainable development agenda, and strengthen our institutions of governance so that they are accountable to future generations of human and nonhuman animals.

Conceptual Innovations in Environmental Public Health: Meeting the Challenge of the "Integration Imperative"

In recent years, concerted attention has been given to what are referred to as the ecological determinants of health (20,34). Now, a proliferation of ideas, frameworks, concepts, and fields of practice can be mapped; these consider the interactions between health, ecosystems, and society more generally, including environmental public health, environmental health justice, ecohealth, One Health, political ecology of health, and planetary health, among others (see Figure 15.1) (35).

It is perhaps of less importance to unpack the differences between these approaches than to identify their overlaps and similarities. Common to these fields are concerns with the equitable distribution of resources and need to redress existing inequalities in health status, attentiveness to systems science, and the interactions between multiple scales of analysis and multiple species, and the need to move knowledge to action (16,36). Each of these approaches operates somewhere on the spectrum of control over nature (i.e., the "traditional" environmental health protection approach) to embracing recommendations from the Ottawa Charter for Health Promotion in promoting healthy and just environments. In summary, these analytic tools, fields, or paradigms signal that we have conceptual tools in our toolkit that can encourage the integration of not only

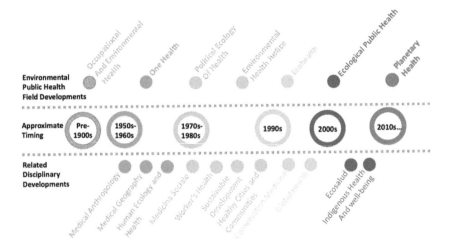

FIGURE 15.1 The emergence of environmental public health fields linking the health of people, ecosystems, and environments.

environmental conditions into public health practice, but also the multi-scalar and systemic implications of changes to those conditions on society and ecological health more generally.

This identified need for "integration" – to bring together the imperatives of multiple fields of practice and focus on the promotion of multiple values, health among them – is a significant shift in public health that has been reiterated at least since the 1978 Alma Ata Declaration. With the rise of so-called "healthy public policy," there is growing recognition from within the public health sphere and beyond that the policy decisions we make today have both direct and indirect impacts to not only human health, but also the health of environments (37). Integrating health in decision making is therefore a considerable benefit for decision makers and the public and a cornerstone of modern public health practice.

The same is true when it comes to making inroads into complex environmental health challenges. We are required to collaborate with a host of likely and unlikely allies and have more than 40 years of experience merging evidence bases, developing common languages, integrating multiple worldviews, and developing solutions. It's no wonder that the "fields" referenced above implicitly and explicitly recognize the value of this broader collaborative approach and the need to integrate new data, values, and perspectives into research, policy, and practice.

Some of the most recent thinking on this topic is what Gillingham et al. (38) refer to as the "integration imperative," which recognizes that the overlapping and cumulative impacts of multiple land uses, climate change, and other forms of ecological degradation have ecological, community (i.e., socioeconomic and sociocultural), and health implications. Merging the imperatives of different tools, such as environmental, social, and health impact assessment, and recognizing the need for not only healthy public policy but also "environment in all policies" is therefore essential to continued progress. The integration imperative is not only about merging diverse land-use values into research and practice, however, but also the imperatives of sectors, research methodologies and disciplines, and our senses (i.e., head, heart and hands) to better understand the "whole" of complex systems (16).

One example of an area where intersectoral action for public health and the environment is perhaps most apparent is the rise of the nature connectedness movement sweeping across the globe. Increasingly, public health institutions are advocating for enhancing greenspace to provide multiple co-benefits, including: reduction of urban-heat islands to mitigate extreme temperature exposure; increasing permeable surfaces to reduce flood risk; improving local air quality and carbon sequestration, especially for traffic-related air pollution; simultaneously, safe places for people to be active, recreate, and connect with nature are provided (39). These actions therefore not only seek to provide ecological and health benefits, but create benefits for other nonhuman animals by providing or restoring critical habitats. In short, problems with multiple domains of impact precipitate a need for interventions that highlight co-benefits across multiple systems.

As Salas and Keyhoe (40) argue, one of the best ways for the health sciences is to generate what they term "evidence-based hope." And the case for improving connectedness to nature is well-reflected in the evidence (41,42), with obvious benefits for not only improving human health and well-being, but also creating regenerative opportunities for the health of nonhuman animals and ecosystems more broadly. This idea has led to a social prescribing trend where doctors are increasingly prescribing connection with nature to improve physical, mental, and social health (43,44). What the overall trajectory of this thinking suggests runs counter to Chandler (29), above; that perhaps the remedy for re-enchantment with nature is a fundamental cognitive switch in which humans find a sense of purpose or meaning in being active stewards of the planet, rather than merely the caretakers of its decline (21,45,46).

PUBLIC HEALTH BREAD AND BUTTER WITH A SIDE OF JAM: (INTEGRATIVE) MONITORING AND SURVEILLANCE

Building on the above logic, one area that public health can have a measurable impact on combatting environmental change is by developing innovative (and integrative) approaches to monitoring and surveillance through transdisciplinary partnerships. Monitoring and surveillance have been key to building research evidence in the public health sciences since John Snow's seminal study of the cholera epidemic in England. Today, basic statistics have been complemented with novel dashboards to track the emergence and mutation of emergent infectious diseases (including COVID-19), satellite measurements are regularly used to remotely assess air pollutants, geospatial modeling tools can use past information to project and visualize changes to landscapes, and real-time surveillance systems are being implemented to follow hospital admissions in times of crisis (e.g., extreme heat events) to better plan for surge-capacity constraints.

Just as wind and solar energies become cheaper and more efficient, so too have our monitoring and surveillance tools, enhancing the power of our predictive capacity and enabling us to merge big datasets that were previously thought to be beyond the purview of our science.

One example of a tool that is capable of integrating a variety of data sources to compute how individual indicators – when combined – can become more than the sum of their parts is the U.S. EPA's Environmental Justice Screening Tool. The tool has been refined several times at smaller levels of analysis, leading to the creation of the CalEnviroScreen (47), the Washington Environmental Health Disparities Map (48), the Maryland Cumulative Effects Map (49), and the Chicago Cumulative Effects map (50), with several pilots being led in the Canadian context. The approach combines indicators of environmental exposures, environmental effects, socioeconomic marginalization, and sensitive populations (including health outcomes) to create meta-indices of "pollution" or "landscape" burden, socioeconomic and health marginalization, and an overall cumulative-impacts score that represents the confluence of all indicators included in the model.

Not only can these tools be ground-truthed with communities to ensure their accuracy and contextual nuance, but they can also be used to screen where issues of environmental health injustice occur and point our attention to which areas may need more or less support in adapting to environmental and corresponding social change. The fact that equity is an underlying theme that animates the tool means it is attentive to the inequitable distribution of both environmental and social harms in their production of human health outcomes (51). Depending on the context, the tools can be adapted based on local needs and priorities to measure and report on other issues of relevance, such as nonhuman animal health, to better draw out connections on how certain ecological impacts play out for multiple species.

And yet, a key drawback of surveillance and monitoring is still reflected in the EnviroScreen tool – that often in public health, we are focused on tracking decline through a focus on detriments rather than assets. Another example that is national in scale incorporates both assets and detriments-based measurement – albeit with more of an anthropocentric focus – is the Lancet Countdown (see: https://www.lancetcountdown.org/), a data-explorer initiative that tracks 41 indicators across five thematic areas: climate change impacts, exposures, and vulnerability; adaptation planning and resilience for health; mitigation actions and health co-benefits; economics and finance; and public and political engagement. By providing users with measures of things that are bad (e.g., the state of the biosphere and human health related to climate change), there are also efforts to track our progress on climate change mitigation and the implementation and funding for robust responses to climate threats that simultaneously produce health co-benefits. However, each of these examples highlights the need for robust planetary health indicators that are relevant across scales (i.e., from the local to global, attending to contextual nuance as needed) that are committed to measuring the good with the bad. Without measuring assets, a sense of hope is less likely to be instilled in data users because of the inherent framing (i.e., relative to reporting on decreases of a negative phenomenon which still focuses and emphasizes on the detriment rather than the asset).

Intersectional and Equity-Informed Environmental Public Health

Similar to the above, we cannot instill a sense of hope through action without being attentive to issues of social, environmental, and health equity. Equity is a long-standing concern within public health practice and differentiates from the unequal distribution of health outcomes (i.e., inequalities) to place emphasis on what is fair, right, and just (52). Equity is also embedded in many of the concepts and tools described in the preceding sections of this chapter; it is a recognition that not all health outcomes are merely unequal, but inequitable, stemming from systemic patterns of oppression, marginalization, or discrimination. In the case of the EnviroScreen tool, it is a potent reminder that the places we live, work, and play may differentially expose us to environmental harms, or worse, those very places we call home are already experiencing such a disproportionate degree of landscape-level change that they have become so-called "sacrifice zones" (53,54).

Increasingly, we've heard calls for "intersectional" analyses that better account for the unique factors that make us who we are (55). Whether they be categories of race/ethnicity, gender, sexual orientation, religion, geographic location, or otherwise, intersectionality posits that multiple social categories can intersect to create conditions for compounding or cumulative disadvantage (56). Attending to these conditions, particularly in the context of climate change, can help to avoid generalizations and victimization by enhancing contextual understandings of assets and detriments for specific populations or communities (57). This approach can in turn lead to the direct consideration of power to promote social justice in the context of climate change (58).

A key dimension of equity that has been instrumental in garnering hopeful action is that of intergenerational equity – the recognition that decisions made today may differentially impact future generations. This point has been made abundantly clear when it comes to climate change, with mass school walkouts, youth-led protests, and climate marches highlighting that our inaction on climate change today means greater impacts for the future generations. Growing social movements around Black Lives Matter and Indigenous issues, whether they be protests against extractive industries on un-ceded traditional territories or against police violence, are further reminders of the potent power of protest. But what of those species and natural systems that cannot speak for themselves? Climate marches have certainly highlighted more-than-human impacts of climate change, but it could be argued that the dominant narrative is still centered on human impacts and the need to adapt for our species over others.

Fortunately, this is another area where we can see how hopeful actions inspire commitments to change, and even transform the very systems and structures that are causing climate change and its inequitable impacts on vulnerable populations, with potential spin-off co-benefits for nonhuman species. More and more governments are taking climate change seriously by making record investments in renewable energy, developing plans and frameworks for the just transition of energy workers away from carbon-intensive fuels for cleaner and greener economies. Support for Green New Deal policies is more popular than ever and forms the foundation of EU and Korean economic and environmental policy.

Erica Chenowith has highlighted how nonviolent protests can bring about many of the changes above. Her research suggests that when as little as 3.5% of a given population engages in mass, peaceful, civil unrest, governments will either ac-commodate the goals of the movement, or in some cases, disintegrate entirely (59). Put simply, the effects of social movements increase as they gain momentum in terms of participants and the number of events over time (60). While the "3.5 rule" has limited validation in western democratic countries, and is as yet untested in a global context, it is still compelling to know that a loud and hopeful minority can make waves of change large enough to begin tackling ecological crises (61).

THE FUTURE OF MULTISPECIES INTERSECTIONAL ANALYSIS AND THE NEED FOR A ROBUST THEORY OF INTERSPECIFIC JUSTICE

Despite many of the efforts listed above, existing accounts of what equity and justice require for public health fail to meaningfully attend to the justice-relevant interests of nonhuman animals and the environment. In other words, even if we were to achieve the desiderata of existing accounts of justice, we would likely still be faced with many of the concerns outlined in this chapter given that such accounts tend to be anthropocentric. If we hope to achieve justice in the Anthropocene, it must be justice among and between humans, nonhuman animals, and the environment. To achieve such a lofty goal, we (as individuals and communities) must commit ourselves to the pursuit of "interspecific justice."

The notion of *justice* has received thousands of years' worth of scholarly attention and social action, but the idea generally refers to people getting their fair share or what they are due. Much 20th and 21st century literature on justice focuses on the just distribution of benefits and burdens among a group of people, often in the context of whole societies (i.e., social justice). Since at least the 1970s, increasing attention has been given to what justice demands of humans as it relates to nonhuman animals. Unsurprisingly, what interspecific justice de-mands of humans varies depending on the author. In 1979, VanDeVeer provided an initial "sketch" of an argument that nonhuman animals have interests that ought to be protected and promoted (62). According to VanDeVeer, balancing of interests between human and nonhuman animals means giving primacy to an animal's basic interests over another animal's peripheral interests, whether that animal is human or not, while accounting for the varying "psychological capa-cities" of different animals and giving primacy to those capable of higher cog-nitive functioning. In more recent times, interspecific theories of justice have argued for various conclusions, for example, that animals ought to have citi-zenship rights in democratic states (63), that animal assent to human actions are a necessary component of interspecific justice (64), and that articulating what rights animals have as laborers (65), or intrinsically otherwise.

The unfortunate reality, however, is that societies are presently far from just, even concerning justice among and between humans. Examples of injustice abound, including profound income inequality globally and within most countries; racism and anti-immigrant sentiment most notable in the present and burgeoning rise of right-wing extremism; and the inability of wealthy nations to share COVID-19 vaccines with low-and-middle-income countries, even when it is in their interest to do so. So, while it may be correct that there exists greater political attention to the rights of animals than in the recent past, it is perhaps a fair to ask: if we can't ensure justice among humans, what realistic chance is there in promoting interspecific justice?

The answer to this question might be, paradoxically, anthropocentric and instrumental. Appealing to the instrumental value that exists for humans to care about nonhuman animals and the environment might be the best shot we have in promoting interspecific justice, at least in the short term. For example, public discussions and debates about the zoonotic genesis of the COVID-19 pandemic – as flawed and overly simplistic as they are – suggest that people do care about animal health, albeit indirectly and self-interestedly. Yet, because we all depend on interventions made in the interests of the environment and nonhuman animals, and because such interventions tend to require collective action, acting in one's self-interest may tend to support actions and behaviors that advance interspecific justice. This support may in turn provide a foundation for broader and more holistic ethical considerations inherent to interspecific justice. Alas, there is hope yet.

CONCLUSION: FINDING HOPE IN A TIME OF DARKNESS

"It's always darkest before the dawn."

Florence + The Machine

This chapter has demonstrated that in the face of significant challenges facing the biosphere in the 21st century, we have a great deal to be hopeful about, that it's not too late to have a meaningful impact on generating creative solutions to the world's most pressing environmental challenges. While collective ecological problems increasingly challenge well-established individually-oriented theories of behavior change (e.g., health literacy), community organization, and innovations within and beyond the public health sector provide ample evidence that hope can inspire action.

Key to achieving continued collective success must include a focus on regenerative actions that heal not only the planet but also our relationship to the nonhuman animals and life forms with which we share our common home. Indigenous knowledge, perspectives, and leadership are often overlooked exemplars in this case, demonstrating rapid and on-going developments linking the ecological, social, and cultural determinants of health in ways that are supportive of nature rather than extractive (66–68). These actions must therefore be attentive to equity challenges within our own species, but also across species and across scales of action, from the local to the global (69). Key to the success of hope in being a motivating factor is its ability to foster tangible action. Unlike fear and despair, which create conditions for inaction, avoidance, and denial, hope *must* motivate action. It must instill a sense of possibility, no matter the odds. We must also be open to celebrating our achievements to further "double-down" on hope, to combat burn-out through demonstrating the art of the possible, and to use those opportunities as fuel for continued engagement. Public health institutions have a significant opportunity moving forward, so long as our collective solutions are integrative, intersectoral, and intersectional.

REFERENCES

1. Climate Change, 2021. The physical science basis. *Contribution of Working Group I to the Sixth Assessment Report of the Intergovernmental Panel on Climate Change.* Cambridge, UK: Cambridge University Press; 2021. (Masson-Delmotte, V., Zhai, P., Pirani, A., Connors, S.L., Péan, C., Berger, S., Caud, N., Chen, Y., Goldfarb, L., Gomis, M.I., Huang, M., Leitzell, K., Lonnoy, E., Matthews, J.B.R., Maycock, T.K., Waterfield, T., Yelekçi, O., Yu, R. and Zhou, B., editor. Working Group I – The Physical Science Basis).

2. Crutzen, P.J., 2006. The "Anthropocene." In: Ehlers, E. and Krafft, T., eds., *Earth System Science in the Anthropocene [Internet].* Berlin, Heidelberg: Springer., pp. 13–18 [cited 2021 Sep 29]. Available from: 10.1007/3-540-26590-2_3

3. Bruininks, P. and Malle, B.F., 2005. Distinguishing hope from optimism and related affective states. *Motivation and Emotion*, 29(4), pp. 324–352.

4. Orr, D.W., 2007. Optimism and hope in a hotter time. *Conservation Biology*, 21(6), pp. 1392–1395.

5. Bury, S.M., Wenzel, M. and Woodyatt, L., 2016. Giving hope a sporting chance: Hope as distinct from optimism when events are possible but not probable. *Motivation and Emotion*, 40(4), pp. 588–601.

6. Coroners. Heat-Related Deaths in B.C. - Province of British Columbia [Internet]. Province of British Columbia; [cited 2021 Oct 4]. Available from: https://www2.gov.bc.ca/gov/content/life-events/death/coroners-service/news-and-updates/heat-related

7. CNN DW. Extreme heat cooked mussels, clams and other shellfish alive on beaches in Western Canada [Internet]. *CNN.* [cited 2021 Sep 27]. Available from: https://www.cnn.com/2021/07/10/weather/heat-sea-life-deaths-trnd-scn/index.html

8. Endangered salmon scorched to death in river during US heatwave | The Independent [Internet]. [cited 2021 Sep 27]. Available from: https://www.independent.co.uk/climate-change/news/us-heatwave-salmon-boiled-alive-b1892230.html

9. Extreme heat triggers mass die-offs and stress for wildlife in the West [Internet]. *Animals.* 2021 [cited 2021 Sep 27]. Available from: https://www.nationalgeographic.com/animals/article/extreme-heat-triggers-mass-die-offs-and-stress-for-wildlife-in-the-west

10. Hochwasser aktuell: 82-Jähriger spendet zehnmal so viel wie FC Bayern - WELT [Internet]. [cited 2021 Oct 4]. Available from: https://www.welt.de/vermischtes/live232509543/Hochwasser-aktuell-82-Jaehriger-spendet-zehnmal-so-viel-wie-FC-Bayern.html

11. Ceballos, G., Ehrlich, P.R., Barnosky, A.D., García, A., Pringle, R.M. and Palmer, T.M., 2015. Accelerated modern human-induced species losses: Entering the sixth mass extinction. *Science Advances*, 1(5), p. e1400253.

12. Orr, J.C., Fabry, V.J., Aumont, O., Bopp, L., Doney, S.C., Feely, R.A., et al., 2005. Anthropogenic ocean acidification over the twenty-first century and its impact on calcifying organisms. *Nature*, 437(7059), pp. 681–686.

13. Cunsolo, A. and Ellis, N.R., 2018. Ecological grief as a mental health response to climate change-related loss. *Nature Climate Change*, 8(4), pp. 275–281.

14. Panu, P., 2020. Anxiety and the ecological crisis: An analysis of eco-anxiety and climate anxiety. *Sustainability*, 12(19), p. 7836.

15. Albrecht, G., 2011. Chronic environmental change: Emerging 'Psychoterratic' Syndromes. In: Weissbecker, I., ed., *Climate Change and Human Well-Being: Global Challenges and Opportunities [Internet]*. New York, NY: Springer, pp. 43–56 [cited 2020 Jul 27]. (International and Cultural Psychology). Available from: 10.1007/978-1-4419-9742-5_3

16. Buse, C.G., Cole, D.C. and Parkes, M.W., 2020. Health security in the context of social-ecological change. In: Lautensach, A. and Lautensach, S., eds., *Human Security in World Affairs: Problems and Opportunities [Internet]*. Prince George, BC: BCCampus and UNBC, 20pp. Available from: https://opentextbc.ca/humansecurity/chapter/social-ecological-change/

17. Herrington, G., 2021. Update to limits to growth: Comparing the World3 model with empirical data. *Journal of Industrial Ecology*, 25(3), pp. 614–626.

18. Meadows, D.H., Rome C. of, Meadows, D.L., Rome, C de, Associates, P., Randers, J., et al., 1972. *The Limits to Growth: A Report for the Club of Rome's Project on the Predicament of Mankind*. Universe Books, 214 p.

19. Steffen, W., Broadgate, W., Deutsch, L., Gaffney, O., Ludwig, C., 2015. The trajectory of the Anthropocene: The great acceleration. *The Anthropocene Review*, 2(1), pp. 81–98.

20. Hancock, T., Spady, D. and Soskolne, C., 2015. Canadian Public Health Association Discussion Document Global Change and Public Health: Addressing the Ecological Determinants of Health [Internet]. Canadian Public Health Association. Available from: http://www.cpha.ca/uploads/policy/edh-discussion_e.pdf

21. Hancock, T., 2019. Beyond science and technology: Creating planetary health needs not just 'Head Stuff', but social engagement and 'Heart, Gut and Spirit' stuff. *Challenges*, 10(1), p. 31.

22. Anon. Only 11 Years Left to Prevent Irreversible Damage from Climate Change, Speakers Warn during General Assembly High-Level Meeting | Meetings Coverage and Press Releases [Internet]. [cited 2021 Oct 4]. Available from: https://www.un.org/press/en/2019/ga12131.doc.htm

23. Whitmee, S., Haines, A., Beyrer, C., Boltz, F., Capon, A.G., de Souza Dias, B.F., et al. Safeguarding human health in the Anthropocene epoch: Report of The Rockefeller Foundation – Lancet Commission on planetary health. *The Lancet [Internet]*. [cited 2015 Sep 23];0(0). Available from: http://www.thelancet.com/article/S0140673615609011/abstract

24. Prüss-Üstün, A., Wolf, J., Corvalán, C., Organization, W.H., Bos, R. and Neira, D.M., 2016. *Preventing Disease through Healthy Environments: A Global Assessment of the Burden of Disease from Environmental Risks*. World Health Organization, 173 p.

25. Buse, C.G., Poland, B., Wong, J. and Haluza-Delay, R., 2019. 'We're all brave pioneers on this road': A Bourdieusian analysis of field creation for public health adaptation to climate change in Ontario, Canada. *Critical Public Health*, 31, pp. 1–11.

26. World Wildlife Foundation. Living Planet Report 2020: Bending the curve of biodiversity loss [Internet]. Gland, Switzerland: World Wildlife Foundation;

2020, p. 160. Available from: https://files.worldwildlife.org/wwfcmsprod/files/ Publication/file/279c656a32_ENGLISH_FULL.pdf?_ga=2.86824408. 1172252376.1638383231-1251308859.1638383231

27. Chapron, G., Epstein, Y. and López-Bao, J.V., 2018. US bill illustrates how conservation triage can lead to extinctions. *Nature*, *554*(7692), p. 300.

28. Wilson, K.A. and Law, E.A., 2016. Ethics of conservation triage. *Frontiers in Ecology and Evolution [Internet]*. [cited 2021 Dec 1]; 4. Available from: http:// journal.frontiersin.org/Article/10.3389/fevo.2016.00112/abstract

29. Chandler, D., 2019. The death of hope? Affirmation in the Anthropocene. *Globalizations*, *16*(5), pp. 695–706.

30. Head, L., 2016. *Hope and Grief in the Anthropocene: Re-conceptualising human–nature relations*. London: Routledge, 194 p.

31. Treanor, B., 2019. Hope in the age of the Anthropocene. *Analecta Hermeneutica [Internet]*, 10(0). Available from: https://journals.library.mun.ca/ojs/index.php/ analecta/article/view/2055

32. Buse, C.G., Gislason, M., Reynolds, A. and Ziolo, M., 2021. Enough tough talk! It's time for the tough action(s) to promote local to global planetary health. *International Journal of Health Promotion and Education*, *59*(5), pp. 271–275.

33. Seneviratne, S.I., Rogelj, J., Séférian, R., Wartenburger, R., Allen, M.R., Cain, M., et al., 2018. The many possible climates from the Paris Agreement's aim of 1.5 °C warming. *Nature.*, *558*(7708), pp. 41–49.

34. Lang, T. and Rayner, G., 2012. Ecological public health: The 21st century's big idea? An essay by Tim Lang and Geof Rayner. *BMJ*, *345*, p. e5466.

35. Buse, C.G., Oestreicher, J.S., Ellis, N.R., Patrick, R., Brisbois, B., Jenkins, A.P., et al., 2018. Public health guide to field developments linking ecosystems, environments and health in the Anthropocene. *Journal of Epidemiology and Community Health*, *72*(5), pp. 420–425.

36. Oestreicher, J.S., Buse, C., Brisbois, B., Patrick, R., Jenkins, A., Kingsley, J., et al., 2018. Onde ecossistemas, pessoas e saúde se encontram: SustDeb, *9*(1), pp. 23–44.

37. Buse, C., 2013. Intersectoral action for health equity as it relates to climate change in Canada: Contributions from critical systems heuristics. *Journal of Evaluation in Clinical Practice*, *19*(6), pp. 1095–1100.

38. Gillingham, M.P., Halseth, G.R., Johnson, C.J. and Parkes, M.W., 2016. *The Integration Imperative: Cumulative Environmental, Community and Health Impacts of Multiple Natural Resource Developments*. Springer International Publishing AG.

39. World Health Organization, 2016. *Urban Green Spaces and Health [Internet]*. Copenhagen: WHO Regional Office for Europe, 91 p. Available from: https:// www.euro.who.int/__data/assets/pdf_file/0005/321971/Urban-green-spaces-and-health-review-evidence.pdf

40. Salas, R.N. and Hayhoe, K., 2021. Climate action for health and hope. *BMJ. 374*, p. n2100.

41. Wolf, K.L., Lam, S.T., McKeen, J.K., Richardson, G.R.A., van den Bosch, M. and Bardekjian, A.C., 2020. Urban trees and human health: A scoping review. *International Journal of Environmental Research and Public Health*, *17*(12), p. 4371.

42. van den Bosch, M. and Ode Sang, Å., 2017. Urban natural environments as nature-based solutions for improved public health – A systematic review of reviews. *Environmental Research*, 158, pp. 373–384.

43. Devissher, T. and van den Bosch, M., 2020. Nature and metnal health: New initiatives in British Columbia, Canada [Internet]. Available from: https://www.greenforcare.eu/news/nature-mental-health-canada/

44. Martin, L., White, M.P., Hunt, A., Richardson, M., Pahl, S. and Burt, J., 2020. Nature contact, nature connectedness and associations with health, wellbeing and pro-environmental behaviours. *Journal of Environmental Psychology*, 68, p. 101389.

45. Logan, A.C., Berman, S.H., Berman, B.M. and Prescott, S.L., 2020. Project earthrise: Inspiring creativity, kindness and imagination in planetary health. *Challenges*, *11*(2), p. 19.

46. Prescott, S.L. and Logan, A.C., 2018. Larger than life: Injecting hope into the planetary health paradigm. *Challenges*, *9*(1), p. 13.

47. Rodriguez, M. and Zeise, L., 2017. *Calenviroscreen 3.0 [Internet]*. California EPA and Office of Environmental Health Hazard Assessment. Available from: https://oehha.ca.gov/media/downloads/calenviroscreen/report/ces3report.pdf

48. Washington Environmental Health Disparities Map [Internet]. [cited 2021 Aug 23]. Available from: https://www.doh.wa.gov/DataandStatisticalReports/Washington TrackingNetworkWTN/InformationbyLocation/WashingtonEnvironmentalHealthDisparitiesMap

49. Driver, A., Mehdizadeh, C., Bara-Garcia, S., Bodenreider, C., Lewis, J., Wilson, S., 2019. Utilization of the Maryland Environmental Justice Screening Tool: A Bladensburg, Maryland Case Study. *IJERPH*, *16*(3), p. 348.

50. October 25, Geertsma 2018 Meleah. New Map Shows Chicago Needs Environmental Justice Reforms [Internet]. *NRDC*. [cited 2021 Mar 30]. Available from: https://www.nrdc.org/experts/meleah-geertsma/new-map-shows-chicago-needs-environmental-justice-reforms

51. Lee, C., 2020. A game changer in the making? Lessons from States advancing environmental justice through mapping and cumulative impact strategies. *Environmental Law Reporter*, *50*(3), pp. 10203–10215.

52. Buse, C.G., 2015. Health equity, population health, and climate change adaptation in Ontario, Canada. *Health Tomorrow: Interdisciplinarity and Internationality [Internet]*. [cited 2018 Feb 18]; *3*(1), pp. 26–51. Available from: https://ht.journals.yorku.ca/index.php/ht/article/view/40177

53. Scott, D. and Smith, A., 2017. "Sacrifice Zones" in the Green Energy Economy: Toward an Environmental Justice Framework. 62:3 McGill LJ 861 [Internet]. Available from: https://digitalcommons.osgoode.yorku.ca/scholarly_works/2691

54. Holifield, R. and Day, M., 2017. A framework for a critical physical geography of 'sacrifice zones': Physical landscapes and discursive spaces of frac sand mining in western Wisconsin. *Geoforum*, 85, pp. 269–279.

55. Kaijser, A. and Kronsell, A., 2014. Climate change through the lens of intersectionality. *Environmental Politics*, *23*(3), pp. 417–433.

56. Crenshaw, K.W., 1990. Mapping the margins: Intersectionality, identity politics, and violence against women of color. *Stanford Law Review*, *43*(6), pp. 1241–1299.

57. Buse, C.G. and Patrick, R., 2020. Climate change glossary for public health practice: From vulnerability to climate justice. *Journal of Epidemiology and Community Health*, *74*(10), pp. 867–871. jech-2020-213889.

58. Osborne, N., 2015. Intersectionality and kyriarchy: A framework for approaching power and social justice in planning and climate change adaptation. *Planning Theory*, *14*(2), pp. 130–151.

59. Chenoweth, E., Stephan, M.J. and Stephan, M., 2011. *Why Civil Resistance Works: The Strategic Logic of Nonviolent Conflict*. Columbia University Press. 316 p.

60. Chenoweth, E. and Belgioioso, M., 2019. The physics of dissent and the effects of movement momentum. *Nature Human Behaviour*, *3*(10), pp. 1088–1095.

61. Matthews, K., 2020. Social movements and the (mis)use of research: Extinction rebellion and the 3.5% rule. *Interface: A Journal for and About Social Movements*, *12*(1), pp. 591–615.

62. VanDeVeer, D., 1979. Interspecific justice. *Inquiry*, *22*(1–4), pp. 55–79.

63. Zoopolis, A., 2011. *Political Theory of Animal Rights*. Oxford, New York: Oxford University Press, 352 p.

64. Healey, R. and Pepper, A., 2021. Interspecies justice: Agency, self-determination, and assent. *Philosophical Studies*, *178*(4), pp. 1223–1243.

65. Blattner, C.E., Coulter, K., & Kymlicka, W. eds., 2020. *Animal Labour: A New Frontier of Interspecies Justice?* Oxford, New York: Oxford University Press, 256 p.

66. Redvers, N., 2018. The value of global indigenous knowledge in planetary health. *Challenges*, *9*(2), p. 30.

67. Redvers, N., Yellow Bird, M., Quinn, D., Yunkaporta, T. and Arabena, K., 2020. Molecular decolonization: An indigenous microcosm perspective of planetary health. *IJERPH*, *17*(12), p. 4586.

68. Robinson, J.M., Gellie, N., MacCarthy, D., Mills, J.G., O'Donnell, K. and Redvers, N. Traditional ecological knowledge in restoration ecology: A call to listen deeply, to engage with, and respect Indigenous voices. *Restoration Ecology* [Internet]. 2021 May [cited 2021 Oct 4]; *29*(4). Online. Available from: https://onlinelibrary.wiley.com/doi/10.1111/rec.13381

69. Buse, C.G., Smith, M. and Silva, D.S., 2019. Attending to scalar ethical issues in emerging approaches to environmental health research and practice. *Monash Bioethics Review*, *37*(1), pp. 4–21.

16 Education to Protect Animal Health in a Changing Climate

Will Sander and Colleen Duncan

CONTENTS

KEY LEARNING OBJECTIVES

- Strengthening sustainability within curricula involves considering emerging issues, tailoring learning experiences to core sustainability competencies, and focusing on problem-based and experiential learning to have solution-oriented outcomes.
- Developing a sustainability workforce entails helping learners to become change agents from local to international levels and pulling on an interdisciplinary approach to enhance collaboration.
- Prioritizing inclusivity and diversity in both faculty and students provides the diverse perspectives needed to address complex sustainability challenges.

IMPLICATIONS FOR ACTION

- Educational institutions must rise to the challenges posed by climate change by providing educational programs that empower learners to act.

DOI: 10.1201/9781003149774-16

- Within animal health, experiential and applied opportunities must be available to learners to put into action sustainability and climate change solutions.
- Further initiatives, commitments, and partnerships must be formed to bridge academic silos to enhance collaboration within and between institutions, as well as other sectors that also interface with societal needs for sustainability and climate change.

"Education is the most powerful weapon which you can use to change the world."

– Nelson Mandela

Our ability to protect the health of animals, now and in the future, depends on an educated and empowered animal health community. Unfortunately, there is little evidence that the topic of climate change has been broadly or consistently taught in animal health programs. For example, veterinarians are considered a key animal health profession, yet survey work exploring the education and interest of veterinary students and veterinarians found that despite a strong belief that anthropogenic climate change has a significant impact on animal and human health, 77% of veterinary medical students and 90% of graduate veterinarians were not aware of, or had no education on, climate change in their veterinary medical curriculum (1,2). To meet this need and address the complexities of the topic, educators must draw on expertise both within and beyond the traditional animal health sector.

The profession of veterinary medicine is not alone in needing to think critically about sustainability and climate change into the future. In 2015, the American Association for the Advancement of Science convened a symposium to discuss how to move sustainability forward in educational institutions (3). Their five key points were the following:

1. Universities must realize that emphasizing solutions is essential when studying sustainability problems.
2. Building effective interdisciplinary collaboration in sustainability programs is challenging but must start early.
3. Academic efforts will be misaligned, misallocated, and mistaken in the absence of productive stakeholder partnerships.
4. The path to solutions requires innovation, risk tolerance, and persistence.
5. Universities must apply their research strengths to examine institutional initiatives and develop evidence-based principles to guide institutional transformations.

In 2020, the U.S. National Academies of Sciences, Engineering, and Medicine (NASEM) released a strong framework on *Strengthening Sustainability Programs and Curricula at the Undergraduate and Graduate Levels* (4). This publication brings together important recommendations on how to better instill sustainability into training programs across post-secondary education. Their report builds off the

Sustainable Development Goal (SDG) 4, "Quality Education," particularly SDG 4.7 to ensure all learners acquire the knowledge and skills needed to promote sustainable development and lifestyles. The following chapter uses veterinary medicine as an example to explore how this framework can be used to prepare animal health professionals to respond to the growing health threats associated with climate change. These recommendations can fall under the three broad areas of the NASEM report: strengthening sustainability within programs, developing a sustainability workforce, and building the academic environment.

STRENGTHENING SUSTAINABILITY WITHIN PROGRAMS

While formal, consistent, and core education on climate change appears to be lacking in many veterinary medical curricula, there are obvious opportunities where the topic could be integrated with or built upon existing material, thereby strengthening sustainability within programs. The NASEM recommendations on how this goal can be achieved are robust and relevant to virtually all curricula (box 1). Where possible, attempts should be made to include climate change and sustainability education in ways that build on core themes and complement current core competencies within veterinary medicine. In their recommendations to strengthen sustainability programs, the NASEM has identified five realms in which sustainability curricula should seek to develop competencies. These recommendations broadly align with the nine domains of competence outlined by the American Association of Veterinary Medical Colleges (AAVMC) in its competency-based learning framework (Figure 16.1).

BOX 1   KEY NASEM (4) RECOMMENDATIONS FOR STRENGTHENING SUSTAINABILITY PROGRAMS INCLUDE

1. Specifically tailoring educational experiences to address core sustainability competencies and capacities
2. Developing programs that encompass key and emerging sustainability content while incorporating problem-based and solution-oriented approaches
3. Training for the complex contexts of sustainability and applying learning in experiential settings

---

These five sustainability realms are:

1. Systems Thinking: ability to analyze complex systems across disciplines, domains, and scales
2. Anticipatory: ability to make decisions in the face of uncertainty by analyzing, evaluating, and assessing sustainability issues
3. Normative: ability to understand, apply, and reconcile sustainability values and goals considering ethical principles and values

4. Strategic: ability to design and implement change through interventions, policies, and strategies at different scales
5. Interpersonal: ability to facilitate collaborative and participatory efforts in sustainability

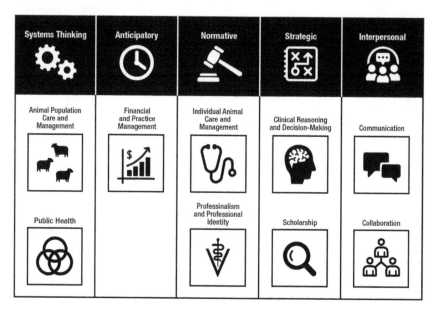

FIGURE 16.1   Intersection of NASEM sustainability competencies and AAVMC nine domains of competence.

Public health is a critical component of veterinary medical education, part of the veterinary professional oath worldwide (5), and a logical area for climate change education to be embedded. In its competency-based learning framework, the AAVMC has included public health as one of the nine competency domains in which "the graduate responds to issues at the interface of animals, humans, and the environment, utilizing a global perspective and sensitivity to local cultures" (6). As described in other chapters, climate change will impact virtually all aspects of veterinary public health, including infectious disease, natural disasters, and food security. Public health education is, therefore, a logical place to include or expand upon the topic of climate change. Climate-associated examples could also be used to highlight other key elements in a veterinary public health class, such as framing an emergency-response module within a natural disaster instead of an infectious disease outbreak. Similarly, the manner in which climate change may alter host-agent relationships or contribute to the development of disease could be infused when discussing the concept of causation or the epidemiological triad. Despite being a foundational component of veterinary medical education, public health is typically only 1/10th of one year in the core, four-year, veterinary curriculum and less than that in some institutions. The lack of knowledge about where public health is taught across veterinary colleges in the U.S. illustrates the large gap in how this teaching is prioritized. The most recent publications in this area that provide any insight are from 2004 and 2008, and many colleges have shifted and changed curriculums since then (7,8). Reinvigoration of public health programs through the inclusion of current topics that transcend species, like climate change, could push the profession to forefront of societal public health problem solving. Climate change education can help to develop a public health workforce that can integrate across disciplines, pulling on the broad training of veterinarians.

The topic of climate change could additionally be infused throughout other areas of the curriculum by using it as a framing to address other veterinary education or skills. Courses such as microbiology, emergency care, internal medicine, and animal production are all appropriate places to introduce key sustainability issues. Such an approach requires commitment by faculty from different backgrounds but may more successfully tie the message to clinical medicine. Examples of this approach may include internists highlighting the impact of wildfire smoke on the athletic performance in horses (9) or associations between indoor air pollution and respiratory disease in dogs and cats (10,11). Critical care teams could highlight the projected increase in heat-associated illnesses in companion animals when teaching how to manage these patients in emergency (8). Range expansion of pathogens carrying vectors, such as those contributing to Lyme disease or Rocky Mountain spotted fever, could be integrated into clinical parasitology and preventive medicine courses (12). Zoo and wildlife medicine can incorporate the discussion of diminishing species and wildlife conflict resulting from lack of natural resources (13). A primary concern of veterinary students and veterinarians regarding the integration of new or additional climate change educational programming was an already full curriculum (1,2). Identification of ways to infuse this content such that it complements other topics will be critical for wider adoption. Restricting the topic to elective courses means the content is only provided to self-selecting students who likely already have an interest and some knowledge of the topic. Additionally, by linking these issues with core medical content, the material is well suited to problem-based learning with a focus on solutions.

Clinical practice and business management education is another route to infuse sustainability topics into the veterinary medical curriculum, particularly the application of learning in experiential situations. As above, the clinical framing of the animal health issues will motivate learners to seek solutions. Not surprisingly, the educational content that veterinary students and practitioners were most interested in was how they can minimize their own deleterious environmental impact (1,2). The provision of healthcare is resource intensive. While no data on veterinary medicine is available, the human healthcare system is estimated to produce ~4.5% of the world's global greenhouse gas emissions (14,15). The NASEM report highlights the importance of experiential learning settings that help learners to become more effective implementers of sustainability action. In veterinary medicine, veterinary teaching hospitals (VTH) are the site of most experiential learning. Unfortunately, a recent survey of faculty, staff, and students at veterinary teaching hospitals around the world found little evidence that sustainable behaviors were being practiced or showcased in this experiential-learning environment, despite the strong belief that clinical sustainability is an important topic. (16) Adopting and implementing more environmentally sustainable practices, in accordance with established physical and behavioral themes, (17) is a logical and feasible way to educate and empower veterinary trainees in this space. Similarly, there are growing economic arguments for integrating sustainability into business. Preliminary work in the companion animal sector suggests that veterinary clients may preferentially select, and pay more for, veterinary services from a clinic that is knowledgeable about climate-associated health impacts and has taken steps to reduce their own environmental footprint (18).

In clinical situations, students also have opportunities to practice client communication. Communication is an AAVMC competency domain, and veterinary graduates are expected to communicate effectively with diverse clients, colleagues, other healthcare professionals, and the public to promote animal, human, and environmental health and well-being (6). Differing political viewpoints were identified as a barrier for veterinarians to communicate with their clients about the animal health impacts of climate change (2), although, interestingly, this result was statistically more common in the USA compared to veterinarians from other countries (1).

Finally, topics of climate change and health can be incorporated into problem- or team-based learning approaches intended to develop skills to address "wicked" problems. Such efforts often include innovative, interprofessional education exercises and ties back to communication. Climate change is related to one of several difficult physical problems that could be discussed in these team-based activities (ex., heat exhaustion in pets, flooding concerns for livestock producers, and increasing disease threats from vector borne diseases), but it also highlights social problems and issues faced by vulnerable populations needing access to veterinary care (19). This situation provides an entry way to discuss the importance of addressing equity around social determinants of health within climate change and how that translates to animal care.

DEVELOPING A SUSTAINABILITY WORKFORCE

Given the monumental and pressing nature of climate change, educational efforts should focus on "doing" rather than "knowing." The NASEM recommendations for the development of a sustainable workforce (box 2) can be used for the development of engaged, action-oriented, animal health professionals who can be leaders on sustainability efforts with the help of professional societies that support these efforts. While the skillsets differ, this concept of developing "change agents" is not unlike existing clinical efforts to graduate "practice ready" or "day 1 competent" veterinarians. The power of "learning by doing" is realized by immersion into applied clinical and professional rotations toward the end of training programs. A similar, experiential-learning approach can be taken by integrating climate action and sustainability activities throughout the curriculum. As an example, studies have shown a constant, strong interest in zoo and wildlife medicine (up to 30% in entering veterinary class years), which draws heavily on the ideas of conservation medicine and ecosystem health (13). With limited post-veterinary opportunities in clinical zoo and wildlife medicine, steering students to the bigger picture both before and during veterinary school provides additional job opportunities that will address ecosystem health, climate change, and sustainability.

BOX 2   NASEM (4) RECOMMENDATIONS FOR DEVELOPING A SUSTAINABILITY WORKFORCE ARE

1. Developing change agents who can take action to meet emerging local, regional, national, and global needs
2. Enhancing collaboration among professional societies and sectors on sustainability and assess programs and jobs in this area

---

Leadership training is another content area that is typical in elective coursework, but it can have significant impacts toward climate change and sustainability. The potential for highly trained and knowledgeable veterinarians to facilitate local actions addressing climate is vast (20). Tied into leadership is the need to influence policy, including knowledge of the legislative process, policy-writing, and understanding how policy shapes global topics of concern. However, these types of opportunities are limited and are not incorporated as core curricular programs. Courses to engage current veterinary students in established public-policy topics (21) should be expanded so that all students understand how their work fits into the broader, global context of climate change and sustainability. By having this background, veterinarians can take action to inform policy decision making in their areas.

After graduation from veterinary school, collaborative and continued growth opportunities around sustainability are numerous. Veterinarians need to build from successes in other professions to enhance their own abilities. One such example from another profession is the Medical Society Consortium on Climate and Health (MSCCH) (22). MSCCH focuses on three simple messages: climate change is harming Americans today, and these harms will increase without action; the way to slow or stop these harms is to decrease the use of fossil fuels and increase energy efficiency and use of clean energy sources; and changes in energy choices will improve air quality and water quality, providing immediate health benefits. The group's focus is on the United States, but it represents over 600,000 health professionals and provides a platform for veterinary collaboration. This is just one example of an organization where national or regional veterinary medical associations (e.g. American Veterinary Medical Association,

Canadian Veterinary Medical Association, state veterinary medical associations), which already have a vested interest in their members, can partner at a larger scale to better develop their members competencies in sustainability.

Partnerships with animal health industries focused on sustainability could provide opportunities for engaged scholarship and experiential learning. For example, a number of pet food companies have strong sustainability commitments under groups like Pet Sustainability Coalition or TerraCycle or independently (e.g. Purina, Mars Petcare, and Hill's Pet Nutrition) (23–27). Veterinarians have been identified as valuable messengers of information and key players for the promotion of sustainability in the companion animal sector (28). As such, the profession can already act as a change agent, and continued coalescence around collaborative partnerships to have unified messaging will advance sustainability in the future. In the livestock sector, the National Cattlemen's Beef Association has a commitment to demonstrate climate neutrality of U.S. cattle production by 2040 (29). Similar initiatives are seen across other species groups. For example, the U.S. National Pork Producers Council reports a continual decrease in land, water, and energy use, as well as a decreasing carbon footprint over the last 50 years (30). These opportunities for collaboration, both within the veterinary profession and between professions, ensure numerous potential jobs and avenues for veterinarians that support sustainability into the future.

Veterinary medicine can also begin to encourage these change agents through their recruiting processes. For students interested in pursuing veterinary medicine, there are pre-veterinary clubs and associations including the American Pre-Veterinary Medical Association, which has over 50 groups as members. With the continual growth of sustainability programs in undergraduate education, there are ample opportunities for veterinary medicine to pull in students that have significant background (31). Veterinary medical programs pre-requisites help drive many undergraduate course selections for those interested in biomedical science (32). Simply adding a course requirement related to sustainability would substantially shift the course backgrounds of many applicant students and increase the potential change within the profession. This "pull" effort by veterinary schools can be married to the "push" efforts, such as those of the United Nations, where over 300 universities internationally have signed on to the Higher Education Sustainability Initiative that interfaces education with science and policy making around sustainability (33).

BUILDING THE ACADEMIC ENVIRONMENT

To educate the change agents of the future, academic institutions must have the resources and capacity to foster this development. Recommendations put forth by NASEM (box 3) are not only applicable to veterinary medicine, but they also are complementary to other professional needs and initiatives within academia. Central to this idea is the bridging of disciplinary silos. While the topic of climate change may not be within the comfort of many veterinary educators, expertise in both subject matter and implementation can undoubtably be found elsewhere on campus or beyond. There are numerous examples of bridging disciplinary silos in veterinary education today. Most notably is through elective or selective courses that students opt into so they can focus on topics as broad as global One Health, policy development, ecosystem health, and business sustainability practices. At institutions where climate change is already embedded in the curriculum, it tends to be in these types of classes (1,2).

BOX 3   NASEM (4) RECOMMENDATIONS FOR BUILDING THE ACADEMIC ENVIRONMENT TO STRENGTHEN SUSTAINABILITY PROGRAMS INCLUDE THE FOLLOWING

1. Bridging disciplinary silos with a commitment to inclusivity and interdisciplinarity
2. Prioritizing diversity, equity, and inclusion through support and recruitment of students and faculty
3. Increased federal support of sustainability programs in higher education, particularly minority serving institutions
4. Conduct research on sustainability education programs to assess effectiveness, employment opportunities, current competencies, and student preparedness

---

Several institutions have implemented interprofessional education components to their curriculum. Some of these are focused on disease scenarios, but they open the door for discussions on climate change, sustainability, and collaborative approaches. A notable example is the Field Experience in One Health and Outbreak Investigation co-hosted by University of Texas Medical Branch, University of Texas Rio Grande Valley, and Texas A&M University, which brings together veterinary, medical, and public health students over three to four weeks to go through a One Health disease field investigation with key components tied to climate change and sustainability (34). As just one successful example of this cross-disciplinary content that has continued for four years, it provides a blueprint for future collaborations during the veterinary curriculum and highlights the need for larger groups to tackle these pressing issues.

Another example that tackles One Health and leadership, two key components of sustainability, and climate change originates from the University of Saskatchewan, which started a One Health Leadership Experience spanning three days (35). This experience was open to students in their first or second year of their health science program to introduce the concept of One Health, stimulate student engagement, and encourage students to be One Health leaders and change agents in their respective professions (36). In a five-year period, 669 participants have been part of this experience. Although most do not go on to be involved in One Health clubs or do additional One Health training, follow up surveys document a significant impact of including One Health in their professions (36).

The human medical education community is a logical partner for the development of curricular content around climate change and health. Gaps, needs, action statements, and policies have been put forth by agencies such as the American Medical Association (37) and the Association for Medical Education in Europe including efforts to align learning domains, objectives, and outcomes for environmentally sustainable healthcare the UN SDGs (38). As with veterinary medicine, efforts to address the health impacts of climate change are often driven by students and can be integrated into both public health and clinical training (39,40). Human medicine has taken the approach of not adding to the already crowded curriculum, but rather interweaving climate into health discussions that are already occurring. The term "eco-medical literacy" has been introduced as "the ability to access, understand, integrate, and use information about the health-related ecological effects of climate change to deliver and improve medical services" (41).

Other human health societies are already working together in the educational space. The Global Consortium on Climate and Health Education is "a global network of health professions schools and programs, including schools and programs of public health, medicine, and nursing" (42). This consortium was launched in 2017, building on a side meeting during the Conference of Parties in December 2015 co-hosted by Mailman School of Public Health and the U.S. White House. At that meeting, 115 health professional schools committed to educate their students about climate and health; these included some low- and middle-income countries (43). This participatory-group vision statement desires "all health professionals trained to prevent, mitigate, and respond to the health impacts of climate change." In a recent survey of these members, over half offer climate health education typically part of a required core course. However, a majority had encountered challenges trying to incorporate climate health into the curriculum, despite overwhelming support from students, faculty, and administration (44).

Public health programs have taken a significant lead in connecting climate change and sustainability to public health domains. The Association of Schools of Public Health in the European Region (ASPHER) has released their Climate and Health Competencies for Public Health Professionals that provides both high level and detailed competencies across four domains: knowledge and analytical skills; communication and advocacy; collaboration and partnerships; and policy (45). Like the profession of public health, these competencies are broad but also applicable across many health disciplines, including veterinary medicine. With the United States, the Association of Schools and Programs of Public Health (ASPPH) has joined the Global Consortium on Climate and Health Education. As the largest association of accredited public health programs, ASPPH plays a strong leadership role in forming future public health professionals. Despite these efforts, more work is still needed to expand educational opportunities around climate change and health (46).

With the variety of engagements going on through climate and health in other health professions, the veterinary academic enterprise has several opportunities it can pursue into the future. One of those is joining in with the Global Consortium on Climate and Health Education so that the veterinary profession is part of the larger health profession discussion. Within current curriculums, additional focus can be placed on public health sections, as well as reframing the animal health perspective around influencing sustainable development goals of food, water, consumption and production, conservation, and climate change. This focus includes nutrition and the increasing reliance on animal protein into the future. These themes can also be incorporated into elective content and business classes. For example, classes could address discussing how to message responsible or green clinics to clients or using it as marketing to appeal to consumers shifting preferences.

Prioritization of diversity, equity, and inclusion efforts (DEI) as part of sustainability education is complementary to ongoing efforts within veterinary medicine. In some countries, like the United States, the need to attract and retain both students and staff from diverse backgrounds is well recognized (47,48). At the federal level, the National Science Foundation has pushed additional funding toward minority serving institutions to integrate science and research, including sustainability (49). The American Veterinary Medical Association (AVMA) and the AAVMC have joined forces to coordinate a comprehensive effort to enhance diversity, equity, and inclusion in the veterinary profession. As part of this effort, recognition that addressing these concerns will also help provide a deeper understand of climate related problems is essential. AAVMC and AVMA's Council on Education are the biggest drivers for curriculum change in U.S. veterinary schools. As these two organizations seek to understand and communicate the complex problems of DEI, they must also push forward the challenges of discussing climate change and health. Cultural competence and cultural humility are essential for effective practice, and these dimensions have huge impacts in shaping the effects of the other complex problems mentioned around climate change. As highlighted by NASEM, ensuring federal support is prioritized for minority serving institutions will be important to address this issue.

Finally, a sustainability education-research agenda would serve to strengthen and inform educational programming. Collaboration between institutions and with national veterinary associations and educational accreditation bodies would help to ensure enough study power and create efficiencies. Graduate student surveys could be expanded to include questions regarding needs and opportunities within the profession, as well as the effectiveness of veterinary curricula in for meeting these needs. This work could be informed by, and conducted in parallel with, educational research in other sectors such as human medicine, public health, and agriculture.

Sustainability initiatives are big drivers of campus-wide initiatives among many major universities. At the same time, cooperative research centers, centers of excellence, and interdisciplinary hubs have become popular in academic environments to not only solve more complex, societal problems but to also seek larger and more diverse extramural funding (50). The current environment is ripe for collaborations and synergies to develop between colleges and departments to link entire animal-education sectors together from animal science to natural resources to veterinary medicine. The majority of veterinary colleges are co-located with these other animal sector related disciplines, and joint graduate and professional courses are possible even at R01 institutions providing decreased workload on each individual faculty member while increasing exposure of perspectives to veterinary students (51).

SUPPORTING ALL ANIMAL HEALTH PROFESSIONALS

While veterinary medicine is a key animal health profession, the number of people who contribute to the health of domestic and wild animals around the world in nonveterinary roles is considerably greater. The Lancet Countdown on Health Climate Change in 2020 highlighted the central role all health professionals have in health system adaptation and mitigation, understanding and maximizing the health benefits of an intervention, and communicating the need for an accelerated response (52). Education has to be central to this key role for health professionals, and all animal health professionals can use the NASEM framework highlighted above. There has been increasing progress on this front, with academia increasing publication on climate change and health by a factor of eight over the last decade, and, from 2018 to 2019, media coverage has increased 96% worldwide on climate change and health, outpacing climate change by itself (52).

Learners are increasingly being exposed to the topic of climate change as early as primary and continuing into secondary education. A public health framing of climate change impacts has been shown to be an effective way to advance climate change action (53,54), so coupling this approach with animal health messaging is a logical action. Presentations on career pathways, including animal health, are common in primary and secondary schools. Animal health professionals could build additional interest, even in young students, by appealing not only to day-to-day components of their jobs but also to the larger impact their profession plays in mitigating and adapting to climate change. These efforts can be in tandem to decade long effort from the U.S. Department of Education to foster Green Ribbon Schools that push primary and secondary public and private schools to commit to reducing environmental impacts and costs and providing effective sustainability education including green career pathways (55).

One of the best opportunities to reach students prior to terminal professional training is through clubs and associations. The Future Farmers of America and Cargill have teamed up recently to advance sustainability curriculum in FFA programs, building on environmental stewardship work being led by farmers and ranchers (56). The focus is on solving sustainable agricultural challenges, particularly with the need to support more people in the future. Another youth development program, 4H, has developed modules focused on environmental stewardship and sustainability to implement across their 100 plus university extension networks (57). These two organizations have a significant influence on the choice of an animal health career. For those going into rural veterinary practice, over a third felt 4H and/or FFA was a significant influence in choosing this path (58).

However, there is still a long way to go in pulling in animal health professionals from across different areas, including farmers, wildlife managers, paraprofessionals, economists, and the larger agricultural sector. It has been over 25 years since the tagline people, planet, and profit sought to shake up the traditional model of capitalism by changing how future leaders were educated and how businesses measured and thought about their impacts. Nothing can be kept within its narrow scope. The SDGs highlight this fact with how animal health impacts food, water, sustainable consumption and production, conservation, and climate change. This approach should be how education of animal health and climate change should be imagined. These concepts can be infused into other challenging problems in the curriculum, into elective content, into public health, into communication, and into business classes.

Educators and educational programs have the potential for huge impact by supporting learners to meet the sustainability needs now and in the future. This work can be done with minimal shift in teaching areas and opportunities that begin in primary school and continue into graduate education. For this shift to be optimally effective and timely, there must be buy in at multiple levels within and outside of academia, starting with individuals and going all the way up to strong leadership and policy creation. But we need not wait to act. Learners, particularly younger generations, are actively seeking opportunities to learn and take action on these issues. While animal health educators may feel the topic of climate change is beyond their typical scope of practice, the solutions and tools to address the health impacts are not. Working together with open minds, we can harness the most powerful weapon, education, to change the health of the world.

REFERENCES

1. Kramer, C.G., McCaw, K.A., Zarestky, J. and Duncan, C.G., 2020. Veterinarians in a changing global climate: Educational disconnect and a path forward. *Frontiers in Veterinary Science*, 7(1029). online journal.
2. Pollard, A.E., Rowlison, D.L., Kohnen, A., McGuffin, K., Geldert, C., Kramer, C., et al., 2021. Preparing veterinarians to address the health impacts of climate change: Student perceptions, knowledge gaps, and opportunities. *Journal of Veterinary Medical Education*, 48(3), pp. 343–350.
3. Hart, D.D., Buizer, J.L., Foley, J.A., Gilbert, L.E., Graumlich, L.J., Kapuscinski, A.R., et al., 2016. Mobilizing the power of higher education to tackle the grand challenge of sustainability: Lessons from novel initiatives. *Elementa: Science of the Anthropocene*, 4. p. 000090. https://doi.org/10.12952/journal.elementa.000090

4. National Academies of Sciences Engineering, and Medicine. 2020. *Strengthening Sustainability Programs and Curricula at the Undergraduate and Graduate Levels.* Washington, DC: The National Academies Press. https://doi.org/10.17226/25821
5. Association, W.V., 2019. World Veterinary Association Model Veterinarians' Oath Brussels, Belgium [Available from: https://www.worldvet.org/uploads/news/docs/wva_model_veterinarians_oath.pdf
6. AAVMC Working Group on Competency-Based Veterinary Education, Molgaard, L.K., Hodgson, J.L., Bok, H.G.J., Chaney, K.P., Ilkiw, J.E., Matthew, S.M., May, S.A., Read, E.K., Rush, B.R. and Salisbury, S.K., 2018. *Competency-Based Veterinary Education: Part 1 – CBVE Framework.* Washington, DC: Association of American Veterinary Medical Colleges. https://www.aavmc.org/wp-content/uploads/2020/10/CBVE-Publication-1-Framework.pdf
7. Riddle, C., Mainzer, H. and Julian, M., 2004. Training the Veterinary Public Health Workforce: A review of educational opportunities in US Veterinary Schools. *Journal of Veterinary Medical Education, 31*(2), pp. 161–167.
8. Wenzel, J.G.W., Nusbaum, K.E., Wright, J.C. and Hall, D.C.A., 2008. Public-Health instruction necessary to supplement the Veterinary Professional Curriculum: The DVM/MPH Coordinated-Degree Program at Auburn University. *Journal of Veterinary Medical Education, 35*(2), pp. 231–234.
9. Bond, S.L., Greco-Otto, P., MacLeod, J., Galezowski, A., Bayly, W. and Léguillette, R., 2020. Efficacy of dexamethasone, salbutamol, and reduced respirable particulate concentration on aerobic capacity in horses with smoke-induced mild asthma. *Journal of veterinary internal medicine, 34*(2), pp. 979–985.
10. Lin, C-H, Lo, P-Y and Wu, H-D., 2020. An observational study of the role of indoor air pollution in pets with naturally acquired bronchial/lung disease. *Veterinary Medicine and Science, 6*(3), pp. 314–320.
11. Lin, C-H, Lo, P-Y, Wu, H-D, Chang, C. and Wang, L-C., 2018. Association between indoor air pollution and respiratory disease in companion dogs and cats. *Journal of Veterinary Internal Medicine, 32*(3), pp. 1259–1267.
12. Sonenshine, D.E., 2018. Range expansion of Tick disease vectors in North America: Implications for spread of Tick-borne disease. *International Journal of Environmental Research and Public Health, 15*(3), p. 478.
13. Aguirre, A.A., 2009. Essential veterinary education in zoological and wildlife medicine: a global perspective. *Revue scientifique et technique, 28*(2), pp. 605–610.
14. Lenzen, M., Malik, A., Li, M., Fry, J., Weisz, H., Pichler, P-P, et al., 2020. The environmental footprint of health care: A global assessment. *The Lancet Planetary Health, 4*(7), pp. e271–e279.
15. Pichler, P-P, Jaccard, I.S., Weisz, U. and Weisz, H., 2019. International comparison of health care carbon footprints. *Environmental Research Letters, 14*(6), p. 064004.
16. Schiavone, S.C.M., Smith, S.M., Mazariegos, I., Salomon, M., Webb, T.L., Carpenter, M.J., et al., 2021. Environmental sustainability in Veterinary Medicine: An opportunity for teaching hospitals. *Journal of Veterinary Medical Education, 49*(2), p. e20200125.
17. Koytcheva, M.K., Sauerwein, L.K., Webb, T.L., Baumgarn, S.A., Skeels, S.A. and Duncan, C.G., 2021. A systematic review of environmental sustainability in veterinary practice. *Topics in Companion Animal Medicine, 44*, p. 100550.
18. Deluty, S.B., Scott, D.M., Waugh, S.C., Martin, V.K., McCaw, K.A., Rupert, J.R., et al., 2020. Client choice may provide an economic incentive for veterinary practices to invest in sustainable infrastructure and climate change education. *Frontiers in Veterinary Science, 7*, p. 622199.

19. Lem, M., 2019. Barriers to accessible veterinary care. *Canadian Veterinary Journal, 60*(8), pp. 891–893.
20. Stephen, C., Carron, M. and Stemshorn, B., 2019. Climate change and veterinary medicine: Action is needed to retain social relevance. *Canadian Veterinary Journal, 60*(12), pp. 1356–1358.
21. Herrmann, J.A., Johnson, Y.J., Troutt, H.F. and Prudhomme, T., 2009. A public-policy practicum to address current issues in human, animal, and ecosystem health. *Journal of Veterinary Medical Education, 36*(4), pp. 397–402.
22. Health TMSCoCa. The Medical Society Consortium on Climate and Health [Available from: https://medsocietiesforclimatehealth.org/
23. Hill's. Hill's Sustainability [Available from: https://www.hillspet.com/about-us/sustainability
24. Terracycle. Partner with us. https://www.terracycle.com/en-US/about-terracycle/partner_with_us
25. Coalition, P.S. Pet Sustainability Coalition [Available from: https://petsustainability.org/
26. Purina. Keeping the Future in Mind Today [Available from: https://www.purina.com/about-purina/sustainability
27. MARS. Today's Commitments to a Sustainable Tomorrow [Available from: https://gbr.mars.com/sustainability-plan
28. Acuff, H.L., Dainton, A.N., Dhakal, J., Kiprotich, S. and Aldrich, G., 2021. Sustainability and pet food: Is there a role for veterinarians? *Veterinary Clinics: Small Animal Practice, 51*(3), pp. 563–581.
29. Association NCsB. Cattle Industry Commits to Climate Neutrality by 2040 2021 [Available from: https://www.ncba.org/ncba-news/news-releases/news/details/27404/cattle-industry-commits-to-climate-neutrality-by-2040
30. Council, N.P.P. Farming for today and tomorrow [Available from: https://nppc.org/issues/issue/farming-today-for-tomorrow/
31. Mintz, K. and Tal, T., 2018. The place of content and pedagogy in shaping sustainability learning outcomes in higher education. *Environmental Education Research, 24*(2), pp. 207–229.
32. Moore, J.N., Cohen, N.D. and Brown, S.A., 2018. Perspectives in professional education: Reassessing courses required for admission to colleges of veterinary medicine in North America and the Caribbean to decrease stress among first-year students. *Journal of the American Veterinary Medical Association, 253*(9), pp. 1133–1139.
33. UNESCO. HESI highlights the role of higher education in building a better world for current and future generations [Available from: https://en.unesco.org/news/hesi-highlights-role-higher-education-building-better-world-current-and-future-generations
34. University, T.A.M. Field experience in One Health and Outbreak Investigation [Available from: https://onehealth.tamu.edu/field-experience-in-one-health-outbreak-investigation/
35. Saskatchewan Uo. One Health Leadership Experience [Available from: https://wcvm.usask.ca/ohle/
36. Uehlinger, F.D., Freeman, D.A. and Waldner, C.L., 2019. The One Health Leadership Experience at the University of Saskatchewan, Canada. *Journal of Veterinary Medical Education, 46*(2), pp. 172–183.
37. AMA. Climate Change Education across the Medical Education Continuum H-135.919 2019 [Available from: https://policysearch.ama-assn.org/policyfinder/detail/%22climate%20change%22?uri=%2FAMADoc%2FHOD.xml-H-135.919.xml

38. Shaw, E., Walpole, S., McLean, M., Alvarez-Nieto, C., Barna, S., Bazin, K., et al., 2021, AMEE Consensus Statement: Planetary health and education for sustainable healthcare. *Medical Teacher*, *43*(3), pp. 272–286.
39. Maxwell, J. and Blashki, G., 2016. Teaching about climate change in medical education: An opportunity. *Journal of Public Health Research*, *5*(1), p. 673.
40. Green, E.I., Blashki, G., Berry, H.L., Harley, D., Horton, G. and Hall, G., 2009. Preparing Australian medical students for climate change. *Australian Family Physician*, *38*(9), pp. 726–729.
41. Bell, E.J., 2010. Climate change: What competencies and which medical education and training approaches? *BMC Medical Education*, *10*(1), p. 31.
42. University, C. Global Consortium on Climate and Health Education [Available from: https://www.publichealth.columbia.edu/research/global-consortium-climate-and-health-education
43. Shaman, J. and Knowlton, K., 2018. The need for climate and health education. *American Journal of Public Health*, *108*(S2), pp. S66–S67.
44. Shea, B., Knowlton, K. and Shaman, J., 2020. Assessment of climate-health curricula at International Health Professions Schools. *JAMA Network Open*, *3*(5), p. e206609-e
45. Region. AoSoPHitE, 2021. *ASPHER's Climate and Health Competencies for Public Health Professionals*, 1st ed. Brussels, Belgium. https://www.aspher.org/download/882/25-10-2021-final_aspher-climate-and-health-competencies-for-public-health-professionals-in-europe.pdf
46. Goshua, A., Gomez, J., Erny, B., Burke, M., Luby, S., Sokolow, S., et al., 2021. Addressing climate change and its effects on human health: A call to action for Medical Schools. *Academic Medicine*, *96*(3), pp. 324–328.
47. Elmore, R.G., 2003. The lack of racial diversity in veterinary medicine. *Journal of the American Veterinary Medical Association*, *222*(1), pp. 24–26.
48. Greenhill, L.M., Nelson, P.D. and Elmore, R.G., 2007, Racial, cultural, and ethnic diversity within US Veterinary Colleges. *Journal of Veterinary Medical Education*, *34*(2), pp. 74–78.
49. Foundation, N.S. Centers of Research Excellence in Science and Technology [Available from: https://beta.nsf.gov/funding/opportunities/centers-research-excellence-science-and-technology
50. Council, N.R., 2013. *Best Practices in State and Regional Innovation Initiatives: Competing in the 21st Century*. In: Wessner, C.W., ed. Washington, DC: The National Academies Press, p. 256. https://nap.nationalacademies.org/catalog/18364/best-practices-in-state-and-regional-innovation-initiatives-competing-in
51. Foundation, National Science Environmental Sustainability [Available from: https://beta.nsf.gov/funding/opportunities/environmental-sustainability-1
52. Watts, N., Amann, M., Arnell, N., Ayeb-Karlsson, S., Beagley, J., Belesova, K., et al., 2021. The 2020 report of The *Lancet* countdown on health and climate change: Responding to converging crises. *The Lancet*, *397*(10269), pp. 129–170.
53. Myers, T.A., Nisbet, M.C., Maibach, E.W. and Leiserowitz, A.A., 2012. A public health frame arouses hopeful emotions about climate change. *Climatic Change*, *113*(3), pp. 1105–1112.
54. Maibach, E.W., Nisbet, M., Baldwin, P., Akerlof, K. and Diao, G., 2010. Reframing climate change as a public health issue: An exploratory study of public reactions. *BMC Public Health*, *10*(1), p. 299.
55. Education USDo. Green Strides: Environment, Health and Facilities at ED [Available from: https://www2.ed.gov/about/inits/ed/green-strides/index.html

56. America FFo. Cargill and FFA Join to Support Sustainable Ag [Available from: https://www.ffa.org/ffa-new-horizons/cargill-and-ffa-join-to-support-sustainable-ag/.

57. 4-H. Exploring Your Environment [Available from: https://4-h.org/parents/curriculum/exploring-your-environment/

58. Villarroel, A., McDonald, S.R., Walker, W.L., Kaiser, L., Dewell, R.D. and Dewell, G.A., 2010. A survey of reasons why veterinarians enter rural veterinary practice in the United States. *Journal of the American Veterinary Medical Association, 236*(8), pp. 849–857.

17 Protecting Animal Health in Our Changing Climate: Key Messages

Craig Stephen and Colleen Duncan

CONTENTS

Climate change impacts on animal health are inevitable and are happening. The Intergovernmental Panel on Climate Change (IPCC) sixth assessment report updated our collective knowledge on impacts, adaptations, and vulnerabilities (1). The panel concluded, with high confidence, that climate change has caused substantial and increasingly irreversible losses that were larger than expected, in terrestrial, freshwater, coastal, and open ocean marine ecosystems. It concluded that half of all assessed species have begun to move toward the poles or to higher elevations, and that hundreds of local species have been lost because of climate changes. Some of these losses are irreversible. Animal diseases and zoonoses are emerging and expanding in range under the influence of climate change. Animal productivity and health are beginning to deteriorate on land and on the sea. The IPCC report made it very clear that there is a rapidly narrowing window of opportunity to enable climate resilience, and time is not on our side. The hope for this book is to empower the reader with knowledge, tools, and perspectives to keep that window open and act now to protect animal health and the value it brings ecosystems, societies, and animal welfare. This chapter presents five messages we saw across all the chapters that might help set the reader onto a course of action to protect animal health in the face of the climate crisis.

DOI: 10.1201/9781003149774-17

MESSAGE 1: YES WE CAN

The relationships between climate change and animal health are many, varied, and complex. Citizens are increasingly pleading for fast and wide-ranging action by governments and businesses, but climate change mitigation and adaption require the participation of individuals. New evidence will not change outcomes unless people apply it in practice and in policy. With rapidly evolving science and experience, and the potential for controversy they bring, it can be intimidating to step up and act. The authors in this book have, however, shown that people can act in many ways. Some undertook research, provided health services, or conducted surveillance to help better understand cause-effect relationships and expose options for action. Others described ways to teach and inspire people to act. Some explored ways to influence animal health policy and practices. One even took direct action to block the production of new carbon pollution through peaceful protest. None started their career as climate and health experts or advocates. All have used the knowledge and tools at their disposal to apply their expertise and passion to influence a part of the climate-health equation, and none tried to take it all upon themselves. The first step to individual climate action is to systematically identify your span of influence within which you can legitimately and effectively make a positive contribution. No matter how big or small, all actions combine to lead to cumulative impacts that come from expertise being mobilized across disciplines, places, and times.

MESSAGE 2: WE MUST WORK TOGETHER

Effective climate change action requires an understanding of the diverse ways in which climate change impacts populations of concern along with the opportunities and obstacles to act in the population's socio-ecological system. A team approach is needed to understand the array of interactions, feedbacks, and pathways through which climate and health interact, even within the span of one problem. Collaborative knowledge networks and social-learning groups can increase access to information and innovations from within and outside of your own discipline or area of responsibility (2). Collaborative, multi- to trans-disciplinary teams are needed to translate what we learn through research and surveillance into tangible actions that are socially acceptable, feasible, and effective. Working together will reduce the likelihood that efforts to reduce risk or promote resilience in one species will unintentionally lessen the capacity for health in other species and future generations to adapt to a new climate regime. Common messaging across species and sectors may more effectively influence social norms and perspectives that influence peoples' perceived empowerment to act (3).

Framing climate change as a health issue can accelerate response (4) to help break the status quo through innovative ways to identify, inspire, and sustain effective actions. Finding and engaging partners needs to happen at the start of a journey into climate action, and these relationships need to be nurtured and leveraged throughout the planning process and implementation of an action plan.

New research on the relationship of climate and animal health has its own merits, but without effectively getting this knowledge to the people who can act to make the necessary changes, little will come of it. For the animal health community to meaningfully participate in climate change planning and action, we need to engage in policy discussions and political action at local, regional, and national levels (5). This goal will be best achieved through alliances with other health, agricultural, conservation, fisheries, and related groups.

MESSAGE 3: WE CAN'T WAIT FOR CERTAINTY

The challenges that face animal health managers and researchers today differ from the past, but there is prevailing uncertainty and ambiguity about those differences. It is not yet possible to predict the exact scale, impacts, places, or times in which these differences will manifest. Research into the impacts of climate change is sparse for animal health and can often result in unclear or conflicting findings. The complex, changing, and wicked problems of climate change and animal health have no "recipe book" of pre-defined management strategies. As there is no root cause of this wickedness, there can be no single solution. Climate-related health challenges are often intractable, complex, and interconnected with other problems that require people to change their mindsets and their behaviors.

There has been a noticeable shift from "a focus on reducing scientific un-certainty towards understanding and managing uncertainty" (6). Uncertainty does not imply ignorance and inaction. We may not be certain about how hot it will get in the Australian outback or how dry in the Canadian prairies, but we know it will get warmer and drier in both places because we can see it already happening, and there is overwhelming scientific agreement that it will change more. Management in the face of uncertainty does not dictate that we wait until we know for sure, but rather, it increases the requirement for ongoing evaluation and adaptive management for continued improvement and refinement of what we do in response to climate extremes and perturbations.

Animal health professionals can expect to face at least three challenges to inspiring action in the face of uncertainty and ambiguity on data, knowledge, and goals for action (adapted from (7)):

   I. how to develop agreement to do something, despite different goals and understanding of the problem;
   II. how to make early interventions or actions "stick" by changing peo-ple's behaviors over long enough time, and covering sufficient places and populations to result in a positive change;
   III. how to show that the small changes being made are leading to transformative changes that address the evolving problem in an ac-ceptable, effective, and sustainable way.

Each of these questions is socially complex. Working on them is fundamentally a social process that requires highly creative solutions.

MESSAGE 4: WE CAN CHALLENGE THE STATUS QUO

"If we keep on doing what we have been doing, we are going to keep on getting what we have been getting" (8). The 21st century has been characterized by its unprecedented rates and scales of change. Climate change serves as a change amplifier because it interacts with other global pressures, such as urbanization, species loss, habitat degradation, and pollution. Many areas of study and policy struggle with the reality that current approaches to pressing health issues are insufficiently translating scientific discovery at the rate and scale needed to inspire or sustain actions against climate change.

It has been posited that the world responds to problems in three stereotypical ways: as "permanent crises, where the only viable responses are decisive commands; or permanent tame problems, where the only viable responses are to keep rolling out the same process that led you into the problem in the first place; or permanent wicked problems, where the only viable response is to delay decision-making while you engage in yet more consultation and collaboration" (9). Effective climate change action will need to challenge this stereotype and unfreeze us from the status quo to inspire and mobilize new types of animal health actions and partnerships.

Change is hard. The outcomes of change are often less certain than the status quo. People resist change if they believe they will lose something they value or they won't be able to adapt to the change. People quickly build psychological defenses against change when uncertainty exists or if the suggestion of change implies they are aren't acting in their own best interests or in the interests of others (10). Change requires patience, persistence, and a process. There are many models and theories of change. None are ideal and suited to all situations and context. Each is subject to limitations and questions. But changes in how we govern collaborations and decision making, who we work with to find solutions, how we characterize animal health, and how we turn knowledge into action is critical for gaining ground on the gap between the pace of climate change and the pace of action.

MESSAGE 5: IT'S ABOUT ANIMAL HEALTH

While it may seem a gross simplification, the core theme of this book is that to keep animal healthy in the face of climate change, you need to keep animals healthy. You do not need to be a climate expert. You need to take sustainable actions to ensure animas have access to their determinants of health, such as safe and secure food and water supplies, access to appropriate health care, security of their physical environments, biodiversity, and protection from intolerable hazards like emerging diseases and extreme weather. Individual and population health is the foundation of resilience to any stressors and especially so for climate change.

Health promotion in the Anthropocene must consider who needs to be empowered and who gets to decide when looking at determinants of animal health that work across species and generations. A first step is to promote environmental awareness and interest in the state of the world among animal health professionals. As society increasingly urbanizes, direct connections with nature are becoming more tenuous for a growing proportion of the population. Empowering people to act for animal health goes beyond documenting and disseminating alarms about the insustainability of the current trajectory of humanity (11). Unless animal health professions highlight the centrality of a convivial climate and environment in animal health they will miss the real point – namely, that climate change action is not just about maintaining the flows from the natural world that sustain the economic engine nor maintaining iconic species and iconic ecosystems. It is about keeping all species and generations healthy.

SUMMARY

Societies now face the problem of facing too many problems. Food insecurity, emerging diseases, habitat loss, pollution, climate change, and social conflict happen concurrently rather than in isolation. The trio of a global pandemic, climate change, and the extinction crisis has caused pundits and politicians to declare that humanity is at a tipping point, where new small changes can result in large, serious consequences. The solutions to modern health problems require us to maintain adaptable functionality so we can adapt to expected, unexpected, and interrelated threats by preserving the social and environmental assets that give us health. Health gains made in the 19th and 20th centuries came largely from advances in knowledge on individual drivers of death and diseases. Preserving those gains and preventing backward momentum of the animal health agenda requires substantial re-imagining of how to inspire and sustain the organizational and individual changes needed to concurrently care for the health of biodiversity, societies, and ecosystems.

Attempts to deal with climate change through a single disciplinary or institutional framework have had histories of inadequacy. No longer is it possible, or advisable, to "just let the science" lead the way. Animal health leaders cannot treat climate change as a new puzzle that can be scientifically or technically solved and expect sustainable and acceptable solutions. They must be able to thrive at the intersection where science meets the human dimensions of interspecies and intergenerational health equity. Rather than finding *the* expert who will provide *the* answers by solving *the* puzzle, new forms of collective action is needed to inspire actions in the face of uncertainty, ambiguity, complexity, and conflict to move us away from crisis management to creating the circumstances for health for all species and generations. This approach will require teams, knowledge and capacities to empower individuals, institutions, and governments to tip critical tipping points in favour for all. When taken as a whole, the chapters in this book show that change is possible, that the willingness and expertise to partner for collective actions to protect animal health exists, and that there are viable pathways to protecting animal health, despite the threat of climate change.

REFERENCES

1. IPCC, 2021. Summary for Policymakers. In: *Climate Change 2021: The Physical Science Basis. Contribution of Working Group I to the Sixth Assessment Report of the Intergovernmental Panel on Climate Change.* In Press. Available from https://www.ipcc.ch/report/ar6/wg1/. Accessed July 7, 2022
2. Hess, J.J., McDowell, J.Z. and Luber, G., 2012. Integrating climate change adaptation into public health practice: Using adaptive management to increase adaptive capacity and build resilience. *Environmental Health Perspectives, 120*(2), pp. 171–179.
3. Stephen, C., Duncan, C. and Pollock, S., 2020. Climate change: The ultimate one health challenge. In: Zinsstag, J., Schelling, E., Crump, L., Whittaker, M., Tanner, M. and Stephen, C., eds., *One Health: The Theory and Practice of Integrated Health Approaches.* CABI.
4. Watts, N., Adger, W.N., Agnolucci, P., Blackstock, J., Byass, P., Cai, W., et al., 2015. Health and climate change: Policy responses to protect public health. *The Lancet*, 386(10006), pp. 1861–1914.
5. Stephen, C., Carron, M. and Stemshorn, B., 2019. Climate change and veterinary medicine: Action is needed to retain social relevance. *Canadian Veterinary Journal*, 60(12), pp. 1356–1358.
6. Mehta, L., Adam, H.N. and Srivastava, S., 2019. Unpacking uncertainty and climate change from 'above' and 'below'. *Regional Environmental Change, 19*(6), pp. 1529–1532.
7. Levin, K., Cashore, B., Bernstein, S. and Auld, G., 2012. Overcoming the tragedy of super wicked problems: Constraining our future selves to ameliorate global climate change. *Policy Sciences, 45*(2), pp. 123–152.
8. Wandersman, A.H., Duffy, J.L., Flaspohler, P.D., Noonan, R.K., Lubell, K.M., Stillman, L., et al., 2008. Bridging the gap between prevention research and practice: The interactive systems framework for dissemination and implementation. *American Journal of Community Psychology, 41*, pp. 171–181.
9. Grint, K., 2010. The cuckoo clock syndrome: Addicted to command, allergic to leadership. *European Management Journal, 28*(4), pp. 306–313.
10. Cohen, G.L. and Sherman, D.K., 2014. The psychology of change: Self-affirmation and social psychological intervention. *Annual Review of Psychology, 65*(1), pp. 333–371.
11. Ansari, W.E. and Stibbe, A., 2009. Public health and the environment: What skills for sustainability literacy – And why? *Sustainability, 1*(3), pp. 425–440.

Index